Hormones and Cell Regulation

Hormones et Régulation Cellulaire

Y0-BYB-128

Colloques INSERM
ISSN 0768-3154

Other *Colloques* published as co-editions by John Libbey Eurotext and INSERM

133 Cardiovascular and Respiratory Physiology in the Fetus and Neonate. Scientific Committee: P. Karlberg, A. Minkowski, W. Oh & L. Stern; Managing Editor: M. Monset-Couchard.
ISBN: John Libbey Eurotext 0 86196 086 6
INSERM 2 85598 282 0

134 Porphyrins and Porphyrias. Edited by Y. Nordmann.
ISBN: John Libbey Eurotext 0 86196 087 4
INSERM 2 85598 281 2

137 Neo-Adjuvant Chemotherapy. Edited by Claude Jacquillat, Marise Weil & David Khayat.
ISBN: John Libbey Eurotext 0 86196 077 7
INSERM 2 85598 283 9

139 Hormones and Cell Regulation. Edited by Jacques Nunez, Jacques E. Dumont & Roger J.B. King.
ISBN: John Libbey Eurotext 0 86196 084 X
INSERM 2 85598 284 7

Hormones and Cell Regulation

Hormones et Régulation Cellulaire

Proceedings of the 10th INSERM European Symposium on Hormones and Cell Regulation, held at Sainte Odile (France), 29 September–3 October, 1985

Sponsored by the Institut National de la Santé et de la Recherche Médicale

Edited by
J. Nunez, J.E. Dumont and R.J.B. King

Scientific Committee

E. Carafoli	B. Hamprecht	F. Morel
R.M. Denton	R.J.B. King	J. Nunez
J.E. Dumont	H.J. van der Molen	G. Schultz

QH604
I 17
1985

LES EDITIONS INSERM

John Libbey
EUROTEXT
LONDON · PARIS

British Library Cataloguing in Publication Data
INSERM European Symposium on Hormones and Cell
    Regulation *(10th: 1985)*
    Hormones and cell regulation: proceedings of the
    10th INSERM European Symposium on Hormones
    and Cell Regulation. — (Colloques INSERM
    ISSN 0768-3154; 139)
    1. Cell metabolism   2. Cellular control mechanisms
    3. Hormones   4. Metabolism—Regulation
    I. Title   II. Nunez, J.   III. Dumont, J.E.
    IV. King, R.J.B.   V. Series
    574.87′61   QH34.5

    ISBN 0 86196 084 X
    ISSN 0768-3154

First published in 1986 by

John Libbey & Company Ltd
80/84 Bondway, London SW8 1SF, England. (01) 582 5266
John Libbey Eurotext Ltd
6 rue Blanche, 92120 Montrouge, France. (1) 47 35 85 52
ISBN 0 86196 084 X

**Institut National de la Santé et de la Recherche
Médicale**
101 rue de Tolbiac, 75654 Paris Cedex 13, France.
(1) 45 84 14 41
ISBN 2 85598 284 7

ISSN 0768-3154

© 1986 Colloques INSERM/John Libbey Eurotext Ltd,
All rights reserved
Unauthorised publication contravenes applicable laws

Typesetting by Gwynnes, Hurstpierpoint, Sussex
Printed in Great Britain by Whitstable Litho Ltd,
Whitstable, Kent

Foreword

The 10th European Symposium on Hormones and Cell Regulation, held at Mont Sainte Odile, Alsace on 30 September–3 October, 1985 concentrated on three topics, hormone action, phosphoinositols and various aspects of protein phosphorylation. The last year has seen major advances in our knowledge of steroid hormone action, largely deriving from the use of immunological, molecular biological and cell biological techniques. It was therefore timely to have an overview of this topic together with a discourse on specificity of steroid action as studied by transfection of mouse mammary tumour virus components into hormone-sensitive cells. In reality, the later paper by Philippa Darbre, together with the one on phosphoinositol metabolism by Bob Michell, were not presented at the meeting as, due to the vagaries of air travel, they could not escape from London Airport. We are therefore grateful that, despite their ordeal, they submitted manuscripts.

Post-receptor events were discussed in two talks, the first concerning steroid regulation of cell proliferation whilst the second outlined the interaction of epithelial and stromal cells in embryonic development of mammary gland. Both these presentations indicated the importance of autocrine and paracrine mechanisms in steroid action. Cell–cell interaction was also the theme of a paper on platelet activating factor. The intricacies of peptide hormone action on the testis were debated after two presentations on that topic, whilst the male–female interface was the subject of an overview of the current status of in-vitro fertilization. In fact, this year saw the wider introduction of review lectures. In broadly based meetings of this type, it is sometimes difficult to convey the principles underlying some of the research topics and it is hoped that review talks will be instructive.

It has long been known that protein function can be regulated by phosphorylation–dephosphorylation cycles but the myriad ways in which this can be achieved are now becoming apparent. The inclusion of tyrosine with the classical phosphorylation substrates serine and threonine was noted some years ago but the inter-relationships of these different phosphorylations is of more current import. This was discussed in the context of insulin and EGF action, whilst three classes of protein kinase, cAMP-dependent, calmodulin-dependent and protein kinase C were each the subject of individual lectures.

Although phosphorylation has been in the limelight of late, sulphation of proteins has been implicated in the regulation of their secretion. This topic led into the role of calcium and inositol phosphates in secretion and in the actions of certain hormones in the liver. The importance of intracellular calcium as a regulator of cell function was underlined in a review lecture and also in two talks on calcium channels.

The enjoyment of this year's meeting was due not only to the contributions of speakers and audience but also to the work put in by the staff at Mont Sainte Odile and the local organizer, Dr Monique Sensenbrenner. As the organizer I would also acknowledge the help provided by Dr Jacques Nunez and his staff for dealing with the French applicants, Ghislaine Maupertuis-Wilmes for handling the arrangements in Brussels and during the actual meeting, and to Margaret Barker in London who did most of the organizing. We also thank INSERM, Paul-Martini Stiftung and the Imperial Cancer Research Fund for financial help.

R.J.B. King

Avant-propos

Le 10ème Symposium Européen "Hormones et Régulation Cellulaire" s'est tenu au Mont Sainte Odile du 30 septembre au 3 octobre 1985 et a été consacré à trois sujets: l'action hormonale, les phosphatidyl inositols et divers aspects de la phosphorylation des protéines. L'année précédente nos connaissances sur l'action des hormones stéroïdes ont fortement progressé. Ceci est du, pour une grande part, à l'utilisation de techniques immunologiques et de biologie cellulaire et moléculaire. Il était temps, par conséquent, de présenter une vue générale sur ce sujet et de commenter simultanément la spécificité de l'action des stéroïdes telle qu'elle est étudiée par transfection de composés de virus de tumeur mammaire de souris dans des cellules sensibles aux hormones. En fait l'article de P. Darbre ainsi que celui sur le métabolisme du phosphoinositol de B. Michell n'ont pas été présentés pendant la réunion car, du fait des incertitudes des voyages aériens ils n'ont pu s'échapper de l'aéroport de Londres. Nous leur savons gré d'avoir soumis leur manuscrit malgré leurs ennuis.

Les événements survenant en aval des récepteurs ont été discutés dans deux présentations, la première concernant la régulation par les stéroïdes de la prolifération cellulaire, alors que la seconde soulignait l'interaction entre les cellules épithéliales et du stroma au cours du développement embryonnaire de la glande mammaire. Ces deux présentations soulignaient l'importance des mécanismes autocrines et paracrines au cours de l'action des stéroïdes. L'interaction cellule-cellule était aussi le thème d'un article sur le facteur activateur plaquettaire. Les dédales de l'action des hormones peptidiques sur les testicules ont été débattus après deux présentations sur ce sujet, alors que l'"interface" mâle-femelle était le sujet d'une revue sur l'état actuel de la fertilisation in vitro. De fait, cette année, davantage de revues d'ensemble ont été présentées. Dans des réunions larges de ce type il est parfois difficile de transmettre les données de base sur certains sujets, et l'on espère que des articles de revue seront instructifs à cet égard. On sait depuis longtemps que la fonction des protéines peut être régulée par des cycles de phosphorylation-déphosphorylation, mais la multiplicité de voies par lesquelles ceci peut être réalisé devient maintenant apparente. L'inclusion de la tyrosine comme substrat de phosphorylation, à côté de la phosphorylation classique de la serine et de la thréonine, a été observée il y a quelques années, mais la compréhension des interrelations existant entre ces diverses phosphorylations est plus récente. Ces points ont été discutés dans le contexte de l'action de l'insuline et de l'EGF, alors que chacune des trois classes de protéine-kinase, cAMP-dépendante, calmoduline dépendante, et protéine kinase C, faisait l'objet d'une conférence.

Alors que la phosphorylation a été sous les feux de la rampe ces dernières années, la sulfatation des protéines a été impliquée dans la régulation de leur secrétion. Ce sujet a conduit à examiner le rôle du calcium et des phosphates d'inositol dans la sécrétion et dans l'action de certaines hormones dans le foie. L'importance du calcium intracellulaire comme régulateur des fonctions cellulaires a été soulignée dans une revue d'ensemble et aussi dans deux exposés sur les canaux calcium.

L'agrément de la réunion de cette année a été dû non seulement aux contributions des orateurs et des participants mais aussi au travail effectué par l'équipe du Mont Sainte Odile et de l'organisateur local, le Dr Monique Sensenbrenner. En tant qu'organisateur, je voudrais aussi remercier le Dr Jacques Nunez et ses collaborateurs qui ont pris en charge les participants français, Ghislaine Maupertuis-Wilmes qui a organisé les opérations à Bruxelles et pendant la réunion et Margaret Barker à Londres qui s'est occupée de l'ensemble de l'organisation. Nous remercions aussi l'INSERM, Paul-Martini Stiftung et l'Imperial Cancer Research Fund pour leur aide financière.

R.J.B. King

List of participants
Liste des participants

Aguilera, G., Physiology Department, Centro de Investigacion y de Estudios Avanzados del IPN, Mexico City, Mexico.

Andrews, T., Department of Medicine, Emanuel Hospital, 2801 N. Gantenbein Avenue, Portland, Oregon 97227, USA.

Assimacopoulos-Jeannet, F., Laboratoires de Recherches Métaboliques, Faculty of Medicine, University of Geneva, 64 avenue de la Roseraie, 1205 Geneva, Switzerland.

Assmann, P., INSERM U 61, ZUP Hautepierre, 3 avenue Molière, 67200 Strasbourg, France.

Auberger, P., INSERM U 145, Faculté de Médecine, Avenue de Vallombrose, 06034 Nice Cedex, France.

Avruch, J., Diabetes Unit, Harvard Medical School, Massachusetts General Hospital, Boston, Massachusetts 02114, USA.

Baudiere, B., ENSC Montpellier, ER CNRS 228, rue de l'Ecole Normale, Montpellier, France.

Beebe, S., Howard Hughes Medical Institute at Vanderbilt Medical Center, Physiology Department, 702 Light Hall, Nashville, Tennessee 37232, USA.

Begeot, M., Laboratoire d'Histologie et d'Embryologie, Faculté de Médecine, B.P. 12, 69600 Oullins, France.

Benveniste, J., INSERM U 200, Université Paris-Sud (Paris XI), 32 rue des Carnets, 92140 Clamart, France.

Billig, H., Department of Physiology, University of Göteborg, PO Box 33031, S-400 33 Göteborg, Sweden.

Blaes, N., INSERM U 63, 22 avenue du Doyen Lépine, 69676 Bron cedex, France.

Bosworth, N., Amersham International plc, Biomedical Department, Cardiff Laboratories, Forest Farm, Whitchurch, Cardiff, Wales CF4 7YT, UK.

Bruni, P., Universita di Firenze, Biochemistry Department, Viale Morgagni 50, Firenze, I-50134, Italy.

Cam, Y., Université Louis Pasteur, Institut de Biologie Médicale, 11 rue Humann, 67085 Strasbourg Cedex, France.

Capony, F., INSERM U 148, 60 rue de Navacelles, 34100 Montpellier, France.

Carafoli, E., Laboratorium für Biochemie, ETH-Zentrum, Universitätstrasse 16, CH-8092 Zurich, Switzerland.

Chabardes, D., Collège de France, Laboratoire de Physiologie, 11 Place marcelin Berthelot, 75231 Paris Cedex 05, France.

Chambaz, E., INSERM U 244, Laboratoire des Régulations endocrines, Département de Recherche fondamentale, Centre d'Etudes Nucléaires – 85 X – 85, avenue des martyrs, 38041 Grenoble Cedex, France.

Cognez, P., IRSC,, Département Membranes et Cancer, 7 rue Guy Moquet, BP no. 8, 94802 Villejuif Cedex, France.

Connolly, E., Wenner-Grens Institute, Department of Metabolic Research, University of Stockholm, S106 91 Stockholm, Sweden.

Cooke, B., Department of Biochemistry, Royal Free Hospital School of Medicine, Rowland Hill Street, London NW3 2PF, UK.

Couchie, D., INSERM U 282, Hôpital Henri Mondor, 51, avenue du Maréchal de Lattre de Tassigny, 94010 Créteil, France.

Darbre, P., Hormone Biochemistry Department, Imperial Cancer Research Fund, PO Box 123, Lincoln's Inn Fields, London WC2A 3PX, UK.

Decoster, C., Institut de Recherche Interdisciplinaire, Campus Hôpital Erasme, Faculté de Médecine – ULB, Route de Lennik 808, B1070 Brussels, Belgium.

Degenhart, H., Pediatric Laboratory, Sophia Childrens Hospital, PO Box 70029, 3000LL Rotterdam, The Netherlands.

Denton, R., Department of Biochemistry, University of Bristol, Medical School, University Walk, Bristol BS8 1TD, UK.

Dokhac, L., CNRS, Laboratoire d'Endocrinologie et Régulations Cellulaires, UA1131, Bt. 432 Université Paris Sud, 91405 Orsay Cedex, France.

Dumont, J., Institut de Recherche Interdisciplinaire, Campus Hôpital Erasme, Route de Lennik 808, B-1070 Brussels, Belgium.

Erneux, C., IRIBHN, C.P.602, Faculté de Médecine, Route de Lennik 808, B1070 Brussels, Belgium.

Falk, H., MHH Haematology, Medizinische Hochschule Hannover, Zentrum Innere Medizin und Dermatologie, Postfach 61 01 80, D-3000 Hannover 61, FRG.

Franklin, T., ICI Pharmaceuticals, Bioscience Department, Alderley Park, Macclesfield, Cheshire, UK.

Friedl, A., Physiologisch-Chemisches Institut, Universität Tübingen, Hoppe-Seyler Strasse 1, 7400 Tübingen, FRG.

Garcia, T., INSERM U 33, Hôpital de Bicêtre, 78 avenue du Général Leclerc, 94270 Le Kremlin-Bicêtre, France.

Guillon, G., INSERM U 264, Centre CNRS INSERM, Pharmacologie et Endocrinologie, rue de la Cardonille, 34033 Montpellier Cedex, France.

Hadjian, A., INSERM U 244, Laboratoire des Régulations endocrines Département de Recherche fondamentale, Centre d'Etudes Nucléaires – 85X, 85 avenue des Martyrs, 38041 Grenoble Cedex, France.

Haller, H., Freie Universität Berlin, Universitätsklinikum Steglitz, Hindenburg Damm 30, D1000 Berlin 45, FRG.

Hamprecht, B., Physiologisch-Chemisches Institüt der Universität, University of Tübingen, Hoppe-Seyler-Strasse 1, 7400 Tübingen, FRG.

Hanson, P., Aston University, Biology Division, Molecular Sciences, Aston Triangle, Birmingham B4 7ET, UK.

Hofmann, F., Physiologische Chemie, Medizinische Fakultät, Universität des Saarlandes, D-6650 Homburg-Saar, FRG.

Huttner, W., Max-Planck-Institute for Psychiatry, Neurochemistry Department, am Klopferspitz 18A, 8033 Martinisried, FRG.

Hyafil, F., Laboratoires Glaxo SA, Département scientifique, 43 rue Vineuse, 75764 Paris Cedex 16, France.

Jakob, F., Medizinische Poliklinik der Universität, Endokrinologie Dept., Klinikstrasse 8, 8700 Würzburg, FRG.

Jung-Testas, I., INSERM U 33, Hôpital de Bicêtre, 78 avenue du Général Leclerc, 94270 Le Kremlin-Bicêtre, France.

Kaiser, C., Laboratoire d'Investigation Clinique, Unité Thyroïde, Hôpital Cantonal, 1211 Geneva 4, Switzerland.

King, R., Hormone Biochemistry Department, Imperial Cancer Research Fund, Lincoln's Inn Fields, London WC2A 3PX, UK.

Klotz, G., Bayer AG, Institute for Chemotherapy, PO Box 101709, 5600 Wuppertal-1, FRG.

Kratochwil, K., Institute of Molecular Biology, Austrian Academy of Sciences, Billrothstrasse 11, A-5020 Salzburg, Austria.

Lamprecht, S., Soroka Medical Centre and Faculty of Health Sciences, Ben Gurion University of the Negev, Gastroenterology Laboratory, Beer Sheva 84101, Israel.

Lamy, F., Institut de Recherche Interdisciplinaire, IRIBHN, Campus Hôpital Erasme, Faculté de Médecine-ULB, Route de Lennik 808, B1070 Brussels, Belgium.

Landron, D., INSERM U 30, Hôpital Necker-Enfants Malades, Tour Technique, 6e étage, 149 rue de Sèvres, 75743 Paris Cedex, France.

Laugier, C., INSERM U 205, Institut National des Sciences Appliquées, Bâtiment 406, 20 avenue Albert Einstein, 69621 Villeurbanne Cedex, France.

Lawen, A., Universität Würzburg, Institut für Physiologische Chemie, Koellikerstrasse 2, D-8700 Würzburg, FRG.

Levy, J., Endocrine Laboratory, Clinical Biochemistry Unit, Soroka Medical Center, University Center for Health Sciences, Beer Sheva 84101, Israel.

Lonchampt, M., Département de Pharmacologie, Institut Henri Beaufour, 72 avenue des Tropiques, 91940 Les Ulis, France.

Ludersdorf, Ms., Freie Universität Berlin, Universitätsklinikum Steglitz, Hindenburg Damm 30, D1000 Berlin 45, FRG.

Magous, R., Ecole Nationale Supérieure de Chimie, Laboratoire de Biochimie Membranes, 8 rue de l'Ecole Normale, 34075 Montpellier Cedex, France.

Marme, D., Godecke Research Institute, Dept. of Biochemical Pharmacology, Mooswaldallee 1-9, 7800 Freiburg, FRG.

Marty, A., Ecole Normale Supérieure, Laboratoire de Neurobiologie, 46 rue d'Ulm, 75005 Paris, France.

Maupertuis-Wilmes, G., Institut de Recherche Interdisciplinaire, Campus Hôpital Erasme, Faculté de Médecine-ULB, Route de Lennik 808, B1070 Brussels, Belgium.

Michell, R., University of Birmingham, Department of Biochemistry, PO Box 363, Birmingham B15 2TT, UK.

Mokhtari, A., Endocrinologie et Régulations Cellulaires, CNRS-UA 1131, Université Paris-Sud, Bat. 432, 91405 Orsay Cedex, France.

Morel, F., Collège de France, Laboratoire de Physiologie Cellulaire, 11 place Marcelin Berthelot, 75231 Paris Cedex 05, France.

Morisset, M., INSERM U 148, 60 rue de Navacelles, 34100 Montpellier, France.

Munari-Silem, Y., INSERM U 197, Laboratoire de Médecine expérimentale, Faculté Alexis Carrel, rue Guillaume Paradin, 69372 Lyon Cedex 08, France.

Nanberg, E., Department of Metabolic Research, The Wenner-Gren Institute, University of Stockholm, Biologi Hus F3, S-106 91 Stockholm, Sweden.

Nicollier, M., INSERM U 198, Bâtiment INSERM, Route de Dole, 25000 Besançon, France.

Nunez, J., INSERM U 282, Hôpital Henri Mondor, 51 avenue du Maréchal de Lattre de Tassigny, 95010 Créteil, France.

Pallast, M., Freie Universität Berlin, Institüt für Pharmakologie, Thielallee 69-73, 1000 Berlin 33, FRG.

Pageaux, J.F., INSERM U 205, Institut National des Sciences Appliquées, Laboratoire de chimie biologique – Bâtiment 406, 20 avenue Albert Einstein, 69621 Villeurbanne Cedex, France.

Pavlovic-Hournac, M., INSERM U 96, Hôpital de Bicêtre, 78 rue du Général Leclerc, 94270 Le Kremlin Bicêtre, France.

Pozzan, T., Institute General Pathology, University of Padova, Via Loredan 16, Padova 35100, Italy.

Probst, I., Georg-August-Universität, Institüt für Biochemie, Humboldtallee 23, 3400 Göttingen, FRG.

Propper, A., INSERM U 198, Bâtiment INSERM, Route de Dole, 25000 Besançon, France.

Rajerison, C., Collège de France, 11 Place Marcelin Berthelot, 75231 Paris Cedex 05, France.

Rebut-Bonneton, C., INSERM U 18, Centre André Lichtwitz, Hôpital Lariboisière, 6 rue Guy Patin, 75010 Paris, France.

Reinhart, P., Physiologisch-Chemisches Institüt, Universität Tübingen, Hoppe-Seyler-Strasse 1, 7400 Tübingen, FRG.

Reiser, G., Physiologisch-Chemisches Institüt, Universität Tübingen, Hoppe-Seyler-Strasse 1, Tübingen, FRG.

Rochefort, H., INSERM U 148, 60 rue de Navacelles, 34100 Montpellier, France.

Roger, P., IRIBHN, Free University of Brussels, Campus Erasme, 808 route de Lennik, B7070 Brussels, Belgium.

Rogers, P., University of Newcastle upon Tyne, Biochemistry Department, Ridley Buildings, Newcastle upon Tyne NE1 7RU, UK.

Rommerts, F., Department of Biochemistry II, Division of Chemical Endocrinology, Postbus 1738, Erasmus University, 3000 DR Rotterdam, The Netherlands.

Rosberg, S., Department of Physiology, University of Göteborg, PO Box 33031, S-400 33 Göteborg, Sweden.

Rudolph, C., Godecke Research Institute, Dept. of Biochemical Pharmacology, Mooswaldallee 1-9, 7800 Freiburg, FRG.

Ruf, H., Physiologische Chemie, Universität des Saarlandes, Gebaude 44, D-6650 Homburg/Saar, FRG.

Saunier, C., INSERM U 14, 10 plateau de Brabois, 54511 Vandoeuvre-les-Nancy, France.

Schachtele, C., Godecke Research Institute, Department of Biochemical Pharmacology, Mooswaldallee 1-9, 7800 Freiburg, FRG.

Schnackerz, K., University of Würzburg, Physiological Chemistry Dept., Koellikerstrasse 2, D-8700 Würzburg, FRG.

Schroder, H., Schering Ag., Endokrinpharmakologie II, Müllerstrasse 170-178, D-1000 Berlin 65, FRG.

Schultz, G., Freie Universitat Berlin, Institüt für Pharmakologie, Thielallee 69/73, D-1000 Berlin 33, FRG.

Sensenbrenner, M., Centre de Neurochimie du CNRS, 5 rue Blaise Pascal, 67084 Strasbourg Cedex, France.

Sladeczek, F., INSERM U 264, Centre CNRS-INSERM de Pharmacologie-Endocrinologie, BP 5055, 34033 Montpellier Cedex, France.

Sobel, A., INSERM U 153, 17 rue du Fer-à-Moulin, 75005 Paris, France.

Soderling, T., Howard Hughes Medical Institute, Vanderbilt University, Nashville, Tennessee 37232, USA.

Tanfin, Z., CNRS-UA1131, Laboratoire d'Endocrinologie et Régulations Cellulaires, Bâtiment 432, Université Paris-Sud, 91405 Orsay Cedex, France.

Tanini, A., Universita de Firenze, Metaboli Research Unit, Clinica Medica III, Viale Morgagni 85, Firenze, I-50134, Italy.

Themmen, A., Medical Faculty, Erasmus University Rotterdam, Biochemistry II, PO Box 1738, 3000 DR Rotterdam, The Netherlands.

Thiard, M.-C., INSERM U 198, Bâtiment INSERM, Route de Dole, 25000 Besançon, France.

Thierauch, K., Schering AG, Biochemische Pharmakologie, Müllerstrasse 170-178, D-1000 Berlin 65, FRG.

Thomas, A., University of Bristol Medical School, Department of Biochemistry, University Walk, Bristol, BS8 1TD, UK.

Thomas, G., Friedrich Miescher-Institüt, Cell Biology Department, PO Box 2543, CH-4002 Basel, Switzerland.

Touitou, I., INSERM U 148, 60 rue de Navacelles, 34100 Montpellier, France.

Treves, S., Universita di Padova, Istituto di Patologia Generale, Via Loredan 16, 35131 Padova, Italy.

Van der Molen, H., Afdeling Biochemie, Faculteit der Geneeskunde, Erasmus University, PO Box 1738, 3000 DR Rotterdam, The Netherlands.

Van der Saag, P., Hubrecht Laboratory, Uppsalalaan 8, 3484 CT Utrecht, The Netherlands.

van Schaftingen, E., Université Catholique de Louvain & International Institute of Cellular and Molecular Pathology, Dept. Physiological Chemistry, UCL 75389, avenue Hippocrate 75, B-1200 Brussels, Belgium.

Ventura, M.-A., INSERM U 282, Hôpital Henri Mondor, 51, avenue du Maréchal de Lattre de Tassigny, 94010 Créteil, France.

Vilgrain, I., INSERM U 244, Laboratoire des Régulations endocrines, Département de Recherche fondamentale, Centre d'Etudes Nucléaires – 85X – 85, avenue des Martyrs, 38041 Grenoble Cedex, France.

Wollheim, C., Institut de Biochimie Clinique, Centre Médical Universitaire, University of Geneva, 9 avenue de Champel, CH-1211 Geneva 4, Switzerland.

Zaninetti, D., Laboratoires de Recherches Métaboliques, Faculty of Medicine, University of Geneva, 64 avenue de la Roseraie, 1205 Geneva, Switzerland.

Zeilmaker, G., Erasmus University, Department of Physiology, PO Box 1738, 3000 DR Rotterdam, The Netherlands.

Contents

Sommaire

PHOSPHOINOSITOL
PHOSPHOINOSITOL

PEPTIDE HORMONE ACTION
ACTION DES HORMONES PEPTIDIQUES

PROTEIN PHOSPHORYLATION AND SULFATION
PHOSPHORYLATION ET SULFATATION DES
PROTEINES

ION CHANNELS
CANAUX IONIQUES

NEW MEDIATORS
NOUVEAUX MEDIATEURS

Introduction

Hormones and cell regulation

The 10th European Symposium on "Hormones and Cell Regulation" has taken place as usual in the monastery of Ste Odile on top of the Vosge mountains (Strasbourg-France) from September 30th to October the 3rd 1985. In four days, it provided an update on recent research results in the control of mammalian cells by extracellular signals. Several types of extracellular signals were considered : hormones, neurotransmitters and growth factors, all acting on receptors on the plasma membrane of their target cells and thus initiating on enzymatic cascade; steroidhormones, on the other hand act on the genome.

The two main regulating circuits studied in the 1980s, were considered : the cyclase-cyclic AMP system, the protein targets of which remain largely unknown, and the now very fashionable agonist activated triphosphoinositide hydrolysis which releases two intracellular signal molecules : diacylglycerol (DIACG) and myosinositol 1-4-5 Phosphate (IP3) - Diacylglycerol activates an ubiquitous protein kinase C discovered by Nishizuka; IP3 activates the release of calcium from intracellular stores.

Thomas (Bristol) elegantly demonstrated the role of PIP2 (triphosphoinositide) hydrolysis and IP3 in the action of vasopressin on liver cells. IP3 causes the release of stored calcium from endoplasmic reticulum, which accounts for the early part of cytosolic calcium rise and of its physiological consequences (eg phosphorylase activation and glycogen breakdown). It was shown, applying the old criteria set up by Sutherland for the demonstration of a cyclic AMP role that the IP3 model as proposed by Irvine, Berridge and Michell for calcium mobilizing hormones, applies to this system. An interesting addition to this model is the fact that IP3 response as such does not desensitize, although Ca^{++} depletion of the endoplasmic reticulum may result in a loss of response.

Wollheim (Geneva) studies the control of insulin secretion on insulinoma cell lines - such cells, contrary to islets, constitute an homogeneous material. However, as they lack hexokinase, the glucose stimulation of insulin secretion has to be mimicked by a metabolite of glucose : glyceraldehyde. The IP3 model is shown to apply to this system. Carbamylcholine increases free intracellular Ca^{++} (as measured by fluorometry of Quinn 2 and fura 2). This effect is preceded by a rise of IP3 (inositol 1-4-5 phosphate) and in permeabilized cells IP3 itself stimulates the release of stored intracellular Ca^{++}. On separated intracellular fractions the effect of IP3 is restricted to microsomes; IP3 does not induce Ca^{++} release from mitochondria or secreting granules.

1

There is a long list of similar systems in which stimulating extracellular
signals enhance calcium influx in cytosol and phosphatidylinositol
turnover : the "PI response". In more and more of these systems the dual model
of Nishizuka, Irvine, Berridge is now demonstrated : activation of
phosphatidylinositol 1-4-5 Phosphate (PIP2) hydrolysis leading to the generation
of 2 intracellular signals : diacylglycerol which activates a specific protein
kinase (C), and myoinositol 1-4-5 (IP3) which activates the release of Ca++
stored in endoplasmic reticulum. Evidence for IP3 generation in response to
extracellular signals was shown in striatal neurones in response to glutamate
(Sladeczek, Montpellier) in bovine adrenal cells in response to angiotensin
(Begeot, Lyon), in brown adipocytes in response to adrenergic agents (Nanberg,
Stockholm) in gastric mucosal cells in response to acetylcholine (Baudière,
Montpellier) and in mammary tumor cell lines through stimulation of V1
vasopressin receptors (Guillon, Montpellier). In the latter system a close
correlation has been demonstrated between the concentrations of vasopressin
analogues acting on PI hydrolysis and binding to the V1 receptors. The effect of
vasopressin is not inhibited by pertussis toxin even though all Ni proteins are
ribosylated under the experimental conditions. Thus, contrary to the platelet
system, Ni does not appear to be involved in vasopressin induced Pi hydrolysis
(Guillon, Montpellier).
The second branch of the PI response is the generation of diacylglycerol, an
activator of protein kinase C. A major boost to our knowledge of this system has
been the discovery by Castagna and Nishizuka that the tumor promoters phorbol
esters are in fact efficient and long acting analogs of diacylglycerol. Protein
kinase C activity had been reported to require Ca^{++}. Pozzan (Padova) showed
convincingly that the action of phorbol esters and in great part of
diacylglycerol does not require this ion. However calcium greatly increases the
sensitivity of protein kinase C to phorbol esters. On the other hand in PC12
pheochromocytoma cells, phorbol esters decrease Ca^{++} influx which suggests a
feedback mechanism of diacylglycerol on the PI - Ca^{++} system activated by the
nicotinic receptors. Similarly phorbol esters inhibit histamine induced gastric
secretion(Hanson - Birmingham) and acetylcholine induced calcium influx in dog
thyroid slices (Vansande, Brussels). Total level of protein kinase C is
correlated to growth in rat thyroid (Omri, Paris).
The "PI" response is coupled to both an influx of extracellular Ca^{++} and to release
of reticulum stored Ca^{++}. Calcium itself by activating directly or through
calmodulin a variety of enzymes, profoundly affects cell metabolism. The control
of calcium fluxes in the cytosol of the cell is therefore of great importance. A
role for mitochondria in this control was considered. The mitochondrion has long
been thought of as an important intracellular store of Ca^{++} which might be
involved in the control of cytosolic free Ca^{++}. However, despite reports to the
contrary, no evidence of modulated mitochondrial Ca^{++} release has been obtained.
Resetting an altered cytosolic Ca^{++} by the mitochondria, by uptake and release,
may be effective, but its role is at a higher than normal Ca^{++} level; i.e. it may
be more a protection than a regulating cell device. On the other hand, cytosolic
Ca^{++} modulates intramitochondrial Ca^{++} level and thereby the metabolism of
mitochondria through the three key enzymes of the tricarboxylic and cycle :
pyruvate, isocitrate and α cetoglutarate dehydrogenases. This is a regulation
operating on supply in contrast to the classical regulation through demand (ADP).
Evidence for the operation of these mechanisms is very strong for heart and liver
cells (Denton, Bristol). In the latter cells, both glucagon and vasopressin
increase in parallel the mitochondrial Ca++ content and the activity of the
dehydrogenases (Assimacopoulos-Jeannet, Genève).

As an intracellular signal Ca++ is also involved in the regulation of ion channels. Lacrymal gland cells have been studied by the patch clamp technique (Marty, Paris) either on whole cells or on membrane fragments. (Carbamylcholine through Ca++ activates a voltage dependent K+ channel, and a Cl- and a monovalent cations channels, which accounts for its effect on fluid secretion.

In the study of the cyclic AMP system, the major remaining black box concerns the natural substrates of cyclic AMP dependent protein kinasesin different tissues. The use of analogs specific for the two sites of protein kinase I and II allows to show that synergistic characteristics of the isolated enzymes can be reproduced in intact cells, which greatly supports the postulate that cyclic AMP acts through the kinases (Beebe, Nashville). On the other hand studies on a variety of systems and end points have until now failed to demonstrate a specificity of these two enzymes. The easy separation of cyclic AMP analogs from cyclic AMP has allowed to demonstrate directly that these analogs decrease basal and enhanced cyclic AMP levels. This confirms the indirect observation of Barber (Houston) that cyclic AMP accelerates its disposal.
Depending on its regulatory network, each intracellular signal (ISM) may modulate the concentration of other ISM in a given type of cells. In cardiac cells ß adrenergic agents enhance calcium influx. This effect is mimicked by cyclic AMP, but also by the catalytic unit of cyclic AMP dependent protein kinase (Hofman-Hombourg-Saar). The phosphorylation of the voltage dependent Ca++ channel increases considerably the probability of its opening. Isolation of the channel protein by use of affinity chromatography on specific channel blockers will soon allow the molecular interpretation of this effect.
In the control of Leydig cells in the testis, it becomes obvious that the previously accepted scheme of action through cAMP does not account for the whole LH effect. LH stimulates steroidogenesis at concentrations where no cAMP increase is observed. Moreover inhibitors of cAMP generation potentiate LH action. A role of the PI pathway is postulated. For instance phorbol esters (i.e. analogs of diacylglycerol) enhance testosterone generation, and other transducing systems than cyclase must be postulated to explain the potentiating effects of LHRH and phospholipase C on this model. Testis interstitial fluid, probably through a protein, greatly enhances LH and LHRH actions (Rommerts , Rotterdam). Besides its presumed action on the PIP2 pathway in Leydig cell, LH also enhances free intracellular calcium levels by a cAMP mediated action (Cooke London). It is therefore apparent again in this model that the different intracellular signals control each other in various ways depending on the cell type. It is evident that although there are only a few actors in the game, there will be many variations in their interrelations. As in electronic circuits, all the possible interrelations will finally be demonstrated in one model or another, but for each cell type there will be a characteristic network. As in many other hormone-target cell interactions, occupied LH receptors desensitize by uncoupling from the Ns transducer; they are also internalized leading to hydrolysis of the hormone and recycling of the receptors.

Calcium as an intracellular signal operates through various effector proteins : through calmodulin it activates several enzymes directly (phosphorylase kinase, myosin light chain kinase, low affinity cyclic nucleotide phosphodiesterase, etc), but it also stimulates 2 calmodulin protein kinases (I, II) which by phosphorylation affect the activity of numerous secondary proteins. One of these kinases (II)is ubiquitous, it is an autophosphorylated hetero-oligomer. In the liver it shares at least one substrate and phosphorylation site with cyclic AMP dependent protein kinase. On the other hand it phosphorylates 2 sites (A and B) of pyruvate kinase while cyclic AMP dependent kinase only phosphorylates one (A). Similar work on other enzymes and systems delineates in

various tissues the partially overlapping phosphorylation sites domains of both
types of kinases (Soderling-Nashville).
Posttranslational modification of proteins is an essential element of the
enzymatic cascades triggered by extracellular signals. However, until now , the
main change involved in such regulations has been protein phosphorylation.
Other posttranslational modifications are involved in traffic signalling of
proteins; for example mannose phosphate directs lysosomal enzymes to their
target. Tyrosine sulfation was shown by Huttner (Munich), to label secreted
proteins. The role of this sulfation is now studied on fibroblasts stably
transfected with cloned cDNA of drosophila vitillogenins. Such sulfation of a
target protein was shown to be hormone regulated (estrone)in uterus (Nicoller-
Besançon).
A very important new intracellular signal modulating glucose metabolism has been
discovered by the group of Hers (Louvain). The properties of the system in the
liver were reviewed by E. Van Shaftingen - Fructose 2-6 phosphate activates
glycolysis and stops gluconeogenesis by direct modulation of phosphofructokinase (positive) and of fructose 1-6 phosphate phosphatase (negative). Its
synthesis and hydrolysis are catalyzed by one bifunctional enzyme. The kinase
activity of the enzyme is stimulated by glucose 6 phosphate and inhibited by
glycerol phosphate and by phosphorylation by a cyclic AMP dependent protein
kinase. The phosphatase activity is regulated conversely. Glycerophosphate
level reflects the NADH/NAD+ ratio of the cytosol; it is increased by ethanol
which thus shuts down glycolysis. Fructose 2-6 phosphate is also an
intracellular signal in yeast, where it also activates F2-6P synthesis and
inhibits its degradation. However, in this system, cAMP activates F 2-6 P
synthesis and inhibits its degradation. These opposite effects of cyclic AMP in
liver and yeast again emphasize that the functional role of such signals, like
that of switches in an electrical circuit depends on the organization of the
particular regulatory network considered.

Several extracellular signals act on receptors in the membrane of their target
cells by other mechanisms than cyclic AMP generation or triphosphoinositide
hydrolysis. Among these, insulin and various growth factors activate the
protein tyrosine kinase activity of their receptors. The primary and quaternary
structure of the insulin receptor have been elucidated in an impressive
technological endeavour by Genentech. It is a tetramer α_2 β_2, with the α chain
completely outside the membrane but bound to the transmembrane ß by ss bonds.
Insulin binds to the α subunits while a protein tyrosine kinase activity is
located inside the cell on the ß subunit. Insulin enhances tyrosine
phosphorylation even in digitonin solubilized receptors. The effect on
exogenous substrates bears on the Vmax; it follows a first autophosphorylation
of insulin receptors own tyrosine residues. The natural substrates of this
kinase are not known. In fact, in intact cells the main tyrosine
phosphorylation observed after insulin treatment is on the insulin receptor !!
One important clue may be that insulin receptor kinase as well as sarckinase
enhances the phosphorylation (but on a serine residue) of ribosomal protein S6,
which has been linked to an activation of translation Avruch, Boston).
Thomas (Basel) showed in 3T3 cells a tight relation between the induction of
proliferation (by serum PGF2α, FGF, etc) the activation of protein synthesis and
the phosphorylation of ribosonal S6 protein. This protein can be heavily
phophorylated (up to 5 phosphates/mole). It is more phosphorylated when bound
to polysomes, which suggests a role of these phosphorylations on the translation
rate. These effects can be related to an activation of S6 kinase presumably due
to phosphorylation of this enzyme on a tyrosyl residue. Purification of this
enzyme from Hela cells was reported (Lawen, Wurzburg).
A stimulation of proliferation in another system, dog thyroid cells in primary
culture, is accompanied by the synthesis of specific proteins (81K for EGF) (39K
for TSH and cyclic AMP). The characteristics of these synthese are compatible
with a role in proliferation induction (Lamy, Brussels).

4

King (London) reviewed the main advances and problems in steroid receptors physiology and biochemistry. On the basis of studies on receptor labeling with tritiated steroids in homogenates, it had been accepted that these receptors are present in the cytosol and only migrate to the nucleus after steroid binding and activation. The use of monoclonal antibodies against these receptors now suggests that a great part of these receptors, even when unoccupied, are localized in the nucleus. However it is not excluded that a small proportion of these receptors could be localized in the cytosol. The relation between structure of steroid receptors and activation remains controversial. It is clear however that their phosphorylation increases the binding of their steroids.

The use of glucocorticoid resistant mutant lymphoma cells allows an elegant approach to the elucidation of the mechanism of action of these steroids : mutants, with receptors unable to bind steroids, with decreased specific binding to DNA promoter regions, or with increased unspecific binding to DNA have been characterized. Studies with glucocorticoid or estrogen receptor binding to promoter DNA of steroid sensitive genes allow to map specific sites. However in each case several such sites are demonstrated, including in the introns of the genes. Transfection of susceptible cells with chimeric genes including part, of the promoter and genes of enzymes such as ß galactosidase allow a better definition of the functional steroid sensitive DNA region.

M. Rochefort (Montpellier) exemplified an example of steroid actions. He uses MCF7 human mammary cell line which proliferates in response to estrogens. Estrogens enhance the formation of several proteins in these cells : some are found in the cell, some in the medium . Such a glycoprotein (C52K) was isolated from the medium of these cells . Monoclonal antibodies allowed to demonstrate its localization in the cytosol of the MCF7 cells. It was also found in human tumors (cancerous or proliferating benign tumors) but not in normal tissue. The purified protein has some mitogenic activity on the MCF7 cells. It is internalized as a lysosomal protein (presumably through mannose phosphate receptors). It could be a marker of proliferation or even an autocrine growth factor.

Benveniste (Paris) summarized our present knowledge on a new very important extracellular signal molecule involved in inflammation, platelet aggregation, etc.: PAF acether (platelet activating factor, an analogue of phosphotidylcholine with an aliphatic alcohol in 1 and an acetic acid in 2). Released by inflammatory cells : leucocytes, macrophages, monocytes, mast cells, and by platelets it acts on these cells but also on many other cell types : endothelial cells, heart cells, bronchial cells. Its level is controlled as much by synthesis (hydrolysis of ether lipid by phospholipase A2, acetylation by an acetyltransferase) as by its degradation. In fact, it is striking, although little appreciated, that in the action of phospholipase A2, for each molecule of fatty acid (i.e. most importantly arachidonic acid) released there is one molecule of the lyso precursor of PAF (a lysophosphatidylcholine) released. The first one is in most studied systems an important extracellular signal; the second causes membrane fusion - Benveniste concluded by suggesting that his newly discovered extracellular signal may be involved in the regulation of many cell types. The tools to investigate the role of this system, i.e. PAF itself and inhibitors of PAF action are available.

Kratochwill (Salzburg) demonstrated the complexity of hormonal action on a tissue composed of different cell populations. In the embryonic mice mammary epithelium induces the differentiation, i.e. the generation of androgen and estrogen receptors, in neighbouring mesenchymac cells. When exposed to androgens (as in males) these will destroy the mammary buds; when exposed to estrogens they will inhibit their further development. Only in the absence of hormones (i.e. in females) will the buds develop.

A growing field of application of the molecular endocrinology of reproduction is the method of human in vitro fertilization. As pointed out by ZEILMAKER (Rotterdam) the demand for this procedure is very important (3000/year in Nederland). However the efficiency (10 % successfull pregnancies) is still very low - Progress is needed at all phases of the process : oocyte maturation, fertilization, development and mostly on the preparation of uterine mucosa for implantation. This synthesis of a very successfull application demonstrated the important medical role of our progression in embryology and molecular endocrinology.

We would like to thank all those who made the Conference a success. Dr R. King organized the meeting. He was helped in Sainte-Odile by Mrs G. Wilmes - Prof. J. Nunez did all the editing of this book.

J. E. DUMONT

The meeting was sponsored and supported by INSERM, Martini/Stiftung, and the Imperial Cancer Research Fund.

Steroid hormone action

Action des hormones stéroïdes

Hormones and cell regulation. Ed J. Nunez *et al.* Colloque INSERM/John Libbey Eurotext Ltd. © 1986. Vol. 139, pp. 9–14.

Hormone action and epithelial-stromal interaction: mutual dependence

K. Kratochwil

Institut für Molekularbiologie der Österreich, Akademie der Wissenschaften, Billrothstraßes 11, A-5020 Salzburg, Austria.

ABSTRACT

The interplay between hormone action and tissue interaction was studied in the embryonic mammary rudiment of the mouse. In male fetuses of day 14, the mammary anlagen are destroyed by the action of testosterone. Utilizing the androgen--insensitive Tfm-mutant in experimental combinations of gland epithelium and mesenchyme it was found that the hormone acts directly only on the mesenchymal component of the organ. Destruction of the epithelium is caused indirectly by testosterone-activated mesenchymal cells. Likewise, inhibition of epithelial outgrowth by estradiol is caused by action of the hormone on the mesenchyme, as suggested by the distribution of estrogen receptors in the gland rudiment. On the other hand, the formation of mesenchymal receptors for both androgens and estrogens depends on inductive interaction with mammary epithelium: Only the few mesenchymal cell closest to the mammary bud form receptors, and experimental combinations have established that the epithelium induces receptor synthesis in the mesenchyme. This induction is organ- and stage-specific, but not species-specific.

KEYWORDS

Steroid action, epithelial-stromal interaction, development of steroid responsiveness, mammary gland.

INTRODUCTION

Development and function of the vertebrate organism depend on continuous inter-action between its cells, tissues and organs. Two principal modes of interaction have evolved: "long range" interaction, as mediated by hormones, and "short range" interaction between closely associated cells and tissues, the mechanism of which is still unknown. The effect of short range interactions is particularly obvious in developmental processes, as seen in the phenomenon of "embryonic induction", or in the complex tissue interactions operating during embryonic organ formation (1, 2).

We are interested in the possible interplay between hormone action and tissue interaction as it can be expected in the development of organs whose morphogenesis is obviously under hormonal control. This is seen in reproductive organs (e.g. the mammary gland or the prostate) and in all other sexually dimorphic organs. Sexual differentiation of the mammalian fetus is accomplished by 2 testicular hormones: A glycoprotein which causes regression of the Müllerian ducts, and the steroid testosterone which is responsible for all other sexual dimorphism from the reproductive tract to secondary sexual characteristics (3). Many of the organs involved form "epithelial-mesenchymal" structures (e.g. glands), and much experimental evidence is available to show that their morphogenesis is controlled by delicate interaction of the two tissues (2, 4). It was therefore logical to ask how this interaction was affected by the hormone.

Our experiments on the embryonic mammary rudiment of the mouse in fact provided evidence for a mutual dependence of hormone action and tissue interaction: The effect of testosterone (and estradiol) is mediated by tissue interaction, and tissue interaction in turn is required for the development of hormone responsiveness in the organ.

HORMONE ACTION MEDIATED BY TISSUE INTERACTION

Unlike the situation prevailing in most mammals, the mammary anlagen of the mouse fetus are subjected to sexual differentiation. Buds form initially (on day 11) in both sexes and for 2 more days develop in identical fashion. On day 14, however, the epithelial buds of male fetuses are surrounded by densely packed mesenchymal cells, the gland epithelium eventually breaks from the epidermis and most or all of its cells become necrotic and disappear (5). This destruction of the mammary anlagen in the male mouse fetus is solely due to the action of testicular androgens, as shown by experiments *in vivo* (6) and *in vitro* (7).

Both tissues of the rudiment are visibly involved in this process: The epithelium is destroyed, the mesenchyme forms the conspicuous condensation. Although the *in vitro* experiments gave evidence for a direct action of testosterone on the gland rudiment, they did not allow the conclusion that both tissues were directly affected by the hormone. We tested the possibility that the steroid acts only on one tissue which then in turn affects its partner by hormone-induced tissue interaction. This was done by preparing "chimaeric" recombinations of gland epithelium and surrounding mesenchyme, one tissue being derived from the androgen-insensitive Tfm mutant (8), the other from a normal embryo. Exposure of these recombination explants to testosterone established the mesenchyme as the sole target tissue for the hormone (9, 10). Epithelial necrosis in the mouse mammary gland, therefore, is not caused by direct, intracellular action of the steroid, the epithelium is destroyed by hormone-induced mesenchymal cells. This system provided the first evidence for a hormone action mediated by tissue interaction, but it was soon shown that the same situation prevails in the androgen-induced development of the urogenital sinus to the prostate (11). A further case may be represented by the action of estrogens on the fetal mammary rudiment. In these early stages, experimentally applied estrogens cause nipple malformation *in vivo* (6) and prevent outgrowth of the primary sprout in organ culture (unpublished). Autoradiographs show estrogen receptors to be present only in the mesenchyme surrounding the gland, suggesting that the effect of the hormone on the gland epithelium is again mediated by tissue interaction (in preparation). In this context it should also be noted that Stumpf and collaborators (12 - 14) localized sex steroid receptors preferentially in the mesenchyme of sexually dimorphic organs of the chick embryo.

TISSUE INTERACTION AND THE DEVELOPMENT OF HORMONE RESPONSIVENESS

The ability to respond to certain hormones is a property acquired by specific cell types at specific stages during ontogenesis. Hormone responsiveness thus is a differentiative function and may in turn be a prerequisite for the expression of other differentiative functions, such as lactation in the mammary gland. Very little is known about the developmental processes controlling the acquisition of hormone responsiveness. One of the prerequisites is the formation of specific hormone receptors, the appearance of which has been described in a number of organs (15 - 17). Nevertheless, none of these systems provided any clues about the processes responsible for the initiation of receptor synthesis. Since the hormone requires preexisting receptors for its action, it cannot induce formation of the first receptors, although it may later regulate receptor levels. At most, one hormone can induce synthesis of receptors for another (e.g. estradiol inducing receptors for progesterone - see ref. 18).

We have studied the development of testosterone responsiveness in the mouse mammary rudiment by exposing explants at various stages to the hormone *in vitro*. It was found that the earliest phase of the gland's development (from day 11 to late on day 13) is not affected by testosterone. The gland then enters a short testosterone-sensitive period of about 30 hrs (essentially day 14 of development), and becomes again insensitive in the first hours of day 15. Interestingly enough, the same precise temporal sequence of acquisition and loss of responsiveness was observed in explanted glands, indicating that all processes responsible for this development must occur within the gland itself and independently of external influence (19).

The appearance of testosterone receptors correlates well with the development of testosterone responsiveness: No receptors were found on day 11, the first could be detected on day 12 and their number increases steadily to the responsive 14-day stage when each gland rudiment possesses about 90 to 100 million androgen--binding sites (20). On the other hand, loss of hormone-sensitivity on day 15 was not reflected in a loss or decline of receptors, or in a change in their affinity for the ligand.

Since the tissue combination experiments mentioned above had established the mesenchyme as the sole target tissue for testosterone, it followed that the binding sites measured in whole glands had to be localized in mesenchymal cells. However, it was puzzling to find that mesenchyme alone, explanted on day 12 without epidermis and mammary buds, never showed any reaction to testosterone (i.e., it did not form the characteristic condensation). Replacing mammary epithelium by epithelia from other organs did not restore mesenchymal hormone responsiveness, whereas mammary epithelia from other species were effective (21). The possibility was ruled out that the mammary bud merely attracted a testosterone--sensitive mesenchymal cell population (21). Instead, the finding that mammary epithelium and mesenchyme must be associated for at least 48 hrs to obtain a hormone reaction indicated that mammary epithelium was somehow involved in the development of mesenchymal testosterone responsiveness.

Eventually, the role played by the epithelium became obvious in ^{3}H-testosterone (and ^{3}H-dihydrotestosterone) autoradiographs (Fig. 1). The sections showed androgen receptors localized in the mesenchyme, thus corroborating the earlier results obtained in tissue combination experiments with the Tfm-mutant, but in addition it was seen that receptors are not present at the same level in the entire mesenchyme (22). As seen in Fig. 1, high receptor levels are present only in a small population of mesenchymal cells in the immediate vicinity of the epithelial mammary bud. This distribution immediately suggested that the mammary bud, in contrast to the epidermis from which it has formed, causes the formation

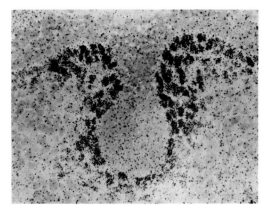

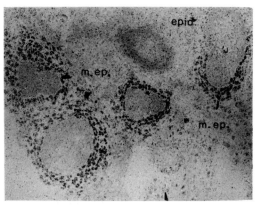

Fig. 1: ³H-5α-Dihydrotestosterone autoradiograph of a mammary rudiment of a
14-day female fetus. Androgen-binding
sites are restricted to a small population of mesenchymal cells surrounding
the epithelial gland bud and its
connection to the epidermis (on top).
Ligand activity: 200 Ci/mmol; concentration: 0.6 nM; exposure time: 7
months. (From reference 22).

Fig. 2: ³H-Testosterone autoradiograph
of a combination explant (12-day mammary
epithelia associated with 12-day mesenchyme, after 3 days in culture). Each
piece of mammary epithelium is surrounded by receptor-containing mesenchymal
cells, no receptors have formed in mesenchyme surrounding pieces of epidermis.
epid = epidermis; m. ep. = mammary epithelium. (From reference 22).

of mesenchymal androgen receptors by a short-range inductive influence. This
hypothesis could be tested in combination explants where mesenchyme-free 12-day
mammary epithelia were associated with 12-day mesenchyme from the mammary region,
but which had not been in contact with a mammary bud before. During subsequent
culture, these epithelia became surrounded by the mesenchyme. After 2 or 3 days
in vitro, the explants were exposed to ³H-testosterone and receptors were either
measured by scintillation counting or visualized in autoradiographs (22).
Addition of the epithelia caused a several-fold increase in the number of testo-
sterone-binding sites in the explants, and autoradiographs showed again that these
receptors were present only in mesenchymal cells surrounding each piece of epithelium (Fig. 2). Epithelia from other organ rudiments, such as the salivary
gland, the pancreas, or the lung, had no or only minimal effects on mesenchymal
receptor synthesis. Mammary epithelia of the 12-day stage were better inductors
than epithelia of older rudiments (14-day, 16-day).

Subsequently we found that estrogen receptors in the mammary rudiment are distributed in the same pattern as androgen receptors, and their induction by mammary
epithelium could be demonstrated by the same type of tissue combination experiment (in preparation).

INTERDEPENDENCE OF SEX STEROID ACTION AND EPITHELIAL-MESENCHYMAL TISSUE
INTERACTION IN THE MAMMARY RUDIMENT

Our experiments show that tissue interaction and hormone action in the mammary
rudiment are mutually dependent processes: In the earliest phase of the organ's
development, beginning with day 12, the newly-formed epithelial mammary bud
induces adjacent mesenchymal cells to form androgen and estrogen receptors.
Short-range inductive tissue interaction is thus responsible for the development
of sex steroid responsiveness in the mammary gland. In the responsive 14-day

stage, testicular androgens of the male fetus (or experimentally administered estrogens) act on this mesenchymal population which then in turn acts on the gland epithelium, causing its destruction in the male (or preventing its outgrowth in the case of estrogen application, respectively). The effect of the hormone on the tissue most affected, the gland epithelium, is not direct but is mediated by hormone-induced tissue interaction. Neither the mechanism of receptor induction, nor that of tissue destruction, is presently understood.

REFERENCES

1. Kratochwil, K. (1983): Embryonic Induction. In Cell Interactions and Development, ed. K.M. Yamada, pp. 99-122. New York: John Wiley & Sons.
2. Grobstein, C. (1967): Mechanisms of organogenetic tissue interaction. Natl. Cancer Inst. Monogr. 26, 279-299.
3. Jost, A. (1953): Problems of fetal endocrinology: the gonadal and hypophyseal hormones. Recent Progr. Horm. Res. 8, 379-418.
4. Fleischmajer,R.,and Billingham R.E. eds. (1968): Epithelial-Mesenchymal Interactions. Baltimore: The Williams & Wilkins Comp.
5. Raynaud, A. (1961): Morphogenesis of the Mammary Gland. In Milk: The Mammary Gland and its Secretion, eds. S.K. Kon and A.T. Cowie, vol.1, pp. 3-46. New York: Acad. Press.
6. Raynaud, A. and Raynaud, J. (1956): La production expérimentale de malformations mammaires chez les foetus de souris, par l'action des hormones sexuelles. Ann. Inst. Pasteur 90, 39-91.
7. Kratochwil, K. (1971): In vitro analysis of the hormonal basis for the sexual dimorphism in the embryonic development of the mouse mammary gland. J. Embryol. Exp. Morphol. 25, 141-153.
8. Lyon, M.F. and Hawkes, S.G. (1970): X-linked gene for testicular feminization in the mouse. Nature 227, 1217-1219.
9. Kratochwil, K. and Schwartz, P. (1976): Tissue interaction in androgen response of embryonic mammary rudiment of mouse: Identification of target tissue for testosterone. Proc. Nat. Acad. Sci. USA 73, 4041-4044.
10. Drews, U. and Drews, U. (1977): Regression of mouse mammary gland anlagen in recombinants of Tfm and wild-type tissues: Testosterone acts via the mesenchyme. Cell 10, 401-404.
11. Cunha, G.R. and Lung, B. (1978): The possible influence of temporal factors in androgenic responsiveness of urogenital tissue recombinants from wild--type and androgen-insensitive (Tfm) mice. J. Exp. Zool. 205, 181-194.
12. Stumpf, W.E., Narbaitz, R. and Sar, M. (1980): Estrogen receptors in the fetal mouse. J. Steroid Biochem. 12, 55-64.
13. Gasc, J.-M. and Stumpf, W.E. (1981): Sexual differentiation of the urogenital tract in the chicken embryo: autoradiographic localization of sex-steroid target cells during development. J.Embryol. Exp. Morphol. 63, 207-223.
14. Gasc, J.-M. and Stumpf, W.E. (1981): The bursa of Fabricius of the chicken embryo: localization and ontogenic evolution of sex-steroid target cells. J.Embryol. Exp. Morphol. 63, 225-231.
15. Clark, J.H. and Gorski, J. (1970): Ontogeny of the estrogen receptor during early uterine development. Science 169, 76-78.
16. Kaye, A.M. (1978): The ontogeny of estrogen receptors. In Biochemical Actions of Hormones, ed. G. Litwack, vol. 5, pp. 149-201. New York: Acad. Press.
17. Teng, C.S. (1980): Ontogeny of the receptor and responsiveness to estrogen in the genital tract of the chick embryo. In The Development of Responsiveness to Steroid Hormones, eds. A.M. Kaye and M. Kaye, pp. 77-94. Oxford: Pergamon Press.
18. Leavitt, W.W., Chen, T.J., Do, Y.S., Carlton, B.D. and Allen, T.C. (1978): Biology of progesterone receptors. In Receptors and Hormone Action, eds. B.W. O'Malley and L. Birnbaumer, pp. 157-188. New York: Acad.Press.

13

19. Kratochwil, K. (1977): Development and loss of androgen responsiveness in the embryonic rudiment of the mouse mammary gland. Dev. Biol. 61, 358-365.
20. Wasner, G., Hennermann I., and Kratochwil, K. (1983): Ontogeny of mesenchymal androgen receptors in the embryonic mouse mammary gland. Endocrinology 113, 1771-1780.
21. Dürnberger, H. and Kratochwil, K. (1980): Specificity of tissue interaction and origin of mesenchymal cells in the androgen response of the embryonic mammary gland. Cell 19, 465-471.
22. Heuberger, B., Fitzka, I., Wasner, G., and Kratochwil, K. (1982): Induction of androgen receptor formation by epithelium-mesenchyme interaction in embryonic mouse mammary gland. Proc. Natl. Acad. Sci. USA 79, 2957-2961.

Résumé

Les effets combinés de l'action hormonale et des interactions tissulaires ont été étudiés dans le rudiment mammaire embryonnaire de la souris. Chez les foetus mâles au 14ème jour, les anlagènes mammaires sont détruits par action de la testosterone. En utilisant le mutant Tfm insensible aux androgènes dans des combinaisons expérimentales de l'épithelium et du mesenchyme de la glande on a trouvé que l'hormone agit uniquement sur le composant mésenchymateux de l'organe. La destruction de l'épithelium est provoquée indirectement par les cellules mésenchymateuses activées par la testostérone. De la même manière, l'inhibition de la croissance épithéliale par l'oestradiol est due à l'action de l'hormone sur le mésenchyme, comme le suggère la distribution des récepteurs des oestrogènes dans le rudiment de la glande. D'autre part, la formation de récepteurs des androgènes et des oestrogènes du mésenchyme dépend de l'interaction inductive avec l'épithelium mammaire: seules les quelques cellules mésenchymateuses qui sont les plus rapprochées du bouton mammaire forment des récepteurs et des combinaisons expérimentales ont établi que l'épithelium induit la synthèse de récepteurs dans le mésenchyme. Cette induction est spécifique de l'organe et du stade mais non de l'espèce.

Hormones and cell regulation. Ed J. Nunez *et al.* Colloque INSERM/John Libbey Eurotext Ltd. © 1986. Vol. 139, pp. 15–26.

A growth-associated protein regulated by estrogens and secreted by human mammary cancer cells

H. Rochefort, F. Capony, G. Cavalié-Barthez, M. Chambon, G. Freiss, M. Garcia, M. Morisset, I. Touitou and F. Vignon

Unité d'Encrinologie Cellulaire et Moléculaire (U 148) INSERM, 60 Rue de Navacelles, 34100 Montpellier et Laboratoire de Biologie Cellulaire, Faculté de Médecine, 2 Rue Ecole de Medecine, 34000 Montpellier, France.

The study of estrogen regulated proteins may be valuable in understanding the mechanism by which estrogens stimulate cell proliferation and mammary carcinogenesis. In estrogen receptor positive human breast cancer cell lines (MCF_7, ZR_{75-1}) estrogens specifically increase the production into the culture medium of a 52,000 dalton (52 K) glycoprotein. The antiestrogen tamoxifen and progestins do not themselves induce the 52 K protein, but inhibit its induction by estrogens.

Several high affinity monoclonal antibodies to the partially purified secretory 52 K protein have been produced. This protein has been detected by immunoperoxydase staining in epithelial cells of some breast cancers and proliferative or cystic benign mastopathia but not in normal mammary gland and other estrogen target tissues such as normal or tumoral endometrium. By immunoaffinity, the 52 K protein has been purified to homogeneity in the medium and in the cell extract. The protein is N–glycosylated with high mannose oligosaccharide chains which are phosphorylated **in vivo**. In the cell, the 52 K protein is processed into a 34 K protein moiety which appears to accumulate in the lysosomes. The uptake and processing of the secreted 52 K protein into MCF_7 cells is inhibited by Mannose 6P suggesting that this protein is directed to lysosomes **via** Mannose 6P receptors.

The biological function of the protein has been investigated and we present results in antiestrogen sensitive and resistant cells which support the concept that the 52 K glycoprotein is a mitogen. Other estrogen regulated proteins are also secreted by breast cancer cells and their respective role as autocrine and/or paracrine growth factors are considered.

KEYWORDS

Breast cancer, estrogens, lysosomal enzyme, autocrine mitogen, monoclonal antibodies.

INTRODUCTION

Steroid hormones in breast cancer cells successively trigger the
synthesis of specific proteins and modulate cell proliferation.
Although the mechanism regulating gene expression is being clarified
due to advances in molecular biology, the regulation of cell
proliferation is much less understood. In hormone-dependent cancer
cells of female target tissues (breast, endometrium), the major
physiological steroid mitogens are estrogens. This assumption is
mostly based on clinical observations of the efficacy of treatment
by ovariectomy and by antiestrogens.

In vivo, estrogens can act indirectly via factors released by other
cells (1), but in the last few years, several laboratories have
clearly shown that estradiol can directly stimulate the growth of
estrogen receptor positive breast cancer cell lines in vitro (2-4).
The mechanism of cell growth regulation by estrogens can therefore
be studied in these hormone-responsive cell lines derived from human
metastatic breast cancer (MCF_7, ZR_{75-1}, $T_{47}D$). These cells contain
receptors for estrogen, progesterone and androgens but their growth
is generally only stimulated by estrogens.

We have more particularly studied the proteins that are secreted by
these breast cancer cells and regulated by estrogens, in an attempt
to find better markers of cell proliferation and hormone dependance
than the classical progesterone receptor (5), and also to analyse
the initial steps of estradiol action leading to its mitogenic
activity. These proteins are generally defined according to their
molecular weight under denaturating conditions by SDS-gel
polyacrylamide gel electrophoresis, e.g. 52,000 (6), 28,000 (7) and
60,000 (8).

The estrogen-regulated proteins secreted by hormone-responsive cells
appear to be particularly attractive for several reasons :
First, they may serve as potential circulating markers of hormone
dependency in breast cancer insofar as they are released into the
blood. Second, some of these proteins or peptides released into the
extracellular medium can modulate the growth of the producing cells
(autocrine) (9), or neighbouring cells (paracrine). Therefore, they
may serve as second messengers of steroid hormones for regulating a
nonspecific mitogenic response. Third, these proteins, being less
numerous than the cellular proteins, are easier to detect and assay
(6).

THE 52 K PROTEIN

Six years ago, we found a protein of Mr = 52,000 (52 K protein)
whose production in culture medium by estrogen-receptor-positive
metastatic breast cancer cell lines was specifically increased by
estrogens and inhibited by antiestrogens (6). The 52 K protein is a
glycoprotein produced in small amounts in culture medium (5 ng/10^6
cells/hour) by estrogen-treated MCF_7 cells, representing 20 to 40 %
of all proteins released in the culture medium. This protein is
specifically regulated by hormones (estrogens and high doses of
androgens) that can bind to and activate the estrogen receptor. The
52 K protein is more closely related to the control of cell
proliferation by estrogens than the progesterone receptor, which is
induced by tamoxifen in cells whose growth is inhibited by this

antiestrogen. However, its release into the medium is not stimulated by the other mitogens that have been tested (insulin, epithelial growth factor and charcoal-treated fetal calf serum (10)).

Due to its rarity, we used a three-step strategy to purify the 52 K glycoprotein and study its structure and function. First, using Concanavalin A sepharose, we partially purified it from 22 l of conditioned medium from MCF$_7$ cells. Second, we obtained several monoclonal antibodies (in collaboration with Clin-Midy Research Laboratory, Pr B. Pau). Third, using these monoclonal antibodies on an immunoaffinity column, we purified the 52 K protein to apparent homogeneity (1,000-fold purification) both in its secreted and cellular form (11).

THE MONOCLONAL ANTIBODIES AND THEIR USE AS MARKERS OF CELL PROLIFERATION IN MASTOPATHIES

Following immunisation of mice and thymocyte myeloma fusion, we detected 16 hybridomas producing monoclonal antibodies to the 52 K protein (12). Seven of them were cloned and purified ; they are all of the IgG1 isotype and their KD range from 0.35 to 2.3 nM. The antibodies specifically recognised the secreted 52 K protein and a cellular 52 K protein as evidenced by double immunoprecipitation and by immunoblotting after electrophoretic separation and transfer. Double-determinant immunoradiometric assay indicated that the 7 MAbs recognised three distinct regions of the Mr = 52,000 protein so that the 52 K protein could be assayed in biological fluids such as culture media conditioned by cells, cytosol from breast cancer and mammary cyst fluid (13).

Using the indirect peroxidase-antiperoxidase staining of frozen sections, the 52 K protein was located in the cytoplasm of certain breast cancers. The staining was heterogeneous, mostly located in epithelial cancer cells and absent from normal resting mammary tissue and other estrogen target tissues such as normal or tumoral endometrium, in all periods of the menstrual cycle (14). In a collaborative study with members of the Cancer Center and pathologists in Montpellier on a series of 114 benign mastopathies, we found most of the positive staining of this protein in proliferative ductal mastopathies and mastopathies with cysts (15). There was no staining in non-proliferative lesions or in lobular hyperplasia. This result indicates that the protein is produced when the ductal mammary cells proliferate. It may therefore be used as a marker of high-risk mastopathies, since proliferative lesions have a higher risk than non-proliferative lesions (15). Another particularity of this protein is its tissue distribution, which appears to be mostly specific for ductal mammary cells, as well as sweat and sebaceous glands.

The 52 K protein has not been identified as a known mammary protein. It is not a major milk protein such as human casein or α -lactalbumin, which have different molecular weights. The relationship of the 52 K protein with other estrogen-regulated proteins, such as the 24 K protein (7), the plasminogen activator (16) or the pS2 protein (17) has been excluded.

17

ESTROGEN–REGULATED AUTOCRINE CONTROL OF CELL PROLIFERATION VIA SECRETED PROTEINS

An indirect mechanism of estrogen action in stimulating cell proliferation was first proposed by Sirbasku (18) who anticipated that estrogens might act via endocrine-secreted estromedin, thus explaining why estradiol is more active in vivo than in vitro.

We then proposed an autocrine-type mechanism (19-20) which was supported by Vignon et al. (21) in studies showing that proteins from serum-free media conditioned by estrogen-stimulated MCF_7 cells increase the growth of resting MCF_7 cells. The mitogenic activity was retained by Con A Sepharose chromatography and eluted with the glycoprotein fraction. By contrast, the conditioned media from estrogen-withdrawn MCF_7 cells were inactive (fig.1). This mitogenic activity of the estrogen-induced conditioned media has now been confirmed by other groups(22-23). In these media however, several estrogen-regulated proteins have been described in addition to the 52 K protein (fig.2). There is a 160 K protein (6) and a 65 K protein recently identified as being the α_1 antichymotrypsin (8). Moreover the pS2 protein of 6-10 K coded by the cloned pS2 mRNA (17) and finally EGF-like peptides able to interact with the EGF receptor of MCF_7 cells have also been described by other laboratories (22-24).

It is important now to know which protein(s) is (are) responsible for the mitogenic activity of conditioned media. We have taken two approaches in an attempt to solve this problem.

One approach was the use of antiestrogen-resistant variants of MCF_7 cells, cloned by their ability to grow in 1 µM tamoxifen. We found that with the R27 and RTx6 clones, kindly given to us by Marc Lippman and Francis Bayard, respectively, tamoxifen became able, like an estrogen, to increase the production of the 52 K protein, whereas it remained unable, as in the wild type MCF_7 cells, to stimulate the production of the estrogen-regulated 160 K secreted protein and of pS2 mRNA (25). In these cell lines, the 52 K protein was therefore a better candidate for being a mitogen than the pS2 or 160 K proteins. A possible mechanism for the resistance to antiestrogens of these clones was that tamoxifen became able to induce the production of autocrine growth factors. In fact conditioned media of tamoxifen-treated R27 cells were able to stimulate the growth of estrogen deprived MCF_7 cells, while conditioned media from tamoxifen-treated MCF_7 cells were not mitogenic (Vignon and Freiss, unpublished). These results indicate that tamoxifen may decrease the production of the estrogen-induced growth factor in antiestrogen-sensitive cells and increase this production in some antiestrogen-resistant cells (26).

18

Fig.1 : Growth-stimulating effects of conditioned media

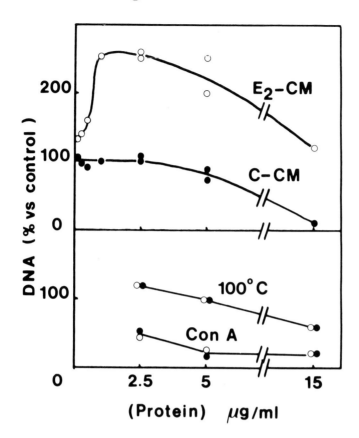

a. Aliquots of conditioned media from control (C–CM) (o) or E_2-treated (E_2-CM) (o) MCF$_7$ cells were added at increasing protein concentrations to recipient MCF$_7$ cells cultured in the presence of 1% FCS/DCC and without insulin. DNA was measured after 10 days of culture on triplicate wells. Results are expressed as percentages of control cells not treated by conditioned media. The effect of estradiol (→) is shown for comparison.

b. The same experiment was performed with C–CM (o) and E_2-CM (o) either preheated at 100° for 3 mn or passed through a Con A Sepharose column (from (21) by permission of the Editor).

Fig.2 : The two types of response to estrogens and their relationship.

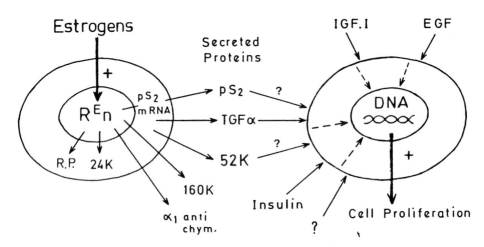

Estrogens induce several specific proteins (left) and stimulate cell proliferation (right). Other growth factors (IGFI, TGF , EGF, insulin) are also mitogens when added in vitro to hormone-responsive cell lines. Increased DNA synthesis and other pleïotypic responses are triggered at the cell surface by the activation of several types of receptors (pleïoreceptor responses). By contrast, estrogen-receptor-specific responses such as the 160 K protein, the progesterone receptor (R.P.), the 24 K protein, and pS2 mRNA, are specifically regulated by activation of the nuclear estrogen receptor (R^En). The best candidates for being (total or partial) mediators of the mitogenic effect of estrogens are the estrogen-induced secreted proteins or peptides. The recipient cells should, however, be able to respond to these signal **via** specific and accessible receptors located at the cell surface. In addition, other growth factors, such as IGF_1, appear to be secreted constitutively but are not estrogen-regulated (22).

The second more direct approach was to purify a biologically active 52 K protein by immunoaffinity and to test it on estrogen-deprived recipient MCF_7 cells (27). The dose-dependent stimulation of cell growth, as evaluated by DNA assay, ranged from 120% to 240%. A mean stimulation of 170% was obtained in the 7 experiments performed. This stimulation represented 40% of the effect obtained by estradiol and was observed at 52 K protein concentrations (1 to 10 nM) similar to those released into the culture medium. The growth stimulation was dose-dependent and not observed with an other control glycoprotein (Fig.3). The 52 K protein was also able like estradiol to stimulate the number and length of microvilli at the cell surface. This mitogenic activity of the purified 52 K protein could be due to a contaminant that we have not yet detected. However we have excluded, by ^{35}S cysteine-labelling experiments, the possibility that it could be peptides containing cysteine residues such as TGFα , pS2 protein or IGFI (27).

Fig.3. Dose-dependent mitogenic effect of purified 52 K protein

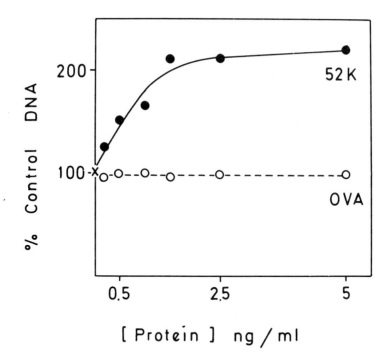

Steroid-deprived MCF$_7$ cells maintained in F12/DEM containing 1% FCS/DCC and 1 mg/ml BSA were either left untreated (X) or treated with the indicated concentrations of purified 52 K protein (52 K-o-) or ovalbumin (ova-o-). After 7 days, DNA was assayed in control and treated cells. Reproduced by permission of the editor (27).

The mitogenic activity of the purified 52 K protein was therefore in agreement with an autocrine mechanism and the next step was to evidence a receptor on the plasma membrane of MCF$_7$ that could specifically bind this secreted protein. The nature of this receptor was not found by classical binding experiments but rather by characterising the co- and post-translational modifications of the protein.

ROUTING TO LYSOSOMES VIA MANNOSE 6 PHOSPHATE RECEPTOR

F. Capony et al. have recently characterised the co and post-translational modifications of the 52 K protein using several types of radio labelling, followed by double immunoprecipitation of cell extracts. A 48 K protein was first N-glycosylated with at least two high-mannose or mixed oligosaccharide chains (28) giving a cellular 52 K protein. Following deglycosylation with endoglycosidase H, an asialo protein of 48 K was still recognised by the antibodies and had the same molecular weight as after Tunicamycin inhibition of glycosylation. Pulse chase experiments indicate that 30% of the cellular 52 K protein was secreted into the medium and 70% was processed into 48 K and a 34 K proteins which were still recognised by the antibodies (11). The 34 K protein (80%

of the total 52 K related proteins in cells) appeared to be mostly accumulated in lysosomes as shown by immuno electron microscopy. The 52 K protein was phosphorylated **in vivo** when MCF$_7$ cells were treated with $|^{32}P|$ H$_3$PO$_4$. However, the majority of the phosphorylation was removed by endoglycosidase H treatment which splits the N-linked high mannose oligosaccharides. Analysis of the detached phosphorylated oligosaccharides indicated that they contained Mannose 6 Phosphate, which was accessible to alkaline phosphatase (29). Moreover, the secreted 52 K protein was able to enter into recipient MCF$_7$ cells and be processed. This was not possible in the presence of Mannose 6 Phosphate, strongly suggesting a competition on Mannose 6 Phosphate receptors located on the plasma membrane. This mechanism has been described in lysosomal hydrolases (30-32) which route **via** intracellular Mannose 6 Phosphate receptor and Golgi vesicles into lysosomes. Under some conditions, these enzymes can be partly secreted and reinternalised through their interactions on Mannose 6 Phosphate receptors located on most of the cell surface. Thus, the 52 K protein can act as an autocrine factor on MCF$_7$ cells since the cells are able a)to produce it, under estrogen stimulation, b)to receive the protein message via Mannose 6 Phosphate receptors located on their cell surface and c)to be growth stimulated by the protein.

We do not yet know the site of action of this mitogen in MCF$_7$ cells and we can not exclude the possibility that the mitogen acts intracellularly even before being secreted or in the culture medium through an enzymatic activity. The mechanism generally described for autocrine growth factors is a stimulation of membrane receptors such as the PDGF receptor on the EGF receptor, leading to a transmembrane activation of a tyrosine protein kinase in the cytosol. In the case of the 52 K protein, the mechanism may be different since no enzymatic activity has been described in the Mannose 6 Phosphate receptor, which appears to function mostly as a carrier protein routing specialised enzymes to lysosomes. The mechanism by which the 52 K protein stimulates cell proliferation is therefore a total mystery. It is unlikely but not totally impossible that the 52 K protein acts by stimulating a trans membrane protein kinase as other growth factors do. It is possible that this protein modulates the action of other growth factors and/or contains an enzymatic activity responsible for its mitogenic effect (such as a lysosomal enzyme) or a growth factor receptor.

It has not yet been demonstrated whether the autocrine mitogenic activity of the 52 K protein observed **in vitro** has any relevance to understanding the effects of estrogens **in vivo**, where other cell types are also putative targets of this protein, such as immuno-competent cells, vascular or connective tissues, etc.. In this respect, since the 52 K protein is secreted, it has the potential to enter, via the Mannose 6 Phosphate receptors, many other cells than mammary cancer cells, and it can therefore act **in vivo** as a paracrine factor in addition to (or instead of) acting as an autocrine factor.

CONCLUSIONS

Considering the recent progress in mammary cancer cell biology, it appears that the studies of estrogen-regulated proteins in breast cancer by endocrinologists are converging with the results obtained by virologists and oncologists, who have described growth factors coded by oncogenes (33-34) and have introduced the concept of autocrine growth factors secreted by cancer cells (39). We propose that during the earlier stages of mammary carcinogenesis and in hormone-dependent breast cancer, estrogens may increase the production of mitogens or growth factors coded or regulated by mammary oncogenes. Although there is some consensus among several laboratories that estrogen-regulated proteins secreted by breast cancer cells are important in regulating their growth (21-24), the nature of the proteins involved is controversial and is currently being studied in these laboratories.

We have reviewed results from our laboratory favouring the description of the 52 K protein as an estrogen-regulated autocrine mitogen. The 52 K glycoprotein, which is secreted under estrogen control in hormone-dependent breast cancer, and constitutively released in other estrogen receptor-negative breast cancers, has been probed by several monoclonal antibodies. Basic studies of the structure and function of the protein indicate that it is a phosphoglycoprotein secreted into the medium, where it can be rebound, internalised, and processed by the same MCF$_7$ cells. The protein is also a mitogen when added to recipient cells that produce little 52 K protein. However, the way by which the 52 K protein stimulates cell proliferation **in vitro** and its exact biological function **in vivo** are not yet understood. The study of the structure of this protein, the cloning of its cDNA and gene, and the selection of antibodies able to prevent its mitogenic activity may provide evidence to clarify these points.

ACKNOWLEDGEMENTS

We are grateful to the CLIN-MIDY/SANOFI Immunodiagnostic Laboratory (B. Pau) which helped us to prepare the monoclonal antibodies, to E. Barrié and M. Egea for their skillful preparation of the manuscript. We thank Drs M. Lippman, M. Rich, I. Keydar, and the Mason Research Insitute, for their gifts of mammary cell lines and Pr P. Chambon for his gift of pS2 cDNA clone. This work was funded by INSERM grant (n°81039), the "Ligue Nationale Française de Lutte contre le Cancer" and the "Association pour le Développement de la Recherche sur le Cancer".

REFERENCES BIBLIOGRAPHIQUES

1. Banbury Report 8 (1981), "Hormones and Breast Cancer", eds. Pike M.C., Siiteri P.K. and Welsch C.W., Cold Spring Harbor Laboratory.
2. Lippman, M.E., Bolan, G. and Huff, K. (1976) : The effects of estrogens and antiestrogens on hormone-responsive human breast cancer in long-term tissue culture. Cancer Res. 36, 4595-4601.
3. Chalbos, D., Vignon, F., Keydar, I. and Rochefort, H. (1982) : Estrogens stimulate cell proliferation and induce secretory proteins in a human breast cancer cell line (T47D). J. Clin. Endocrin. Metab. 55, 276-283.
4. Darbre, P., Yates, J., Curtis, S. and King, R.J.B. (1983) : Effect of estradiol on human breast cancer cells in culture. Cancer Res. 43, 349-354.
5. Mc Guire, W.L. (1980) Steroid hormone receptors in breast cancer treatment strategy. In Recent Progress in Hormone Research, vol. 36, ed Greep R.O., Academic Press, Inc., pp 135-156.

6. Westley, B. and Rochefort, H. (1980) : A secreted glycoprotein induced by estrogen in human breast cancer cell lines. Cell 20, 71-85.
7. Edwards, D.P., Adams, D.J., Savage, N. and McGuire, W.L. (1980) : Estrogen induced synthesis of specific proteins in human breast cancer cells. Biochem. Biophys. Res. Commun. 93, 805-812.
8. Massot, O., Baskevitch, P.P., Capony, F., Garcia, M. and Rochefort, H. (1985) : Estradiol increases the production of α_1 antichymotrypsin in MCF 7 and T47D human breast cancer cell lines. Mol. Cell. Endocrin. 42, 207-214.
9. DeLarco, J.E. and Todaro, G.J. (1978) : Growth factors from murine sarcoma virus-transformed cells. Proc. Natl. Acad. Sci USA 75, 4001-4005.
10. Vignon, F., Capony, F., Chalbos, D., Garcia, M., Veith, F., Westley, B. and Rochefort, H. (1984) : Estrogen regulated 52 K protein and control of cell proliferation of human breast cancer cells. In Hormones and Cancer 2, Progress in Cancer Research and Therapy, vol. 31, eds Gurpide E., Calandra R., Levy C. and Soto R.J., Alan R. Liss, Inc., New York, pp 147-160.
11. Capony, F., Morisset, M., Garcia, M. and Rochefort, H. (1985-86) : Purification and characterization of the 52-kDa protein regulated by estrogen in human breast cancer cells. Submitted for publication.
12. Garcia, M., Capony, F., Derocq, D., Simon, D., Pau, B. and Rochefort, H. (1985) : Monoclonal antibodies to the estrogen-regulated Mr 52,000 glycoprotein : Characterization and immunodetection in MCF7 cells. Cancer Res. 45, 709-716.
13. Rochefort, H., Capony, F., Garcia, M., Morisset, M., Touitou, I, and Vignon, F. (1985) : The 52 K protein : An estrogen regulated marker of cell proliferation in human mammary cells. In Tumors Markers and Their Significance in the Management of Breast Cancer, eds Ip C. et al., Alan R. Liss, Inc., New York, in press.
14. Garcia, M., Salazar-Retana, G., Richer, G., Domergue, J., Capony, F., Pujol, H., Pau, B. and Rochefort, H. (1984) : Immunohistochemical detection of the estrogen-regulated Mr 52,000 protein in primary breast cancers but not in normal breast and uterus. J. Clin. Endocrin. Metab. 59, 564-566.
15. Garcia, M. et al. (1985-86) Specific tissue distribution of the 52 K estrogen-regulated protein in proliferative and cystic benign mastopathies. Submitted for publication.
16. Massot, O., Capony, F., Garcia, M. and Rochefort, H. (1984) : The estrogen-regulated 52 K protein and plasminogen activators released by MCF 7 cells are different. Mol. Cell. Endocrin. 35, 167-175.
17. Chambon, P., Dierich, A., Gaub, M.P., Jakowley, S., Jongstra, J., Krust, A., Lepennec, J.P., Oudet, P. and Reudelhuber, T. (1984) : Promoter elements of genes coding for proteins and modulation of transcription by estrogens and progesterone. In Recent Progress in Hormone Research, vol.40, ed Greep R.O., Academic Press, Inc., pp.1-42.
18. Sirbasku, D.A. and Benson, R.H. (1979) : Estrogen-inducible growth factors that may act as mediators (estromedins) of estrogen-promoted tumor cell growth. In Hormones and Cell Culture, vol. 6, Book A, eds Sato G.H. and Ross R., Cold Spring Harbor Laboratory, pp 477-497.

24

19. Rochefort, H., Coezy, E., Joly, E., Westley, B. and Vignon, F. (1980) : Hormonal control of breast cancer in cell culture. In Hormones and Cancer, vol 14, eds Iacobelli S. et al, Raven Press, New York, pp 21-29.
20. Rochefort, H., Chalbos, D., Capony, F., Garcia, M., Veith, F., Vignon, F. and Westley, B. (1984) : Effect of estrogen in breast cancer cells in culture : Released proteins and control of cell proliferation. in Hormones and Cancer, vol.142, eds Gurpide E. et al., Alan R. Liss, Inc., New York, pp 37-51.
21. Vignon, F., Derocq, D., Chambon, M. and Rochefort, H. (1983) : Les protéines oestrogéno-induites sécrétées par les cellules mammaires cancéreuses humaines MCF 7 stimulent leur proliferation. C. R. Acad. Sci. Paris 296, 151-156.
22. Dickson, R.B., Huff, K.K., Spencer, E.M. and Lippman, M.E. (1985) : Induction of EGF-related polypeptides by 17ß estradiol in MCF 7 human breast cancer cells. Endocrinology, in press.
23. Dembinski, T.C. and Green, C.D. (1984) : Estrogen-induced proteins secreted by ZR 75-1 human breast cancer cells stimulate their proliferation. Lymphokine Res. 3, 84 abst.
24. Salomon, D.S., Zwiebel, J.A., Bano, M., Losonczy, L., Fehnel, P., Kidwell, W.R. (1984) : Presence of transforming growth factors in human breast cancer cells. Cancer Res. 44, 4069-4077.
25. Westley, B., May, F.E.B., Brown, A.M.C., Krust, A., Chambon, P., Lippman, M.E. and Rochefort, H. (1984) : Effects of antiestrogens on the estrogen regulated pS2 RNA, 52-kDa and 160-kDa proteins in MCF 7 cells and two tamoxifen resistant sublines. J. Biol. Chem. 259, 10030-10035.
26. Rochefort, H., Vignon, F., Bardon, S., May, F.E.B. and Westley, B. (1985) : Different efficacy of antiestrogens on estrogen regulated proteins in human breast cancer cell lines. In Estrogen and Antiestrogen Action : Basic and Clinical Aspects, ed Jordan V.C., University of Wisconsin Press, Madison, in press.
27. Vignon, F., Capony, F., Chambon, M., Garcia, M. and Rochefort, H. (1985-86) : Autocrine growth stimulation of the MCF 7 breast cancer cells by the estrogen-regulated 52 K protein. Endocrinology, in press.
28. Touitou, I., Garcia, M., Westley, B., Capony, F. and Rochefort, H. (1985) : Effect of tunicamycin and endoglycosidase H and F on the estrogen regulated 52000-Mr protein secreted by breast cancer cells. Biochimie, in press.
29. Capony, F., Vignon, F. and Rochefort, H. (1985) In preparation/
30. Neufeld, E.G. and Cantz, M.J. (1971) : Ann. N. Y. Acad. Sci. 179, 580.
31. Kaplan, A., Achord, D.I. and Sly, W. (1977) : Phosphohexoxyl components of a lysosomal enzyme are recognized by pinocytosis receptors on human fibroblasts. Proc. Natl. Acad. Sci. USA 74, 2026-2030.
32. Kornfeld, S., Reitman, M.L., Varki, A., Goldberg, D.E. and Gabel, C.A. (1982) : Steps in the phosphorylation of the high mannose oligosaccharides of lysosomal enzymes. In Membrane Recycling, Pitman Books Ltd, London (Ciba Foundation Symposium 92), pp 138-156.
33. Heldin, C.H. and Westermark, B. (1984) : Growth factors : Mechanism of action and relation to oncogenes. Cell 37, 9-20.
34. Bishop, J.M. (1985) : Viral oncogenes. Cell 42, 23-38.

Résumé

Le mécanisme de stimulation de la croissance des cancers du sein par les oestrogènes a été étudié sur des cellules métastatiques de cancers du sein (MCF$_7$, ZR$_{75-1}$) en caractérisant certaines protéines induites par les oestrogènes. Parmi celles-ci, la glycoprotéine 52 K est sécrétée mais son induction est bloquée par les antioestrogènes. Des anticorps monoclonaux ont été développés contre cette protéine qui permettent de la purifier, de la doser dans les extraits cellulaires et les milieux de culture, et de la localiser dans les cellules et divers tissus humains. Elle a été détectée par immunocytochimie dans les tissus mammaires en prolifération ou cancéreux, et dans les glandes sudoripares mais pas dans l'endomètre. L'étude biochimique de ses modifications co- et post-traductionnelles a établi que cette protéine contient au moins 2 chaines riches en mannose porteuses à leur extrémité d'un signal Mannose 6 Phosphate. Ce signal permet le triage de la protéine intracellulaire vers les lysosomes où elle est protéolysée et accumulée en protéines de Mr 48,000 et 34,000. Une partie est sécrétée et peut être réinternalisée par des récepteurs Mannose 6 Phosphate localisés à la surface des cellules MCF$_7$. Après purification, la protéine 52 K peut stimuler la prolifération de cellules MCF$_7$ dépourvues d'oestrogènes à des concentrations nanomolaires, similaires aux concentrations de cette protéine dans les milieux de culture.

Ces résultats indiquent que la protéine 52 K peut agir **in vitro** comme un mitogène autocrine sur les cellules MCF$_7$ et suggère qu'**in vivo**, elle agit comme médiateur de l'action mitogène des oestrogènes. Son action autocrine ou paracrine **in vivo** reste cependant à démontrer de même que le mécanisme de son effet observé en culture de cellules.

Hormones and cell regulation. Ed J. Nunez *et al.* Colloque INSERM/John Libbey Eurotext Ltd. © 1986. Vol. 139, pp. 27–42.

Molecular studies of steroid hormone action by means of gene cloning and transfection

P.D. Darbre, M.J. Page* and R.J.B. King

*Hormone Biochemistry Department, Imperial Cancer Research Fund, Lincoln's Inn Fields, London WC2A 3PX and *Wellcome Research Laboratories, Beckenham, Kent, UK.*

ABSTRACT

The technique of gene transfection is a very powerful method to investigate the biological importance of regions of DNA and has been useful in studies of the regulation of gene expression by steroid hormones. In particular, use of this technique has revealed much about glucocorticoid regulation of the long terminal repeat (LTR) of mouse mammary tumour virus (MMTV). Our own data, however, have demonstrated that androgens as well as glucocorticoids can regulate MMTV RNA accumulation in the S115 mouse mammary tumour cell line. This discrepancy from previous work could be explained on the basis that the MMTV LTR can respond to several classes of steroid if the appropriate receptors are present in the cells. In view of the importance of the LTR in the mechanism of carcinogenesis and tumour growth, it is necessary to know which hormones can regulate its function. Recently, we have used transfection experiments to show that androgens act through the LTR in S115 cells and that, in addition, hormonal regulation is also conferred by the LTR on expression of an adjacent gene. The importance of receptor status of the cell in determining steroid hormone action at the DNA level is discussed. Also, the possibility that hormonal regulation of an LTR can be transmitted to other adjacent promoters by an enhancer insertion mechanism is suggested.

KEYWORDS

Androgen, mouse mammary tumour virus, transfection, steroid-regulated genes, enhancers, glucocorticoid

INTRODUCTION: MECHANISM OF STEROID HORMONE ACTION

a) General outline
It has long been the dream of steroid biochemists to understand the molecular events involved in steroid hormone action both in normal and in cancer cells. Whilst we are still a long way off full comprehension, considerable progress is being made.

Steroid hormones produce their physiological response by regulating the expression of specific genes in target cells. Early events begin with transport of the hormone into the cell and this is followed by binding of the steroid to its own specific receptor (1,2). The steroid-receptor complex then binds to discrete sequences in the DNA in the vicinity of the induced genes (3-13). This receptor-DNA interaction stimulates increased or de novo RNA transcription and then protein synthesis. Steroid hormones regulate many body functions and in some cases this involves simply production and secretion of a specific protein (e.g. progesterone regulates production of chicken egg-white proteins). However, the more complex question of how steroid hormones regulate cell growth remains unsolved. A diagram of a general model for steroid action is shown in Fig. 1.

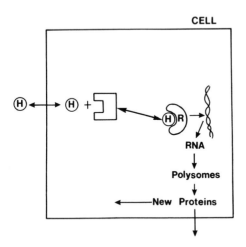

Fig. 1. Mechanism of steroid hormone action

b) Steroid action at the DNA level
Several steroid-regulated genes have now been cloned and their nucleotide sequences have been determined. In some cases, this includes sequences to both the protein coding region and to regions 5' and 3' to the gene. Table 1 lists the steroid-regulated genes for which DNA sequences have been worked out.

Table 1. Steroid-regulated genes for which nucleotide sequences are now known

Gene	Steroid Regulation	Reference
Mouse mammary tumour virus	Glucocorticoid/Androgen	14-20
Lysozyme	Glucocorticoid	21
Metallothionein IIA	Glucocorticoid	22
Rat prostate steroid binding protein	Androgen	23,24
Ovalbumin gene family	Oestrogen/Glucocorticoid/Progesterone	25-28
Rat α2U globulin	Androgen/Glucocorticoid/Oestrogen	29
Growth hormone – human	Glucocorticoid	30,31
– rat		32,33
– bovine		34

Cloning of these genes has allowed more detailed studies on the molecular mechanisms of steroid action at the DNA level. Most of our current understanding rests on the results of experiments involving receptor-DNA binding studies or gene transfection. It is now clear that there are specific regions in the DNA to which steroid-receptor complexes can bind (3-13). Further studies have revealed the existence of multiple binding sites within the mouse mammary tumour virus (MMTV) genes (6-8), the metallothionein IIA promoter sequences (10) and the lysozyme gene (11). Comparisons of the nucleotide sequences required for binding of the different steroid receptors have revealed some structural similarities and indeed consensus sequences for receptor binding have been put forward (13). However, it may well be that the quality of the DNA-receptor interaction also determines its functional consequences and this could be influenced by factors such as DNA methylation or chromatin structure.

Considerable effort is now being made to identify all the transcriptional control elements in the DNA for the steroid-regulated genes. Most effort has been focused on the DNA sequences upstream of the coding sequences, but recent reports of other genes suggest that crucial control elements may also reside downstream of the transcription start site (35-39). The upstream control elements can be broadly classified into three types based on their function, their sequence characteristics and their position relative to the start site of RNA transcription. The first type, the TATA box element (because the sequence is often TATA) fixes the start site of transcription to a point 30 base pairs downstream from its own position. The second type is called the upstream element and encompasses a broad class of sequences at a variable distance from the transcription start site which determine the level of transcription. These elements have been referred to also as G-C rich elements or CCAAT boxes. The third type is the enhancer elements, which are able to stimulate gene transcription from long distances and appear to be very flexible in their requirements for spacing and orientation relative to the start site of transcription. It has been suggested that it is through enhancer-type elements that the steroid hormones exert their effects (40).

c)   Model systems for studying steroid regulation of cell growth
Studies with mammary tumour cell lines have demonstrated that steroids can affect both cell proliferation and cell morphology and thus provide model systems for studying steroid hormone action on cell growth. We have been studying the S115 mouse mammary tumour cell line, which is androgen-responsive. Removal of testosterone from these cells results in a reduced growth rate (41), an increased density regulation (42) and a dramatic change from fibroblastic to epithelial morphology (43). The cells become more sensitive to certain growth stimuli (42) and develop an increased dependence on anchorage to the substrate for growth (44) including loss of ability to grow in suspension culture (43). In addition, marked changes occur in the cell membrane and cytoskeletal structure (45). On the basis of these behavioural differences in vitro, it seems that androgen action in these cells results in a change from a normal (-hormone) to a transformed (+hormone) phenotype. We have thus been searching for any increased products from transforming genes in the presence of hormone and in particular from mouse mammary tumour virus (MMTV) since this virus has a well-known involvement in the development of mammary tumours in the mouse. We have shown that androgens regulate MMTV RNA production in S115 cells (46-48) and that in addition, prolonged culture in the absence of testosterone, which results in loss of proliferative response to androgen and loss of MMTV RNA, results also in increased methylation of MMTV sequences (46, 49). It has been known for a long time that glucocorticoids regulate MMTV RNA production and that such action is through hormone-regulatory regions in the long terminal repeat (LTR) region (5). However, regulation of MMTV RNA by androgen also has raised many questions, beginning with whether the androgen acts directly through the LTR and whether the nucleotide sequences

required are the same as for glucocorticoid action. Models of the mechanism
of carcinogenesis by MMTV stress the importance of the regulatory regions of
the LTR in that they could control expression levels of adjacent oncogenes
(50-53) or other cellular genes (54,55). Given the importance of what compounds
can regulate the LTR function, we have been studying the possibility of multiple
steroid regulation of the LTR by gene transfection experiments using cells
containing the appropriate receptors.

GENE TRANSFECTION IN THE STUDY OF STEROID-REGULATED GENES

The success of nucleotide sequencing of eucaryotic genes has been enormous, but
it cannot lead alone to a direct correlation between particular nucleotide
sequences and defined biological functions. Confirmation of exact nucleotide
sequences required for initiation of transcription, for polyadenylation and for
hormonal regulation of genes needs some sort of functional assay, and one
powerful tool in this analysis has been the technique of gene transfection.

a) Principles of transfection
By this procedure, cloned genes can be introduced into cultured mammalian cells
where they have been shown to be accurately expressed. It is then possible to
introduce changes in the cloned nucleotide sequences by recombinant DNA techniques
and to analyse the consequence of such alterations on expression after transfer,
thereby identifying regulatory sequences in the DNA for each step in gene
expression.

Such transfection experiments can be carried out in one of two ways: gene
expression can be analysed either shortly after transfection when the gene is
expressed transiently, or at a later date after stable integration of the gene
into the host genome. Transient expression assays offer the advantage that
experiments can be performed within one week and, because of the speed, enable
use of primary cell cultures as well as cell lines. However, the disadvantages
are that if transfection efficiencies are low, alterations in gene expression may
be too low to measure, and that experiments are carried out on a whole population
of cells where differing responses in each cell makes interpretation of results
difficult. The use of stable transfection involves much more time and labour,
but does allow repeated studies of hormone effects on clones of transfected cells.

Whichever method is chosen, the gene of interest is inserted first into a DNA
vector. A variety of vectors have been constructed (56,57) all containing plasmid
sequences for growing them in E. coli, certain viral sequences needed to maximise
expression for transient transfection assays and a selectable gene required for
cloning transfected cells in stable transfection experiments. Several selectable
genes have been used. One method involves use of the thymidine kinase (tk) gene,
introducing the DNA into cells lacking the tk gene (tk⁻ cells) and selecting for
transfected cells by growth in medium containing hypoxanthine, aminopterin and
thymidine (HAT) where tk⁻ cells alone could not grow (58). Another method uses
a vector with the gene for xanthine guanine phosphoribosyltransferase (gpt) (59).
This bacterial enzyme converts xanthine into xanthine monophosphate in the purine
biosynthetic pathway. If the mammalian enzyme, which uses hypoxanthine in purine
biosynthesis, is inhibited with mycophenolic acid, only those cells which have
taken up the bacterial gpt gene will survive in HAT medium. Other dominant
selectable markers include dihydrofolate reductase to confer methotrexate
resistance to cells (60), the neo gene to confer G418 resistance to cells (61)
and the hygromycin resistance gene (62).

Methods for introduction of the DNA into mammalian cells include protoplast
fusion (63), DEAE-dextran (64), and co-precipitation with calcium phosphate (58).
This latter method is the one most widely used and the most successful.

Transfection experiments designed to study the mechanism of steroid hormone action always require the use of cells containing the appropriate receptors. Since many cell lines contain glucocorticoid receptors, much work has been done on genes regulated by glucocorticoids, and in particular on the glucocorticoid regulation of mouse mammary tumour virus (MMTV). A review of the use of gene transfection in the study of steroid-responsive genes in general has been written recently (65) and so the rest of this article will discuss more specifically studies on steroid regulation of MMTV.

b) Gene transfection in the study of glucocorticoid regulation of MMTV
Mouse mammary tumour virus is a non-acutely oncogenic retrovirus which replicates specifically in the epithelial tissue of the mammary gland. The DNA proviral form has three genes, the gag, pol and env genes, which are bounded at both ends by the long terminal repeat (LTR) regions (Fig. 2).

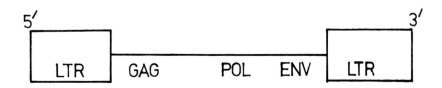

Fig. 2. Structure of proviral DNA of mouse mammary tumour virus

The regulation of transcription of these genes by glucocorticoids is a well-established fact and transfection experiments have played a large part in elucidation of the mechanism involved. Initial experiments showed that cloned MMTV was expressed accurately in transfected cells and that expression was regulated by glucocorticoids (66,67). Further transfections located the site of glucocorticoid action to the LTR region. This was done by ligating the LTR to either dihydrofolate reductase c-DNA (60) or the transforming gene, p21, of Harvey murine sarcoma virus (68) and showing that gene expression from both chimaeric genes was regulated by the synthetic glucocorticoid, dexamethasone. By use of fragments of the LTR, this approach has now been extended to map the hormone responsive regions more precisely (69-72). Current conclusions as to the mechanism of glucocorticoid regulation on the LTR are summarized in Fig. 3.

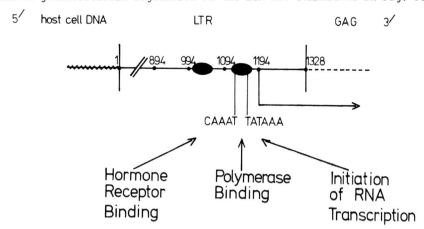

Fig. 3. A schematic model indicating regions of the MMTV LTR important in the control of gene expression

The LTR is 1328 nucleotides long. The initiation of RNA transcription is at
position 1194 with TATAAA and CAAAT sequences 0-100 nucleotides upstream. The
main glucocorticoid-sensitive region is distinct from this promoter region and is
found at a position between 100 and 200 base pairs upstream of the site of
transcription initiation. The hormone regulatory regions appear able to function
irrespective of orientation (73) and this has led to the suggestion that they
resemble enhancer elements in other viruses (40,74).

c) Gene transfection in the study of androgen regulation of MMTV
Our own studies on the S115 mouse mammary tumour cell line have suggested that
MMTV RNA can be regulated by androgens as well as glucocorticoids in these cells,
and thus that the LTR region might be capable of responding to more than one class
of steroid hormone. The reason why multiple steroid regulation had not been
suspected earlier was that the appropriate receptors had never been present in the
cells used. So, we have been trying to confirm whether, given the correct cell
receptors, testosterone can act directly on the LTR of MMTV and to ensure that the
action was not through adjacent host DNA at the site of MMTV DNA integration. To
answer this question, we have used also gene transfection.

We have joined the LTR of MMTV to the coding sequence of a gene not normally
expressed in S115 cells so that the expression of the foreign gene could be
assayed for steroid regulation following transfection into S115 cells. The foreign
gene chosen was the genomic rat C3(1) gene for prostatic steroid binding protein
(23). All 5' sequences were removed from the gene so that the promoter and any
other upstream control elements needed for transcription were provided solely by
the MMTV LTR. The chimaeric sequence was then inserted into the pSV2gpt
transfection vector (Fig. 4).

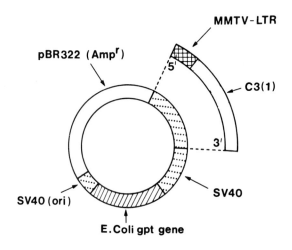

Fig. 4.   Construction of the LTR-C3-pSV2gpt transfection vector to study steroid
hormone regulation of the LTR of mouse mammary tumour virus. The LTR-C3
chimaeric gene was inserted into the pSV2gpt vector DNA at the position indicated

This LTR-C3-pSV2gpt DNA was transfected into S115 cells growing in monolayer
culture by the standard calcium phosphate precipitation method and clones of
transfected cells selected as containing the gpt gene by growth in HAT/myco-
phenolic acid medium. Expression of the C3(1) gene was assayed in several clones
for steroid regulation by measuring C3 RNA on Northern blots of whole cell RNA.
Proliferation of S115 cells is androgen-responsive (41) and the cells contain

receptors for both androgen and glucocorticoid, although not for oestrogen or progesterone, thus providing an assay for possible androgen regulation on the LTR (see Fig. 5).

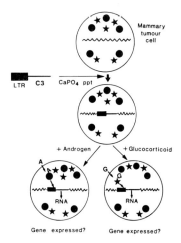

Fig. 5. Principles of gene transfection in the study of steroid hormone regulated genes

We have studied eight different clones of transfected cells in which C3 RNA was expressed from the incorporated DNA. In each case, C3 RNA made in the presence of testosterone was reduced after 6-7 days in the absence of hormone but was recovered again following readdition of either testosterone or dexamethasone for 24 hours. Recovery of C3 RNA was not observed, however, on readdition of either oestrogen or progesterone. This followed the receptor status of the cells since the cells possess only receptors for androgen and glucocorticoid. The data for one clone of cells is shown in Fig. 6.

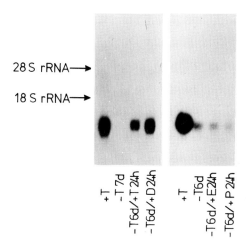

Fig. 6. Steroid hormone regulation of expression of the transfected rat C3(1) gene in S115 cells. Northern blot of C3 RNA from one clone of cells transfected with LTR-C3-pSV2gpt DNA to show regulation by testosterone (T), dexamethasone (D), oestrogen (E) and progesterone (P). (d = days)

Previous studies have shown that androgens do not alter total RNA levels in S115 cells (75) but to confirm the specificity of the steroid effects on the C3 transcripts, similar hormone manipulations have been shown not to affect actin levels in the cells (Fig. 7).

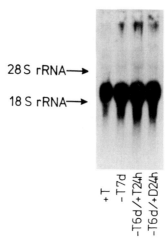

Fig. 7. Steroid hormone regulation of C3 RNA is a specific effect in S115 cells. Northern blot of actin RNA from the same clone of cells as in Fig. 6 showed no regulation by testosterone (T) or dexamethasone (D). (d = days)

The effect of hormone manipulations on levels of RNA within a cell can result either from increased transcription or from stabilization of existing RNA. However, after removal of testosterone for 7 days, actinomycin D blocked the recovery of the C3 RNA normally seen on readdition of testosterone or dexamethasone (data not shown), demonstrating that the effects were on transcription in this case. We then tested the effect of an inhibitor of protein synthesis, cycloheximide, on the recovery of the C3 RNA (see Fig. 8). Cells were pretreated for 1 h with cyclo-heximide (2 μl/ml) such that protein synthesis was inhibited by greater than 75%.

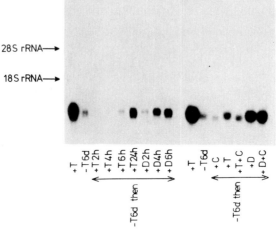

Fig. 8. Kinetics of the steroid regulation of recovery of C3 RNA after removal of hormone and dependence of the process on protein synthesis. Cells were grown for 6 days without hormone followed by readdition of testosterone (T) or dexamethasone (D) with or without cycloheximide (C)

34

Cycloheximide alone (7 h) weakly inhibits the basal levels of C3 RNA in S115 cells, which makes interpretation of these experiments difficult. However, it does appear that cycloheximide partially inhibited the recovery of the C3 RNA (after 6 h) in response to androgen but not to glucocorticoid. It is thus possible that the androgen effects on the LTR are dependent to some degree on protein synthesis and may be less direct than for glucocorticoid. This is further supported by the kinetics of recovery of the C3 RNA which takes longer in the presence of androgen than glucocorticoid (see Fig. 8).

It has recently been suggested that regulatory sequences for eucaryotic genes may reside in intervening sequences and not solely 5' to the gene (35-39). In our first set of experiments, we used a genomic clone of the C3 gene with two intervening sequences present (24) and in vivo this gene is also regulated by androgen (76). To confirm that androgen regulation on the chimaeric gene was due solely to the LTR and not to an intervening sequence in the C3 genomic clone, a second set of analogous transfections of S115 cells was carried out but using in the vector DNA, a c-DNA sequence for human β-interferon rather than the genomic rat C3(1) gene. Construction of this LTR-interferon chimaeric gene is shown in Fig. 9.

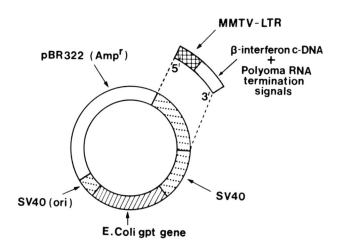

Fig. 9. Construction of the LTR-interferon-pSV2gpt transfection vector. The LTR-interferon chimaeric gene was inserted into the pSV2gpt vector DNA at the position indicated

We studied four clones of transfected cells in which interferon RNA was produced. In each case, interferon RNA was increased if the cells were grown for 24 hours in the presence of either androgen or glucocorticoid. The data for one clone of cells is given in Fig. 10. Thus androgen regulation was also observed for the chimaeric LTR-interferon gene.

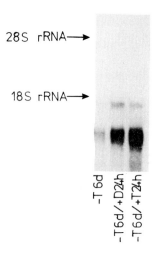

28S rRNA→

18S rRNA→

-T 6d

-T6d/+D24h

-T6d/+T24h

Fig. 10. Regulation of interferon RNA by testosterone (T) and dexamethasone (D) in one clone of S115 cells transfected with LTR-interferon-pSV2gpt DNA. (d = days)

An interesting observation in the transfected clones of cells was found when expression of the cotransfected gpt gene from the vector was studied. Since in these vectors, transcriptional regulation for the gpt gene is provided by SV40 DNA, no hormonal regulation of the gpt transcripts would be expected. However, in clones of cells transfected with LTR-C3-pSV2gpt DNA, the gpt RNA was also regulated by androgen and glucocorticoid. Fig. 11 shows that in these cells removal of hormone for 7 days resulted in loss of gpt RNA which could be recovered after 24 hours by readdition of testosterone or dexamethasone. The regulation is less than for the C3 (Fig. 6) or interferon (Fig. 10) RNA but is nonetheless still there.

To demonstrate that this hormonal regulation on the gpt gene was solely due to the presence of the LTR in the vector and was not merely an internal function of S115 cells, a third set of transfections were carried out but using this time vector pSV2 gpt DNA alone without an inserted chimaeric gene (Figs. 5 and 9). The cells then produced gpt RNA which was no longer affected by hormonal manipulation (Fig. 11). The lack of hormonal control could not be explained by the cells having become unresponsive to steroids since both cell proliferation and MMTV LTR RNA were still regulated by hormones (data not shown).

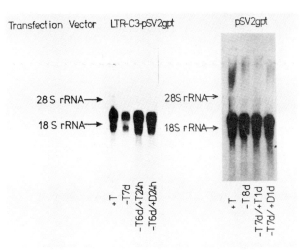

Fig. 11. Steroid hormone regulation of expression of the cotransfected gpt gene in S115 cells. Northern blots of gpt RNA from cells transfected with A) LTR-C3-pSV2gpt DNA, or B) vector pSV2gpt DNA alone to show regulation by testosterone (T) and dexamethasone (D). (d = days)

CONCLUSIONS

From gene transfection experiments, we conclude that the MMTV LTR is capable of responding to multiple steroid regulation. When appropriate receptors are present in the cells, androgens as well as glucocorticoids can act through the LTR of MMTV to increase RNA production from any linked gene (see Fig. 12).

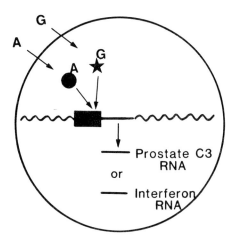

Fig. 12. Conclusions: when appropriate receptors are present in the cells, androgens as well as glucocorticoids can act through the LTR of MMTV

Since S115 cells possess only receptors for androgen and glucocorticoid and not for oestrogen or progesterone, the hormonal regulation of our transfected chimaeric genes follows the receptor status of the cells. The hypothesis that

37

receptor status or even receptor levels are critical in determining the hormonal
regulation of genes in a cell is further supported by recent data suggesting that
the regulatory elements in DNA for different steroid hormones may share some
structural similarities (13). However, in view of the partial blockage of
androgen effects by cycloheximide, it remains in question whether the androgen
action involves direct binding of the steroid receptor or an intermediate protein
to the LTR region. The fact that the response to dexamethasone is faster than the
response to testosterone could be due to lower androgen receptor levels or lower
affinity of the androgen receptor for the LTR region and does not necessarily
imply a separate mode of action. Similarly, the cycloheximide results could be
explained as instability of the androgen receptor itself in the absence of protein
synthesis. It is now of importance to determine whether oestrogen or progesterone
can also act on the LTR if cells are used with the relevant receptors.

Our second conclusion is that the hormonal regulation through the LTR may not be
confined only to the sequences controlled by the LTR promoter but that other
adjacent promoters can also be affected. The hormonal regulation of the gpt
transcripts suggests that the LTR can confer hormonal control not only on its own
promoter (regulating C3 or interferon RNA) but also on the neighbouring gpt gene.
Since SV40 promoter and enhancer sequences are present in the vector to control
transcription of the gpt gene, this suggests that hormonal effects through the
LTR must be able to operate over and above not only the adjacent viral promoter
but also the viral enhancer sequences. It has already been shown that the
glucocorticoid regulatory element in the MMTV LTR can act as an enhancer by virtue
of its flexibility in spacing and orientation relative to the promoter (73). Our
data would imply further that hormonal influences of an LTR can stretch to
neighbouring genes by means of an enhancer insertion mechanism. Such a hypothesis
is consistent with other published data demonstrating effects of the MMTV LTR on
the function of exogenous enhancers (51,70) although it remains to be seen how
far through the genome such effects might operate.

REFERENCES

1. DeSombre ER, Greene GL, King WJ, Jensen EV (1984) Prog Clin Biol Res 142:1-21

2. King RJB (1985) In: Dumont J, Nunez J (eds) Hormones and Cell Regulation,
    Vol.10. John Libbey & Company Limited, London, this volume

3. Govindan MV, Spiess E, Majors J (1982) Proc Natl Acad Sci USA 79:5157-5161

4. Pfahl M (1982) Cell 31:475-482

5. Ringold GM (1983) Curr Top Microbiol & Immunol 106:79-103

6. Scheidereit C, Geisse S, Westphal HM, Beato M (1983) Nature 304:749-752

7. Payvar F, DeFranco D, Firestone GL, Edgar B, Wrange O, Okret S,
    Gustafsson JA, Yamamoto KR (1983) Cell 35:381-392

8. Pfahl M, McGinnis D, Hendricks M, Groner B, Hynes NE (1983) Science
    222:1341-1343

9. Dean DC, Knoll BJ, Riser ME, O'Malley BW (1983) Nature 305:551-554

10. Karin M, Haslinger A, Holtgreve M, Richards RI, Krauter P, Westphal HM,
    Beato M (1984) Nature 308:513-519

11. Renkawitz R, Schutz G, von der Ahe D, Beato M (1984) Cell 37:503-510

12. Jost JP, Seldran M, Geiser M (1984) Proc Natl Acad Sci USA 81:429-433

13. Von der Ahe D, Janich S, Scheidereit C, Renkawitz R, Schutz G, Beato M (1985) Nature 313:706-709

14. Donehower LA, Huang AL, Hager GL (1981) J Virol 37:226-238

15. Fasel N, Pearson K, Buetti E, Diggelmann H (1982) EMBO J 1:3-7

16. Kennedy N, Knedlitschek G, Groner B, Hynes NE, Herrlich P, Michalides R, van Ooyen AJJ (1982) Nature 295:622-624

17. Donehower LA, Fleurdelys B, Hager GL (1983) J Virol 45:941-949

18. Fasel N, Buetti E, Firzlaff J, Pearson K, Diggelmann H (1983) Nucleic Acids Res 11:6943-6955

19. Majors JE, Varmus HE (1983) J Virol 47:495-504

20. Redmond SMS, Dickson C (1983) EMBO J 2:125-131

21. Grez M, Land H, Giesecke K, Schutz G, Jung A, Sippel AE (1981) Cell 25:743-752

22. Karin M, Richards RI (1982) Nature 299:797-802

23. Hurst HC, Parker MG (1983) EMBO J 2:769-774

24. Parker MG, White R, Hurst H, Needham M, Tilly R (1983) J Biol Chem 258:12-15

25. Cochet M, Gannon F, Hen R, Maroteaux L, Perrin F, Chambon P (1979) Nature 282:567-574

26. Gannon F, O'Hare K, Perrin F, LePennec JP, Benoist C, Cochet M, Breathnach R, Royal A, Cami B, Chambon P (1979) Nature 278:428-434

27. Heilig R, Perrin F, Gannon F, Mandel JL, Chambon P (1980) Cell 20:625-637

28. Heilig R, Muraskowski R, Mandel JL (1982) J Mol Biol 156:1-19

29. Kurtz DT, McCullough L, Bishop DK, Manos MM (1982) In: Cold Spring Harbor Symposia on Quantitative Biology, Vol. XLVII, pp 985-988

30. DeNoto FM, Moore DD, Goodman HM (1981) Nucleic Acids Res 9:3719-3730

31. Seeburg PH (1982) DNA 1:239-249

32. Barta A, Richards RI, Baxter JD, Shine J (1981) Proc Natl Acad Sci USA 78:4867-4871

33. Page GS, Smith S, Goodman HM (1981) Nucleic Acids Res 9:2087-2104

34. Gordon DF, Quick DP, Erwin CR, Donelson JE, Maurer RA (1983) Mol Cell Endocrinol 33:81-95

35. Banerji J, Olson L, Schaffner W (1983) Cell 33:729-740

36. Gillies SD, Morrison SL, Oi VT, Tonegawa S (1983) Cell 33:717-728

37.   Schlissel MS, Brown DD (1984) Cell 37:903-913

38.   Wright S, Rosenthal A, Flavell R, Grosveld F (1984) Cell 38:265-273

39.   Charnay P, Treisman R, Melton P, Chao M, Axel R, Maniatis T (1984)
      Cell 38:251-263

40.   Parker M (1983) Nature 304:687-688

41.   King RJB, Cambray GJ, Robinson JH (1976) J Steroid Biochem 7:869-873

42.   Yates J, King RJB (1978) Cancer Res 38:4135-4137

43.   Yates J, King RJB (1981) Cancer Res 41:258-262

44.   Yates J, Couchman JR, King RJB (1980) In: Iacobelli S, King RJB, Lindner HR,
      Lippman ME (eds) Hormones and Cancer I. Raven Press, New York, pp 31-39

45.   Couchman JR, Yates J, King RJB, Badley RA (1981) Cancer Res 41:263-269

46.   Darbre P, Dickson C, Peters G, Page M, Curtis S, King RJB (1983)
      Nature 303 431-433

47.   Darbre PD, Curtis SA, King RJB (1984) In: Bresciani F, King RJB, Lippman ME,
      Namer M, Raynaud JP (eds) Hormones and Cancer II. Raven Press, New York,
      pp 261-268

48.   Darbre P D, Moriarty A, Curtis SA, King RJB (1985) J Steroid Biochem
      23:379-384

49.   Darbre P, King RJB (1984) J Cell Biol 99:1410-1415

50.   Jakobovits EB, Majors JE, Varmus HE (1984) Cell 38:757-765

51.   Ostrowski MC, Huang AL, Kessel M, Wolford RG, Hager GL (1984)
      EMBO J 3:1891-1899

52.   Papkoff J, Ringold GM (1984) J Virol 52:420-430

53.   Stewart TA, Pattengale PK, Leder P (1984) Cell 38:627-637

54.   Dickson C, Smith R, Brooks S, Peters G (1984) Cell 37:529-536

55.   Nusse R, van Ooyen A, Cox D, Fung YKT, Varmus H (1984) Nature 307:131-136

56.   Rigby PWJ (1982) In: Williamson R (ed) Genetic Engineering, Vol. 3.
      Academic Press, New York, pp 83-141

57.   Rigby PWJ (1983) J gen Virol 64:255-266

58.   Wigler M, Pellicer A, Silverstein S, Axel R, Urlaub G, Chasin L (1979)
      Proc Natl Acad Sci USA 76:1373-1376

59.   Mulligan RC, Berg P (1980) Science 209:1422-1427

60.   Lee F, Mulligan P, Berg P, Ringold G (1981) Nature 294:228-232

61. Southern PJ, Berg P (1982) J Mol Appl Genet 1:327-341

62. Gritz L, DAvies J (1983) Gene 25:179-188

63. Rassoulzadegan M, Binetruy B, Cuzin F (1982) Nature 295:257-259

64. Sompayrac LM, Danna KJ (1981) Proc Natl Acad Sci USA 78:7575-7578

65. Parker MG, Page MJ (1984) Mol Cell Endocrinol 34:159-168

66. Buetti E, Diggelmann H (1981) Cell 23:335-345

67. Hynes NE, Kennedy N, Rahmsdorf V, Groner B (1981) Proc Natl Acad Sci USA 78:2038-2042

68. Huang AL, Ostrowski MC, Berard D, Hager GL (1981) Cell 27:245-255

69. Buetti E, Diggelmann H (1983) EMBO J 2:1423-1429

70. Hynes NE, van Ooyen AJJ, Kennedy N, Herrlich P, Ponta M, Groner B (1983) Proc Natl Acad Sci USA 80:3637-3641

71. Majors J, Varmus HE (1983) Proc Natl Acad Sci USA 80:5866-5870

72. Lee F, Hall CV, Ringold GM, Dobson DE, Luh J, Jacob PE (1984) Nucleic Acids Res 12:4191-4206

73. Chandler VL, Maler BA, Yamamoto KR (1983) Cell 33:489-499

74. Khoury G, Gruss P (1983) Cell 33:313-314

75. Jagus R (1979) Expl Cell Res 118:115-125

76. Parker MG, Scrace GT, Mainwaring WIP (1978) Biochem J 170:115-121

Résumé

Bien que l'élucidation des séquences nucléotidiques des gènes régulés par les stéroïdes aient progressé d'une manière impressionnante, nous sommes encore loin de comprendre la régulation de l'expression génique. Une méthode puissante permettant d'évaluer l'importance biologique des régions du DNA est de transfecter des gènes clonés dans des cellules de mammifères et ensuite d'étudier le contrôle des gènes exprimés dans les cellules transfectées.

Le virus de tumeur mammaire de souris (MMTV) est un virus ARN associé avec le cancer du sein chez la souris. Une de ses caractéristiques bien connues est que la transcription de l'ARN à partir de sa forme d'ADN provirale est régulée par les hormones glucocorticoïdes. On a montré, par des expériences de transfection, que le mécanisme de cette induction d'ARN intervient par l'intermédiaire de la longue région terminale répétée (LTR). Elle implique également une interaction directe du complexe récepteur-stéroïde avec des séquences proches du promoteur de la LTR. Cette région régulatrice des glucocorticoïdes peut agir comme un élément hormone - dépendant de type stimulateur ("enhancer"); en outre, si elle est liée aux oncogènes, elle exercera un contrôle de l'expression de l'oncogène à la fois in vitro et in vivo. Etant donné que ces faits pourraient être pertinents en ce qui concerne le mécanisme de la cancérogénèse et de la croissance tumorale il est important de savoir si d'autres hormones peuvent réguler la fonction MMTV-LTR.

Nos données ont démontré que les autres gènes peuvent réguler comme les glucocorticoïdes, l'accumulation d'ARN de MMTV dans la lignée cellulaire S115 de tumeur de sein de souris. Cette différence avec des travaux antérieurs pourrait être expliquée par le fait que la MMTV-LTR peut répondre à plusieurs classes de stéroïdes si les récepteurs appropriés sont présents dans les cellules. Plus récemment nous avons utilisé des expériences de transfection pour montrer que les androgènes agissent par l'intermédiaire de la LTR dans les cellules S115. Des vecteurs de transfection ont été construits dans lesquels la MMTV-LTR a été liée à la séquence du gène non normalement exprimée dans les cellules S115. Après transfection des gènes chimeriques dans les cellules S115 l'expression génique est régulée à la fois par les androgènes et les glucocorticoïdes. De plus la régulation hormonale est aussi conférée par la LTR sur l'expression d'un gène adjacent. L'importance de l'état récepteur de la cellule pour déterminer l'action des hormones stéroïdes au niveau de l'ADN est discutée. De même, la possibilité que la régulation hormonale d'un LTR peut être transmise à un autre promoteur adjacent par un mécanisme d'insertion d'un stimulateur ("enhancer") est suggérée.

Hormones and cell regulation. Ed J. Nunez *et al.* Colloque INSERM/John Libbey Eurotext Ltd. © 1986. Vol. 139, pp. 43–54.

The laboratory procedure of in-vitro fertilization; some observations and conclusions

G.H. Zeilmaker

Department of Physiology, Erasmus University, Rotterdam, The Netherlands

ABSTRACT

A laboratory procedure of in vitro fertilization is described in detail. With this method 60 ongoing pregnancies have been induced in the programme of the Erasmus University at Rotterdam. The observations have led to the following conclusions:
a) Oocytes from undersized follicles can lead to embryo formation and pregnancy.
b) Oocytes can safely be transported in the aspirate from a distant hospital to a central IVF laboratory.
c) The fertilization rate per patient is nearly 90% in the case of tubal pathology. Absence of tubal pathology diminishes the chance of fertilization to about 50%, probably due to male infertility.
d) Screening for male infertility with the hamster test led to the conclusion that fertilization in vitro of the human oocyte and the hamster oocyte are closely correlated.
e) The hamster test is indicated when there is no known cause of infertility and when the first IVF attempt did not lead to fertilization in vitro.

Keywords: In vitro fertilization, hamster test, laboratory procedures.

INTRODUCTION

In vitro fertilization (IVF) has become a well-established method of treatment for infertile couples with a wide range of indications, tubal infertility being by far the most common. Since the first success with the method was reported in 1980 (Edwards et al.) many papers have been published dealing with various aspects of the procedure.
The method consists of the following parts:

        preoperative treatment and monitoring
        oocyte aspiration
        laboratory phase
        embryo transfer

In this paper the emphasis will be on the laboratory procedure as it is employed in our institute. Moreover some observations will be reported and conclusions will be formulated.

Preoperative treatment and monitoring

Patients are treated with HMG, 225 U/day starting on day 3,4 or 5. Follicular development is monitored by ultrasound from day 9 onwards. HCG treatment (10000 IU) is given on the day the largest follicle has a diameter of 18-19 mm and the oestradiol level in plasma is at least 1100 pMol/l. HCG is administered 36 hours after the last HMG injection.

Oocyte aspiration

The operation is performed 34 hours after the HCG injection. The techniques employed are laparoscopic puncture under complete anaesthesia and ultrasound guided follicle aspiration (transvesical and transvaginal). The transvaginal aspiration route is considered to be the method of choice: painless,no sedation required, good visibility of the follicles and a high yield of oocytes.

THE LABORATORY PROCEDURE

This part of the IVF method consists of the following parts:

1. oocyte intake
2. fertilization
3. embryo culture
4. embryo transfer

Oocyte intake

The aspirates arrive from the operation theatre in the adjacent clinic (5´ walking) or from peripheral hospitals (1 hour transport).

The aspirates are kept in an insulated box at 35°C or against the body of the husband in a compartimentalized cotton sac. Upon arrival the vials are placed upon a heated plate (35°C) in a laminar flow unit.

Oocyte recovery

Upon arrival in the laboratory the vials are decanted in a plastic petridish (1) one by one and the search for an oocyte is started with the naked eye. In 90% of the cases this results in the identification of a clear clot of about 1 mm in the slightly bloody aspirate or of a misty clot of the same size in a clear yellow aspirate. Then 2 drops of equilibrated medium are prepared in a petri-dish (150 µ l each).

The clot is taken in a fire polished pasteur-pipette (2) controlled by suction (3) and placed in the first drop of medium and subsequently to the second drop, in order to get rid of the blood cells. Thereafter the cumulus mass is placed in the final equilibrated culture drop which is covered with mineral oil.

If an oocyte is not seen with the naked eye or after it has been picked up with the pipette, a thorough microscopic (5) scanning is done before the aspirate volume is finally measured.

The data and name of the patient are written in the laboratory book and, if possible, the underlying pathology (tubal, endometriosis, antibodies, idiopathic, male factor) is added.

The volume of the aspirate and a note on the aspect of the cumulus is made (radiating, dispersed, dense, with clumps, very small or absent).

The culture medium (3) is maintained in a loosely capped vial at $37^{\circ}C$ and 5% CO_2 and air for one night in the laminar flow unit. With the aid of a ballooned pasteur pipette the above mentioned drops of medium can be prepared after spotting the cumulus mass. For each oocyte spotted fresh drops, are taken from the bottom of the vial in order to minimize pH changes.

The culture drop is prepared 1-3 days before oocyte aspiration.

Equilibrated medium (3) is taken up in a 1 or 2 ml syringe subsequently fitted with a $0,45\mu$ Millipore filter unit (4). Five drops from the filter opening (150μl) are allowed to fall on the bottom of the dish, which is subsequently covered with mineral oil (5). The use of mineral oil has several advantages: protection against evaporation, CO_2 dissipation and contamination. The oil should not be sterilized.

The name of the patient is written on the lid and on the periphery of the bottom of the dish. For each patient 6 dishes are prepared and placed in the incubator for overnight equilibration.

> Problems: blood clotting in the aspirate
> coagulation of the whole aspirate
> blood in the cumulus
> transfer of cumulus mass to clean drop

Blood clots in the aspirate are common and require careful scanning under the microscope whereby an occasional cumulus mass is separated from the clot by means of 2 syringes fitted with a needle. The two needles can act like a pair of scissors and cut through the cumulus and through the blood clot. Two drops of heparin (6) added immediately after aspiration do not inhibit fertilization and development.

Coagulation of the whole aspirate, occurs infrequently and its cause is unknown. Sometimes the cumulus mass can still be spotted and recovered with 2 needles.

Blood in the cumulus cannot be removed by rinsing. In this case the oocyte is transferred to a clean drop 3 hours after sperm is added. By that time most of the cumulus is digested and the oocyte with surrounding cumulus is picked up with a 300μ pipette (7). This procedure is also carried out when too many blood cells are present in the culture drop.

Transfer of the cumulus mass to a clean drop before insemination is risky because the cumulus tends to stick to the bottom of the dish in the presence of fibrin material from coagulated blood. Therefore transfer to a clean drop is only carried out after digestion of the cumulus mass by spermatozoa.

Sperm intake

The husband or partner of the patient arrives in the lab at the same time as the oocyte since he is present at the operation or brings the oocytes from another hospital.

He waites outside the lab until one or more oocytes are found. He is asked to produce an ejaculate in a sterile plastic vial (8). This is allowed to liquefy for 15 minutes. The concentration of living and dead spermatozoa is determined with a Makler counting chamber (9). The volume of the ejaculate and the sperm concentrations (alive and dead) are written in the laboratory book. In order to achieve fertilizing ability the spermatozoa have to be separated from the seminal plasma. This can be done by a swim-up procedure into medium carefully layered upon the ejaculate or by centrifugation and washing with medium.

The ejaculate is transferred to a 15ml tube (10) and mixed with 8 ml equilibrated medium (3) and inspected for hyalinic masses. When not completely free of coarse materials the mixture is filtered through sterile Kleenex in a small funnel into a new tube (with name).

The mixture is centrifuged during 5 minutes at 200g in a clinical centrifuge. The supernatant is carefully decanted and the pellet is mixed with 5 ml equilibrated medium and again centrifuged and decanted. The pellet and some back-flow are placed in the incubator.

> Problems: failure to masturbate
>   liquefaction problems
>   too few spermatozoa

Failure to masturbate, is a problem which can largely be prevented when the husband is requested to produce an ejaculate several weeks (days) earlier during the IVF-intake, in order to verify the sperm parameters. When problems arise at that time, a sample can be frozen before the actual IVF operation. In case there is no sperm sample available on the morning of the operation, husband and wife should be brought together in privacy.

Liquefaction problems are rare and should lead to a preliminary sperm investigation ("IVF without egg") in order to see whether spermatozoa can be isolated and whether survival is good (>24 hrs). Spermatozoa can be obtained from a non-liquefied ejaculate by shaking for some minutes in equilibrated medium.

Oligospermia is not necessarily a problem since good sperm suspensions and pregnancies can be obtained when the ejaculate contains less than 5.10^6 motile spermatozoa.

Insemination of the cultures

Five to six hours after the operation sperm is added to the oocytes. Recent work in Paris has shown that this interval can be shortened safely to two hours.

The tube with the sperm pellet is taken from the incubator and placed in the flow-cabinet. A mouth-blown micropipette of 300 μ diameter (11) is lowered into the tube and some of the misty suspension above the pellet is aspirated. When there is no misty suspension above the pellet the sperm failed to swim-up. This is recorded and the pellet has to be mixed with the supernatant in order to obtain enough spermatozoa.

The dishes with the oocytes are taken from the incubator one by one (to spread the risk of dropping on the floor) and the oocyte is inspected with the microscope. With the mouth-blown pipette 3 "clouds" of spermatozoa are deposited around the oocyte. We do not count the numbers added to the oocyte. The swimming pattern is observed and abnormal behaviour such as extreme lateral movements with the head, "laziness" and impurities are recorded.

When erythrocytes are on the bottom of the dish the oocyte is transferred to a clean culture drop 3 hours after insemination. At that time the adherence of the cumulus to the bottom has disappeared and transfer can be done without risk with a 300μpipette.

Inspection of the IVF cultures

One day after operation the culture can be inspected and the following items can be recorded:
1. survival of sperm
2. sedimentation of the cumulus cells on the bottom of the dish
3. the presence of 2 (or rarely more) pronuclei.

The pronuclei become visible after complete removal of the cumulus cells. This can be done with two injection needles, fitted to a 1 ml syringe. Some very careful dissection should be done here! We prefer to denude the oocyte with the aid of a fire-polished mouth-blown 160μpipette (12). First the oocyte and surrounding cumulus cells are carefully dissected from the bottom, where it may have settled firmly. Then the mouth-blown pipette is lowered into the drop, distant from the oocyte and fluid is allowed to enter by capillary force. Subsequently the pipette is brought near the oocyte and careful suction brings the oocyte complex at the tip of the pipette. Further suction will separate the oocyte from the ovarian cells.

We only inspect for pronuclei in patients with idiopathic infertility or when the transfer time has to be fixed before a weekend. In other cases we leave the culture undisturbed for another 24 or 48 hours and dissect at the 4- or 8-cell stage.

The disadvantage is that the pronuclei are not always seen but the advantage is that the oocyte is not subjected to stress at a vulnerable stage. Opinions may differ about this strategy however.

        Problems: 1. absence of pronuclei
                  2. no sperm motility next morning
                  3. abnormal cumulus reaction next morning

Absence of fertilization can be seen in some or all of the oocytes. In the latter case male infertility is a possibility especially when there are no technical problems (e.g. other cultures doing well) and when the appearance of the oocyte was normal after operation.

In case of suspected male infertility a second IVF attempt should preferably be preceeded by a hamster test. A repeated negative hamster test gives less than 5% fertilization chance in IVF and the use of donor sperm should be recommended for at least half of the oocytes at a second IVF attempt.

<u>Absence of motile sperm the next morning</u> is rare and indicates a male problem which in most cases is associated with fertilization failure.

<u>Abnormal cumulus reaction</u> i.e. the cumulus cells have not settled on the bottom of the dish but have formed a sticky mass around the oocyte. This abnormality does not exclude a good embryonic development. Careful dissection may still yield an embryo.

Inspection of embryonic development

The culture is inspected (again) 48 hrs or 72 hrs after the operation. Separation of the embryo from the surrounding ovarian cells occurs as described in the previous section on fertilization diagnosis. When the oocyte was dissected already for pronucleus inspection, no further manipulation is required.

The observation is recorded and may show:

 On day 2: a) a regular 2- or 4-cell embryo with equally
 sized blastomeres
 b) slightly unequal blastomeres
 c) blastomeres and tiny fragments
 d) irregular blastomere-like fragments with or
 without tiny fragments

 On day 3: a) an embryo with (4,5,6,7 or) 8 blastomeres
 b) blastomeres and fragments
 c) irregular fragments

Embryo transfer can be performed on day 2 or on day 3 with similar pregnancy rate. Transfers due on day 2 are postponed to day 3 to avoid the Sunday and when the most advanced stage on day 2 is the 2-cell stage (9 o´clock), the transfer is postponed one day in order to have more information about the growth of the embryo.

Embryo transfer

On the day of transfer, after dissection of the embryos, up to four of the best looking embryos are transferred to one culture drop.

On the day <u>before</u> transfer 3ml 50% patient serum in culture medium is put in the incubator for equilibration.

When the time of transfer has arrived
1) a 2ml syringe is filled with the 50% patient serum, fitted with a

Millipore filter and a capped needle
2) all embryos to be transferred are introduced into a 300 μ pipette
   (13) and placed in a polyurethane foam box for thermal insulation.
   In this capillary the embryos are safe since CO_2 cannot escape
   from the medium.

The embryos are transported to the transfer site in this container
together with
   a 2 ml syringe with 50% serum,
   a 1ml sealed syringe,
   a tube with 5 ml equilibrated medium,
   and the tubing used for blowing the pipette.

At the transfer site (at a distance of 10 minutes from the lab) the transfer
catheter (14) is connected to the 1 ml syringe filled with medium. The syringe
is emptied through the sterile catheter to rinse it. A drop of 50% patient
serum is filtered into a depression slide (12) and the catheter (keep end
sterile) is filled with a column of 5 cm 50% serum.

In a second fresh drop of serum, prepared in order to minimize pH changes, the
embryos are collected by blowing the embryo pipette carefully into the serum
under visual control with the inverted microscope. The embryos tend to float
but this can be prevented by sucking them back and forth into the pipette with
surrounding serum.

When all embryos are on the bottom they are taken into a 3 cm fluid column in
the catheter between 2 small air bubbles followed by 2 cm 50% serum (as safety
plug). Next comes the go-ahead signal from the gynaecologist.

The gynaecologist inserts the catheter guide into the cervical opening and the
embryologist inserts the catheter into the opening of the catheter guide taking
great care not to touch the sterile end of the catheter.

The final position of the catheter tip, 0,5 cm from the fundus is achieved by
inserting the catheter until a marking reaches the opening of the guide. The
catheter guide, in its turn is inserted into the cervical opening until a ring
prevents further entrance.

Some observations made in the IVF-laboratory

The results obtained with this procedure in 1985 are mentioned in Table 1

Table 1

IVF Results 1985 Erasmus University

	Pat.oper.	Pregn. (%)	Ongoing preg (%)
Dijkzigt + tubal pathology	159	27 (17%)	20 (12%)
- tubal pathology	31	2 +4D[1](6%)	2 + 4D(6-19%)
Egg transport:			
+ tubal pathology	58	9 (16%)	7 (12%)
- tubal pathology	18	1 (5%)	1 (5%)
	266	43 (16%)	33 (12%)
Since October 1985:			
Dijkzigt	66	16 (24%)	13 (20)
egg transport	32	7 (22%)	5 (16%)

[1]) 4 patients pregnant after IVF with donor sperm

Since the beginning of the programme 60 children have been born or are soon to be born (1 triplet, 7 twins)

This table indicates that in patients without tubal pathology the success rate is lower. From the data presented in Table 2 it appears clearly that this lower pregnancy rate is due to a diminished fertilization rate in this category of patients. It can furthermore be seen that oocyte transport from peripheral hospitals (up to one hour of transportation) does not appreciably influence the IVF results. In all cases follicular volumes were recorded and it was found that two pregnancies were obtained with one single embryo each, originating from a follicle with a volume of 1 ml or less (Zeilmaker et al. 1984).

50

Table 2

Effect of tubal pathology on IVF results (Erasmus University 1985)

	No. patients	No. Patients with at least one fertilized oocyte
with tubal pathology	217	195 (90%)
without tubal pathology	49	25 (51%)

The hamster test was applied in a series of IVF cases, in order to see whether a correlation could be found between the two procedures. The results are presented in Table 3.

Table 3

Correlation between hamster test and IVF-test (1985 results)

Patient category	No. of positive hamster tests/total
IVF positive	47/48
IVF negative	28/42
Andrology Clinic	13/30
Young fathers (1983-1985)	72/75

Two fathers negative in two repeated hamster tests. Fatherhood was not investigated.

The procedure of the test is described in Zeilmaker & Verhamme 1986.

The following conclusions are drawn from these observations:

1. After a negative IVF result (no fertilization) a hamster test should be performed in order to identify possible male infertility.
2. When the hamster test is negative (preferably twice) the use of donor sperm for at least some of the oocytes should be proposed to the couple, well in advance of the second IVF attempt.
3. The hamster test should be performed before an IVF attempt is made in couples with idiopathic infertility, regardless of the sperm concentrations in the ejaculate.

These observations show that there is a definite place for the hamster test in fertility diagnosis, particularly in the IVF laboratory. This conclusion confirms earlier observations of Margalioth et al. (1983); Junca et al.(1982) and Foreman et al. (1984). Good correlations with clinical fertility have also been described by several authors. It is clear from the data that a positive test gives no information about infertility. A negative test, preferably repeated, indicates that there is a severe fertility problem.

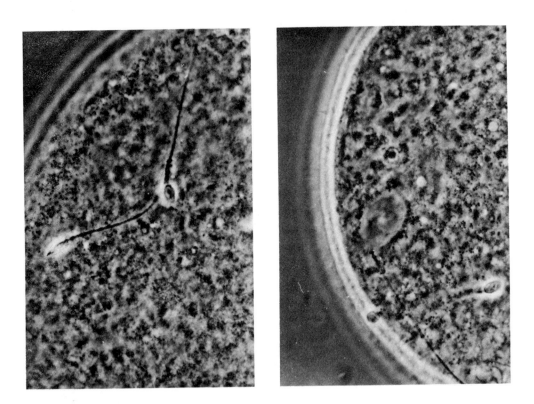

Fig. 1 (L)
Zona free hamster oocyte with two intact human spermatozoa
adhered to the egg surface. The diameter of the hamster egg is
about 100μ.

Fig. 2 (R)
Zona-free hamster oocyte with decondensed human sperm (3 hours after
insemination, 18 hours of preincubation, in order to induce acrosome
reaction).

1. Corning dish, diameter 6 cm.
2. A pasteur pipette (long model) is fire-polished in a hot gas flame. This procedure should be verified by microscopic inspection, the lumen should not be narrowed and the end should be completely smooth.
3. The pipet is connected to a piece of silicone tubing and the movement of fluid and the cumulus can be controlled by gentle suction and blowing of the tubing. Before picking up the oocyte, medium should be allowed to enter by capillary force spotaneously.
4. Inverted Olympus microscope (CK) objective 4x, oculars 10x.
5. Squibb mineral oil (intestinal lubricant), equilibrated in the incubator but not sterilized.
6. Two drops of heparin (Thromboliquin, Organon, 500 IU/ml) are added to the follicular aspirate in the clinic when it appears very bloody.
7. The drawn end of a pasteur pipet is removed and a 10 cm piece of capillary glass tubing is mounted in the shaft of the pipet with laquer. The capillary is hand made from soft soda-glass tubing or with a capillary drawing machine (Shimadzu Seisakusho Ltd. Kyoto). The diameters in use in this laboratory are:
   160μ for denudation of zygotes and embryos.
   200μ for routine handling of embryos
   300μ for insemination, transfer to other culture
          drops and transport purposes.
   The end of these capillaries is broken straight and fire-polished with a micro-flame. The lumen is preserved, the edge is made entirely smooth.
8. A plastic beaker with lid of 125 ml (Greiner) sterilized by radiation.
9. Makler counting chamber (The Israel Electro-Optical Industry Ltd., P.O. Box 1165, Rehovot 76110, Israel).
10. The culture medium has the following composition:

         17 parts Earle (modified)
         3 parts Ham F10 modified
         3 parts inactivated patient serum.

250 ml Earle solution: 1690 mg NaCl; 100 mg KCl; 50 mg Mg SO_4.$7H_2O$;
                       504 mg $NaHCO_3$; 39,5 mg NaH_2PO_4.
                       $2H_2O$; 250 mg glucose; 5 mg pyruvate;
                       160 mg Ca-L-Lactate; penicillin 3U/ml/streptomycin
                       $3\mu g$/ml
                       Osmolarity 280-285 mOsm, pH equilibrated and
                       cooled to 20°C: 7,42.

250 ml Ham F10 (1,2 g $NaHCO_3$/liter, Flow powder) with the addition
               of 49 mg Ca-L-Lactate and penicillin 3U/ml/streptomycin 3 g/ml
               and osmolarity adjusted to 285 mOsm.

Solutions are prepared weekly. Ca-lactate is added last.

11. The dimensions of the catheter are:
    The guide: Portex flexible grade Nylon no. 11 (800/200/250).
               Inner diameter 1,5 mm. Outer diameter 2,1 mm.
               Length: 20 cm.
    Inner catheter: Portex Stand grade Nylon no. 6 (800/200/175).

Inner diameter 0,75 mm. Outer diameter 0,94 mm.
Length: 58 cm, mounted on needle base.
12. A glass plate (L 7,5 cm x W 2,5 cm), thick 5 mm with 3 depressions
    (diameter 22 mm, depth 4 mm), custom made.

REFERENCES

Edwards, R.G.; Steptoe, P.C., Purdy, J.M. (1980): Establishing full
    term human pregnancies using cleaving embryos in vitro.
    Brit. J. Obstet. Gynec. 87, 737-756.

Foreman, R., Cohen, J., Fehilly, C.B., Fishel,S. Edwards, R.G.
    (1984): The application of the zona-free hamster egg test for the
    prognosis of human in vitro fertilization. J. in Vitro Fert. Embryo
    Transfer 1, 166-170.

Junca, A.M.; Mandelbaum, J., Plachot, M. & de Grouchy (1982):
    Evaluation· de la Fecondite du sperme humain par la fecondation in
    vitro interspecifique (homme-hamster). Annales de Genetique 25, 92-95.

Margalioth, E.J., Navot, D., Laufer, N., Yosef, S.M., Rabinowitz, R.,
    Yarkoni, S. Schenker, J.G.(1983): Zona-free hamster ovum
    penetration assay as a screening procedure for in vitro fertilization.
    Fertil. Steril. 40, 386-388.

Plachot, M., Mandelbaum, J., Cohen, J., Salat-Baroux, J. Junca, A.M.
    (1984) In: Recent progress in human in vitro fertilization,
    Ed. W. Feichtinger & Kemeter,P. p.249-252.

Zeilmaker, G.H. & Rijkmans, C.M.P.M.(1986): Experience with the
    hamstertest in an IVF laboratory. Int. J. Androl. Suppl. WHO symposion
    on evaluation of the hamster test. (in press)

Zeilmaker, G.H., Alberda, A.T., Gent van I., Rijkmans, C.M.P.M.&
    Drogendyk, A.(1984): Two pregnancies following transfer of intact
    frozen-thawed embryos. Fertil. Steril. 42, 293-296.

Résumé

Une méthode de fertilisation in vitro est décrite en détail. Avec
cette méthode 60 grossesses continues ont été induites dans le programme
de l'Université Erasme à Rotterdam. Ces observations ont conduit aux
conclusions suivantes:
a) Les oocytes des follicules de taille inférieure peuvent conduire à la
   formation d'enbryon et à la grossesse.
b) Les oocytes peuvent être facilement transportés dans le milieu
   d'aspiration d'un laboratoire lointain jusqu'au laboratoire central
   d'IVF.
c) Le taux de fertilisation par patient est proche de 90% dans les cas de
   pathologie tubaire. L'absence de pathologie tubaire diminue les
   chances de fertilisation jusqu'à 50%, et cela est dû probablement à
   une infertilité mâle.
d) Le criblage de l'infertilité mâle avec le test du hamster conduit à la
   conclusion que la fertilisation in vitro de l'oocyte humain et de
   l'oocyte de hamster sont étroitement corrélées.
e) Le test du hamster est indiqué lorsqu'il n'existe pas de cause connue
   d'infertilité et quand la première tentative d'IVF n'a pas conduit à
   la fertilisation in vitro.

Hormones and cell regulation. Ed J. Nunez *et al.* Colloque INSERM/John Libbey Eurotext Ltd. © 1986. Vol. 139, pp. 55–67.

An overview of the steroid receptor machinery

R.J.B. King

Hormone Biochemistry Department, Imperial Cancer Research Fund, PO Box 123, Lincoln's Inn Fields, London WC2A 3PX, UK.

ABSTRACT

The structure and intracellular loci of steroid receptors are reviewed in the light of recent data derived in large part from work with monoclonal antibodies against glucocorticoid, progesterone and oestradiol receptors and with gene probes for the glucocorticoid and oestradiol receptors. It now seems possible that most unoccupied receptor may be in the nucleus and not the cytosol as previously thought. The biological relevance of data obtained with cytosol receptors is discussed in relation to this suggested nuclear location.

Results on the characterization of mRNA's for the glucocorticoid and oestradiol receptors have markedly increased our knowledge of the molecular aspects of receptor structure particularly in relation to the mechanisms by which lymphoma cells become resistant to glucocorticoids. The cloning of both structural and regulatory regions of several steroid-sensitive genes and transfection experiments with chaemeric genes are providing new insights into the ways by which steroid receptors influence gene function.

KEYWORDS

Steroid receptors, review, monoclonal antibodies, steroid insensitivity, DNA binding

INTRODUCTION

Steroid hormones are simple molecules and yet they influence many cell functions in different ways. Thus, physiological levels of androgen (male sex hormones) induce proliferation of epithelial cells in the prostate gland but not in uterus whereas oestrogens (female sex hormones) can have the converse specificity; neither type of sex hormone changes proliferation of stomach epithelium. Within one class of hormones, a spectrum of responses can be elicited depending on the cell type. For example, oestrogens increase egg white proteins in chick oviduct and frog liver but control neurone development in certain regions of the rat hypothalamus. How do these simple molecules elicit such complex responses? Most, but not all of the answers lie in the specificities of the intracellular receptor machinery.

The synthesis of high specific activity tritiated steroids and related compounds
in the early 1960's enabled the fate of physiological concentrations of these
hormones to be studied. Target cells could retain hormone, a process that
unresponsive cells lacked. This led to the identification of receptor proteins,
the specificities of which explained some of the biological phenomena outlined
above. The main feature not explained by receptor specificity is the different
responses induced by a single steroid. This is probably due to specificity of
chromatin acceptor sites within the chromatin.

The identification of receptor proteins led to the two stage model of intracellular binding in which the steroid first formed a non-covalent attachment to
a cytosol receptor protein which then translocated to the nucleus. A great deal
of biochemical analysis has been conducted on all the facets of this process
(early literature reviewed in 1). This article will consider these data in the
light of new technologies that suggest revision of some of our earlier ideas.

RECEPTOR MODELS

A considerable body of information has been published on steroid receptor
structure and function that has given rise to the general model presented in
Fig. 1A. Steroid, on entering a cell containing the appropriate receptor, forms
a non-covalent complex in the cytoplasm. After a series of ill-defined changes
termed activation, the steroid receptor complex binds tightly to specific
chromatin acceptor sites and initiates the response(s) characteristic of that
steroid and cell. Little is known of the subsequent fate of the receptor and
the term processing has been used to describe this area of ignorance. With
certain differences mentioned below, this model is applicable to oestrogen
(2,3,4), androgen (5), progestin (6,7) and glucocorticoids (8,9). The data on
which this model is based have predominantly been derived from techniques of
tissue homogenization and fractionation of relevant macromolecules. More limited
information has been obtained with autoradiography after labelling with tritiated
steroids. The latter reagents have been the only means of studying receptors so
that, in effect, only ligand binding sites have been investigated and it has long
been realized that alternative methods were required. With the advent of
monoclonal antibodies against these proteins, such alternative methods became
feasible and produced data that conflicted with the model shown in Fig. 1A and
which were more compatible with that presented in Fig. 1B.

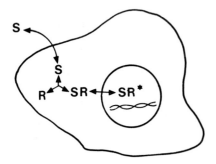 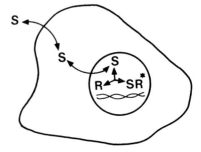

Fig. 1. Models of intracellular localization of steroid receptors.
    A. Classical model   B. Revised model
    S - steroid;   R - receptor;   R* - activated receptor

In this alternative picture, unliganded receptor is in the nucleus before it binds steroid. The revised nuclear model is predominantly based on the immunohistochemical localization within the nucleus of ligand-free oestradiol (10,11), progesterone (12,13) and glucocorticoid (14,15) receptor. Supportive evidence has been presented for oestradiol receptor (RE) by cytochalasin enucleation of cultured cells (16). Proponents of the revised model explain the earlier data as being due to artefactual release of unliganded receptor from the nucleus into the soluble fraction during homogenization. The earlier autoradiographic results, showing cytoplasmic localization with time- and temperature-dependent nuclear transfer, could be explained either by leakage of nuclear receptor during the preparation of the frozen sections or to the non-attainment of equilibrium conditions (17,18).

The revised model must receive serious consideration, but it would be premature to accept its complete veracity particularly with respect to the absence of cytoplasmic receptor. Receptor is presumably synthesized in the cytoplasm and the inability to detect extranuclear receptor by immunohisto-chemistry in some but not all experiments may reflect either the relative insensitivity of the technique or artefactural loss of epitope.

'8S' CYTOSOL RECEPTOR AND RECEPTOR ACTIVATION

If one accepts the possibility that a major fraction of free receptor is within the nucleus one is led to question the biological relevance of data obtained on cytosol receptor after cell disruption.

Since its first identification in the oestradiol/rat uterus system, the '8S' receptor has been a cornerstone of respectability with all classes of steroid as has its salt dissociation into smaller subunits: despite 20 years effort, its structure remains uncertain. From the points made above, the relevance of this structure is also in doubt. Furthermore, if one questions the biological significance of the 8S aggregate, then the literature dealing with activation must also be discussed, activation being defined as the ability to bind to DNA. In essence, the data show that treatments as diverse as salt, heat or cell fractions convert receptor into a form with increased affinity for DNA or chromatin; sodium molybdate is inhibitory. The data on activation are strongest for oestradiol (2,3) and glucocorticoid receptor (19) but are less so for the other classes of steroid and one major group do not include an activation step in their model of progesterone action (6,20).

In simplistic terms, a number of scenarios can be envisaged (Fig. 2). In this cartoon, the receptor is depicted as having dissimilar subunits but this is unknown. Probably the commonest format says that the basic 8S form is activated by a multistep process (Fig. 2.1). Thus for RE, dissociation to a 4S (65K) and subsequent dimerisation to a 5S (130K) form has been proposed (21). Molybdate inhibits activation by maintaining the 8S structure (22). Alternatively, if the biologically relevant receptor is in the nucleus, its release into the soluble fraction during homogenization may result in artefactual interaction with macromolecules with which it never usually comes in contact (Fig. 2.2). It is known that steroid receptors are capable of complexing with a wide range of molecules. This scheme assumes an essential requirement for an activation step.

A third possibility is that the simple act of a steroid binding to the
unoccupied nuclear receptor, induces conformational changes that result in
higher affinity for DNA (Fig.2.3); no additional changes may be required.
A good example that some molecular changes are required comes from the
experiments showing changes in ionic properties of the glucocorticoid
receptor within minutes of intact cell exposure to steroid (23). According
to this model, the activation studies with salt, temperature, molybdate,
etc. simply represent changes in entities that have little meaning outside
of the test-tube. A final consideration concerns the physiological relevance
of the putative soluble component (Fig. 2.4). The only data on this point
concerns the 90 Kd protein that is present in all classes of steroid receptor
aggregate ('8S' form) but whose function is unknown (see below).

I would suggest that none of the schemes outlined in Fig. 2 can be said to be
wrong and that some of the more unconventional models require serious
consideration.

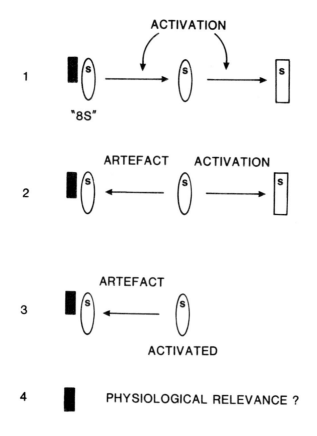

Fig. 2. Models of receptor structure

Three techniques that are increasingly being applied to receptor analysis are affinity labelling (24), immunodetection (25) and the use of dense amino acids to follow protein turnover (26). In the case of RE, the three techniques have been combined to provide evidence in favour of a homodimer structure for nuclear RE (21) whilst the antibody data for RP (27) throws some doubt on the idea of dissimilar A and B subunits (20,28). A simplistic explanation for the close similarity of physical properties of soluble and nuclear RE (Table 1) would be that the two are identical but this must be tempered by the fact that the detergent, sodium dodecyl sulphate, has been used in the gel work which would dissociate subunit structure. However, the similar half-lives of the soluble and nuclear proteins would not have been anticipated if heterosubunit structures were important. Nevertheless, it should be stressed that none of the data provide unequivocal support for one particular model.

Table 1. Properties of cytosol and nuclear oestradiol receptor

Property	Assay method	Cytoplasm	Nucleus	Reference
Half-life	Dense amino acids	4 h	3 h	26
	Affinity label	–	4 h	22
Molecular weight	Affinity label	63 Kd	63 Kd	22
	Antibody	65 Kd	65 Kd	21
	Denature/renature	65 Kd	65 Kd	29
Isoelectric point	Affinity label	5.7	5.7	22

COMPONENTS THAT DO NOT BIND STEROID

These fall into two categories: those that can be converted from non-binding to binding forms and non-binding units that may be integral parts of the receptor machinery.

Interconversion of steroid binding and non-binding forms is receiving current attention particularly in relation to the topic of receptor phosphorylation. This was first described for glucocorticoid receptor (RG) in which phosphorylation of soluble receptor increased glucocorticoid binding: molybdate as a putative phosphatase inhibitor protected against this loss. The molybdate effect is seen to varying degrees with all classes of steroid receptor but it is doubtful if all its effects are mediated by phosphatase inhibition. It has been suggested that molybdate affects RG via interaction with sulphydryl groups (30).

For RE, dephosphorylation in the nucleus has been suggested as a factor in nuclear processing which can be reversed by a calmodulin-dependent protein kinase (31,32). A somewhat different role for phosphorylation has been described for the RE/chick oviduct system involving the interconversion of high and low affinity, oestradiol binding forms with a non-binding component (33).

59

Clearly, the possibility of interconversion of steroid binding and non-binding forms is pertinent to the topic of receptor processing (see above) and further information will be awaited with interest.

Monoclonal antibodies against the steroid binding unit of the glucocorticoid receptor will recognize a similar sized protein that does not bind ligand in receptor-minus and nuclear transfer-increased lymphoma cells. These are clearly variants of the ligand binding component of RG (see below). Other examples of monoclonal antibodies, supposedly raised against receptor proteins, have been described that recognize epitopes on proteins that have little if any homology with the hormone binding units (Table 2).

Table 2. Antibodies against components that do not bind steroids

Immunogen	Size of antigen	Other receptors	Non-receptor cells	Reference
Oviduct RP	90 Kd	Yes	Yes	34
	108 Kd	Yes	Yes	35, 36
Liver RG	94 Kd	No	?	37, 38
Myometrium RE	29 Kd	No	No	39, 40

The antigens recognized by these antibodies are in considerable excess of receptor protein and the 90 Kd and 108 Kd 'RP' antigens are present in receptor negative as well as positive cells. It has been suggested that they may be heat shock proteins. On the other hand, the 29 Kd antigen recognized by the 'RE' antibody shows good qualitative and quantitative relationship to the oestradiol binding unit although the former is exclusively located in the cytoplasm (40,41).

The place of the two 'RP' and one 'RE' proteins in our understanding of the receptor machinery remains to be established.

RECEPTOR GENES

Antibodies against receptors have been used to isolate mRNA and thereby DNA clones for both the glucocorticoid and oestradiol receptors. Two independent isolates of the glucocorticoid receptor gene have been made, one utilising monoclonal (37) and the other a polyclonal antiserum (42). Both monoclonal antibodies and protein sequence data were employed to obtain the oestradiol receptor gene (43). The results are pictorially represented in Fig. 3. Although the receptor protein for oestradiol has a lower molecular weight than that for glucocorticoids, their mRNA's are of comparable size. If the entire mRNA sequence were translated, a protein of about 300 Kd would be produced which is clearly not the case. Hence, these two genes must be complex with the mRNA's having extensive intervening sequences.

GLUCOCORTICOID RECEPTOR

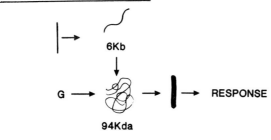

6Kb

G → → RESPONSE

94Kda

OESTRADIOL RECEPTOR

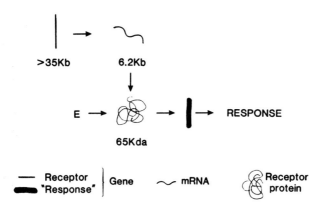

>35Kb   6.2Kb

E → → RESPONSE

65Kda

—— Receptor | Gene   ∿ mRNA   Receptor
■■ "Response" | protein

Fig. 3.   Cartoon of pathway of receptor synthesis
          G - glucocorticoid; E - oestrogen. Data derived from refs. 37, 38, 43

MOLECULAR BASIS OF STEROID RESISTANCE

Steroid-sensitive tumours frequently progress to the insensitive state. Work
predominantly with glucocorticoid-sensitive lymphoma cells, has identified three
types of receptor change from the wild type (wt), determined by ligand binding
that results in the insensitive phenotype (44). Most common is the r⁻ phenotype
in which ligand binding is lost but nt⁻ and nt^i cells can also be identified in
which the glucocorticoid can bind to the receptor but nuclear transfer is either
decreased (nt⁻) or increased (nt^i). A fourth category has been described in
which the receptor machinery appears to be normal and in which a post-receptory
defect has been suggested. A combination of immunoprecipitation (37,38,45),
mRNA (37) analysis and the investigation of receptor-DNA interactions (46,47,48),
has been used to identify the molecular basis of these receptor defects (Table 3).
A number of interesting points can be made about these data. The r⁻ cells
contain reduced levels of normal-sized protein and mRNA although the former does
not bind glucocorticoid. There is no method of monitoring DNA binding in the
absence of ^{3}H steroid binding but even if it does occur it must be devoid of
biological effect as the cells grow; the normal response to receptor-specific
DNA interaction in these cells is death.

61

The nt⁻ cells are qualitatively and quantitatively similar to the wild type cells with the exception of decreased affinity for specific DNA. The nt^i cells exhibit a complex phenotype with some of the features of r⁻ cells (94 Kda, non-steroid binding protein) but with a smaller protein that does bind glucocorticoids. The increased, but biologically inactive, DNA binding is due to elevated binding to non-specific regions of the DNA. These mouse cells are functionally diploid and it has been proposed that one allele has a point mutation resulting in loss of ligand binding, whilst the second allele has undergone a frame shift mutation, thereby producing a truncated form of receptor that will still bind glucocorticoids (38,45). Additional examples of glucocorticoid insensitivity also exist (49).

Other examples of receptor modifications resulting in insensitive cells have been partially described for androgens (50), and antioestrogens (51).

Table 3. Receptor phenotypes of glucocorticoid-sensitive and -resistant lymphoma variant cell lines

Class[+]	Receptor			Receptor binding to DNA	
	mRNA (Kb)	Protein (Kd)	Steroid binding	Specific	Non-specific
wt	6	94	Yes	+	±
r⁻	6*	94*	No	?	?
nt⁻	6	94	Yes	±	±
nt^i	6*	94*	No	?	?
	5	40	Yes	±	++

+ see text for definitions; * reduced amounts

For greater detail, consult refs. 37, 38, 45

RECEPTOR BINDING TO DNA AND CHROMATIN

Genetic engineering techniques have allowed the isolation of specific fragments of several hormone sensitive genes which has facilitated studies on receptor DNA interactions; several novel observations have been made. It has been part of established dogma that a major effect of steroid hormones is to increase transcription by interaction with regulatory regions of the genome. DNA-receptor binding studies have been published for both glucocorticoid (46,47,48) and progesterone receptor (52,53,54) that are compatible with this view. What has yet to be established is the nature of the regulatory element(s). The idea, that enhancer elements are important is being widely canvassed (55,56,57) but other components must also be involved (55). A feature of the receptor-DNA experiments is that additional binding sites have been detected within the structural genes themselves (46,58). The biological relevance of these additional sites has not been resolved. In contrast, the upstream elements have been convincingly allocated a biological function by experiments in which chaemeric genes have been transfected into receptor positive cells (55,56,59,60,61).

Such transfection experiments provide a powerful method for testing the biological function of specific regions of DNA. This type of experiment has also indicated that the regulatory sequences in the promoter may be less specific than at first supposed. There is a well-established literature that proviral mouse mammary tumour viral DNA is regulated by glucocorticoids but not other classes of steroid (62). All these data were obtained with cells that only possessed glucocorticoid receptors. When chaemeric genes containing the regulatory regions within the long terminal repeat (LTR) of mouse mammary tumour virus are transfected into cells containing other classes of receptor the LTR is able to respond to other steroids (63). However, it is not yet clear whether the identical nucleotide sequences are used by the different receptors.

If all cell types contain similar complements of genes, why should the response of different cells to a given steroid be so varied? It is noteworthy that rat liver glucocorticoid receptor was the starting reagent in many of the experiments described above with the mouse lymphoma variants. The implication of this lack of cross-tissue specificity is that the type of response to a given steroid is determined by the acceptor site and not the receptor. However, our knowledge of the structure of receptors from different cell types is not sufficiently detailed to be able to rule out some variation in receptor. The most likely explanation for the different responses is that certain DNA acceptor sites for receptor are in an open state in one cell type but that different acceptors are available in other cells. DNA methylation might influence receptor binding (48) but a more important factor in determining the availability of acceptor sites is the chromosomal proteins. Tissue-specific, non-histone chromosomal proteins may determine the availability of acceptor sites (64,65) and furthermore, some of these sites may be in the nuclear matrix (66).

CONCLUSIONS

Our ideas about the steroid receptor machinery are in a state of flux. I think it probable that most of the unoccupied receptor is within the nucleus but would not rule out a cytoplasmic component. The physiological relevance of large, cytosol receptor aggregates must be re-examined as must the data on receptor activation. It is however clear that some change in receptor structure must accompany steroid binding to explain the enhanced affinity for DNA.

Three types of protein that do not bind steroid have been described as components of the receptor machinery. One type is capable of changing to a steroid binding form and receptor phosphorylation/dephosphorylation may be involved. The second is a mutant form of RG, whilst the third type has been identified with monoclonal antibodies and is of unknown provenance.

Genes and mRNA's for RG and RE have been identified and the reagents so generated used in combination with monoclonal antibodies to elucidate mechanisms of steroid resistance.

The physiological importance of receptor DNA interactions has been affirmed by transfection experiments. Receptor binding sites within both promoter and structural regions of several steroid-sensitive genes have been described.

REFERENCES

1. King RJB, Mainwaring WIP (1974) Steroid-Cell Interactions,
    Butterworths, London

2. Clark JH, Peck EH (1979) Monographs on Endocrinology, vol.14.
    Springer-Verlag, Berlin

3. Knowler JT, Beaumont JM (1985) Essays in Biochemistry 20:1-39

4. Rochefort H, Westley B (1984) In: Biochemical Actions of Hormones,
    vol.21. Academic Press, London, pp.241-266

5. Mainwaring WIP (1977) Monographs on Endocrinology, vol.10.
    Springer-Verlag, Berlin

6. Birnbaumer M, Weigel NL, Grody WW, Minghetti PP, Schrader WT, O'Malley BW
    (1983) In: Progesterone and Progestins. Raven Press, New York, pp.19-32

7. Horwitz KB, Wei LL, Sedlacek SM, d'Arville CN (1985) In: Recent Progress
    in Hormone Research, vol.41. Academic Press, New York, pp.249-316.

8. Rousseau GG (1984) Biochem J 224:1-12

9. Sherman MR (1984) Ann Rev Physiol 46:83-105

10. King WJ, Greene GL (1984) Nature 307:745-747

11. McClennan MC, West NB, Tacha DE, Greene GL, Brenner RM (1984)
    Endocrinology 114:2002-2014

12. Gasc, J-M, Renoir J-M, Radanyi C, Joab I, Tuohimaa P, Baulieu E-E (1984)
    J Cell Biol 99:1193-1201

13. Perrot-Applanat M, Logeat F, Groyer-Picard MT, Milgrom E (1985)
    Endocrinology 116:1473-1484

14. Antakly T, Eisen HJ (1984) Endocrinology 115:1984-1989

15. Gustafsson J (1985) J Steroid Biochem in press

16. Welshons WV, Lieberman ME, Gorski J (1984) Nature 307:747-749

17. Sheridan PJ, Buchanan JM, Anselmo VC (1979) Nature 282:579-582

18. Stumpf WE, Sar M (1976) In: Receptors and Mechanism of Action of Steroid
    Hormones, part 1. Dekker, New York, pp.41-84

19. Housley PR, Grippo JF, Dahmer MK, Pratt WB (1984) In: Biochemical Actions
    of Hormones, vol. 21. Academic Press, London, pp.347-376.

20. Schrader WT, Birnbaumer ME, Hughes MR, Weigel NL, Grody WW, O'Malley BW
    (1981) In: Recent Progress in Hormone Research, vol. 37. Academic Press,
    New York, pp.583-629.

21. Miller MA, Mullick A, Greene GL, Katzenellenbogen BS (1985)
    Endocrinology 117:515-522

22. Monsma FJ, Katzenellenbogen BS, Miller MA, Ziegler YS, Katzenellenbogen JA (1984) Endocrinology 115:143-153

23. Munck A, Foley R (1979) Nature 278:752-754

24. Katzenellenbogen JA, Katzenellenbogen BS (1984) In: Vitamins and Hormones, vol. 41. Academic Press, New York, pp.213-274

25. Greene GL (1984) In: Biochemical Actions of Hormones, vol. 21. Academic Press, London, pp.208-239

26. Eckert RL, Mullick A, Rorke EA, Katzenellenbogen BS (1984) Endocrinology 114:629-637

27. Logeat F, Pamphile R, Loosfelt H, Jolivet A, Fournier A, Milgrom E (1985) Biochemistry 24:1029-1035

28. Birnbaumer M, Schrader WT, O'Malley BW (1983) J Biol Chem 258:7331-7337

29. Sakai D, Gorski J (1984) Endocrinology 115:2379-2383

30. Carter-Su C, Pratt WB (1984) In: Conn PM (ed) The Receptors, vol. 1. Academic Press, New York, pp.541-585

31. Auricchio F, Migliaccio A, Castoria G, Lastoria S, Rotondi A (1984) In: Progress in Cancer Research and Therapy, vol. 31. Raven Press, New York, pp.49-62

32. Migliaccio A, Rotondi A, Auricchio F (1984) Proc Natl Acad Sci USA 81: 5921-5925

33. Raymoure WJ, McNaught RW, Smith RG (1985) Nature 314:745-747

34. Radanyi C, Joab I, Renoir J-M, Richard-Foy H, Baulieu E-E (1983) Proc Natl Acad Sci USA 80:2854-2858

35. Edwards DP, Weigel NL, Schrader WT, O'Malley BW, McGuire WL (1984) Biochemistry 23:4427-4435

36. Peleg S, Schrader WT, Edwards DP, McGuire WL, O'Malley BW (1985) J Biol Chem 260:8492-8501

37. Miesfeld R, Okret S, Wikstrom A-C, Wrange O, Gustafsson J-A, Yamamoto KR (1984) Nature 312:779-781

38. Westphal HM, Mugele K, Beato M, Gehring U (1984) EMBO J 3:1493-1498

39. Coffer AI, Lewis KM, Brockas AJ, King RJB (1985) Cancer Res 45:3686-3693

40. Coffer AI, Spiller GH, Lewis KM, King RJB (1985) Cancer Res 45:3694-3698

41. King RJB, Coffer AI, Gilbert J, Lewis K, Nash R, Millis R, Raju S, Taylor RW (1985) Cancer Res 45:5728-5733

42. Weinberger C, Holleberg SM, Ong ES, Harmon JM, Brower ST, Cidlowski J, Thompson EB, Rosenfeld MG, Evans RM (1985) Science 228:740-742

43. Walter P, Green S, Greene G, Krust A, Bornert J-M, Jeltsch J-M, Staub A, Jensen E, Scrace G, Waterfield M, Chambon P (1985) Proc Natl Acad Sci USA in press

44. Sibley CH, Yamamoto KR (1979) Monographs on Endocrinology, vol. 12. Springer-Verlag, Berlin, pp.357-376

45. Northrop JP, Gametchu B, Harrison RW, Ringold GM (1985) J Biol Chem 260:6398-6403

46. Payvar F, DeFranco D, Firestone GL, Edgar B, Wrange O, Okret S, Gustafsson J-A, Yamamoto KR (1983) Cell 35:381-392

47. Pfahl M, McGinnis D, Hendricks M, Groner B, Hynes NE (1983) Science 222:1341-1343

48. Scheidereit C, Beato M (1984) Proc Natl Acad Sci USA 81:3029-3033

49. Lipsett MB, Chrousos GP, Tomita M, Brandon DD, Loriaux DL (1985) In: Recent Progress in Hormone Research, vol. 41. Academic Press, New York, pp.199-241

50. Griffin JE, Kovacs WJ, Wilson JD (1985) In: Bruchovsky N, Chapdelaine A, Neumann F (eds) Regulation of Androgen Action, Congressdruck R. Brückner, Berlin, pp.127-131

51. Nawata H, Ta Chong M, Bronzert D, Lippman ME (1981) J Biol Chem 256:6895-6902

52. Bailly A, Atger M, Atger P, Cerbon M-A, Alizon M, Vu Hai MT, Logeat F, Milgrom F (1983) J Biol Chem 258:10384-10389

53. Compton JG, Schrader WT, O'Malley BW (1983) Proc Natl Acad Sci USA 80:16-20

54. Mulvihill ER, LePennec J-P, Chambon P (1982) Cell 24:621-632

55. Chambon P, Dierich A, Gaub M-P, Jakowley S, Jongstra J, Krust A, LePennec J-P, Oudet P, Reudelhuber T (1984) In: Recent Progress in Hormone Research, vol. 40. Academic Press, New York, pp.1-39

56. Chandler VL, Maler BA, Yamamoto KR (1983) Cell 33:489-499

57. Parker M (1983) Nature 304:687-688

58. Slater EP, Rabenau O, Karin M, Baxter JD, Beato M (1985) Mol Cell Biol 5:2984-2992

59. Parker MG, Page MJ (1984) Mol Cell Endocrinol 34:159-168

60. Renkawitz R, Schutz G, von der Ahe D, Beato M (1984) Cell 37:503-510

61. Ringold G, Costello M, Dobson D, Frankel F, Hall C, Lee F (1984) In: Progress in Cancer Research and Therapy, vol. 31. Raven Press, New York, pp.7-18

62. Ringold GM, Dobson DE, Grove JR, Hall CV, Lee F, Vannice JL (1983) In: Recent Progress in Hormone Research, vol. 39. Academic Press, New York, pp.387-424

63. Darbre PD, Page MJ, King RJB (1985) In: Dumont J, Nunez J (eds)
Hormones and Cell Regulation, vol. 10. John Libbey & Company Limited,
London, this volume

64. Spelsberg TC, Littlefield BA, Seelke R, Dani GM, Toyoda H, Boyd-Leinen P,
Thrall C, Lian Kon O (1983) In: Recent Progress in Hormone Research,
vol. 39. Academic Press, New York, pp.463-517

65. Royoda H, Seelke RW, Littlefield BA, Spelsberg TC (1985) Proc Natl
Acad Sci USA 82:4722-4726

66. Barrack ER, Coffey DS (1983) In: Biochemical Actions of Hormones, vol. 20.
Academic Press, New York, pp.23-90

Résumé

La structure et la localisation intracellulaire des récepteurs
des stéroïdes sont analysés à la lumière des données récentes, qui
découlent en grande partie de travaux effectués avec des anticorps
monoclonaux dirigés contre les récepteurs des corticostéroïdes, de la
progestérone et de l'oestradiol et des sondes géniques pour les
récepteurs des glucocorticoïdes et de l'oestradiol. Il semble maintenant
possible que la plus grande partie du récepteur non occupé soit présent
dans le noyau et non dans le cytosol comme on le pensait auparavant. La
signification biologique des données obtenues avec les récepteurs
cytosoliques est discutée en relation avec cette localisation nucléaire
ainsi suggérée.
Les résultats sur la caractérisation des mARN des récepteurs des
glucocorticoïdes et de l'oestradiol ont considérablement augmenté nos
connaissances sur les aspects moléculaires de la structure du récepteur
en particulier en ce qui concerne les mécanismes par lesquels les
cellules de lymphome deviennent résistantes aux glucocorticoïdes. Le
clonage des régions structurales et régulatrices de plusieurs gènes
sensibles aux stéroïdes et des expériences de transfection avec des gènes
chimériques sont en train de fournir des vues nouvelles sur les voies par
lesquelles les récepteurs des stéroïdes influent sur la fonction génique.

Phosphoinositol

Phosphoinositol

Hormones and cell regulation. Ed J. Nunez *et al.* Colloque INSERM/John Libbey Eurotext Ltd. © 1986. Vol. 139, pp. 71–79.

Phosphatases in human erythrocytes and rat liver that degrade inositol polyphosphates

Robert H. Michell, Andrew B. Cubitt, C. Peter Downes, Philip T. Hawkins, Christopher E. King, Christopher J. Kirk, Andrew J. Morris, Marc C. Mussat, Stephen B. Shears and Diane J. Storey

Department of Biochemistry, University of Birmingham, PO Box 363, Birmingham B15 2TT, UK.

ABSTRACT

Inositol 1,4,5-trisphosphate, a hormonally generated second messenger that mobilizes Ca^{2+} from an intracellular store, is degraded in human erythrocytes and rat liver by a Mg^{2+}-dependent phosphomonoesterase that removes the 5-phosphate and is present at the inner surface of the plasma membrane and in the cytosol. Inositol 1,3,4-trisphosphate, which also accumulates in stimulated tissues, is degraded in liver by a phosphatase that is severalfold less active. Inositol 1,4-bisphosphate is degraded, more slowly than inositol 1,4,5-trisphosphate, by cytosolic phosphatases that remove the 1- and 4-phosphate groups. Inositol 1-phosphate phosphatase is cytosolic, and much more active than inositol 4-phosphate phosphatase, which is particle-bound. Of these activities, only inositol 1- phosphatase and inositol 1,4-bisphosphate 4-phosphatase are inhibited by millimolar concentrations of Li^+. It thus seems that inositol 1,4,5-trisphosphate 5-phosphatase catalyses a rapid reaction that inactivates inositol 1,4,5-trisphosphate, and that the inositol derived therefrom can return to the free inositol pool of the cell by two routes, only one of which is blocked by Li^+.

KEY WORDS

Inositol 1,4,5-trisphosphate, Ca^{2+}, phosphatases, liver, erythrocytes, plasma membranes.

BACKGROUND

It has been known for many years that inositol bisphosphate ($InsP_2$) and trisphosphate ($InsP_3$) can be produced and may accumulate in cells exposed to receptor-directed stimuli or to a rise in cytoplasmic $[Ca^{2+}]$ (1-4). However, general interest in these water-soluble metabolites produced by phospho-inositidase C-catalysed hydrolysis of phosphatidylinositol 4-phosphate (PtdIns4P) and phosphatidylinositol 4,5-bisphosphate ($PtdIns(4,5)P_2$) has only developed since it was realised that receptor-activated hydrolysis of $PtdIns(4,5)P_2$, and possibly also of PtdIns4P, is a widespread and important mechanism by which cells respond to external signals such as hormones, neurotransmitters and growth stimuli (reviewed in 5-9). In particular, it is now known that inositol 1,4,5-

trisphosphate (Ins(1,4,5)P3) liberated from PtdIns(4,5)P2 serves as an intracellular second messenger responsible for mobilization of Ca^{2+} from a membrane-associated store, probably in the endoplasmic reticulum (reviewed in 7,9). Initially, it was naturally assumed that all of the InsP3 produced in stimulated cells would be the 1,4,5-isomer derived from PtdIns(4,5)P2 (e.g. ref. 10), so the discovery that most is the novel 1,3,4-isomer (Ins(1,3,4)P3), at least in the parotid gland and liver, came as a considerable surprise (11-13). Ins(1,3,4)P3 has no known cellular function as yet. Initially its origin was also unknown, but Batty, Nahorski and Irvine (14) have recently detected the very rapid formation of inositol 1,3,4,5-tetrakisphosphate (Ins(1,3,4,5)P4) in cholinergically stimulated brain cortex slices. This novel inositol polyphosphate is almost certainly the precursor of the Ins(1,3,4)P3 that accumulates in stimulated tissues (14). It is not known whether some, or even all, of the accumulated Ins(1,4,5)P3 might arise from Ins(1,3,4,5)P4 rather than from PtdIns(4,5)P2. The origin and function of Ins(1,3,4,5)P4 remain unknown. Its most likely origin would appear to be as a product of phosphoinositidase C (9) attack on Ptd(3,4,5)P3, an as yet undetected inositol lipid. It also seems probable that it will have an important intracellular messenger function.

Given the rapidity with which Ins(1,3,4,5)P4, Ins(1,4,5)P3 and Ins(1,3,4)P3 are produced in stimulated cells, together with proven or anticipated functions as intracellular messenger molecules (at least for Ins(1,4,5)P3 and Ins(1,3,4,5)P4), it is to be expected that they will be rapidly degraded, and thus inactivated, by specific enzymic mechanisms. This expectation is borne out by the rapidity with which accumulated Ins(1,4,5)P3, and to a lesser extent Ins(1,3,4)P3, disappear from cells when a stimulus is terminated (12,13). Some years ago, Akhtar and Abdel-Latif (4) showed that InsP3 (isomeric form undefined) is rapidly degraded by iris smooth muscle homogenates, and Downes and Michell (15) then detected rapid dephosphorylation of Ins(1,4,5)P3 by human erythrocyte membranes. The latter observation, made before the Ca^{2+}-mobilizing function of Ins(1,4,5)P3 was recognised, led to detailed description of an active Ins(1,4,5)P3 5-phosphatase in human erythrocytes (16,17).

In this paper, we shall first summarise the characteristics of the Ins(1,4,5)P3 5-phosphatase of human erythrocytes: this remains one of the best-described activities and it seems unlikely that it is markedly different from the activities present elsewhere. We shall then briefly describe our more recent work on a family of phosphatases, present in rat liver, that achieve the total dephosphorylation of Ins(1,4,5)P3. In addition, a new technique has been developed for the quantitative precipitation of trace quantities of inositol polyphosphates, and this has allowed us to initiate studies on the degradation of Ins(1,3,4)P3. Finally, we shall compare our results with those published recently by other laboratories.

THE INOSITOL 1,4,5-TRISPHOSPHATE 5-PHOSPHATASE OF HUMAN ERYTHROCYTES

Human erythrocyte membranes possess a phosphoinositidase C that hydrolyses 40-70% of the PtdIns4P and PtdIns(4,5)P2 of the membranes, yielding stoichimetrically appropriate quantities of Ins(1,4)P2 and Ins(1,4,5)P3, during incubation at low ionic strength in the presence of Ca^{2+} at above micromolar concentrations (15). However, on subsequent addition of Mg^{2+} the accumulated Ins(1,4,5)P3 is quickly degraded: this observation provided the first evidence for an active Ins(1,4,5)P3 phosphatase (15). The characteristics of this membrane-associated enzyme activity can be summarised as follows:

1. It removes only the 5-phosphate from Ins(1,4,5)P3 or GroPIns(4,5)P2 (16,17);

2. The Km for Ins(1,4,5)P3 is approximately 25 µM (16), and under

conditions of low substrate concentration the first-order rate constant for attack upon $Ins(1,4,5)P_3$ is about 25-fold greater than for $GroPIns(4,5)P_2$ (17). $GroPIns(4,5)P_2$ hydrolysis shows little deviation from first order kinetics until substrate concentrations of 15 µM or more are achieved (17, Fig 3), suggesting that the major reason for slow hydrolysis of this ester may be a substantially lower affinity than for $Ins(1,4,5)P_3$. $PtdIns(4,5)P_2$ is attacked very slowly, if at all (16).

3. Neither $Ins(1,4)P_2$ nor GroPIns4P is attacked (16). In addition, $Ins(1,3,4)P_3$ is resistant to this enzyme. This was one of the first features by which $Ins(1,3,4)P_3$ was distinguished from $Ins(1,4,5)P_3$, and it facilitates the isolation of $Ins(1,3,4)P_3$ (11, see below).

4. The pH optimum is around 7. Mg^{2+} is the best activating divalent cation that was tested, with Mn^{2+} less active and Ca^{2+} inactive (16). Zn^{2+} and Cd^{2+} are inhibitory (16, unpublished data).

5. 2,3-diphosphoglycerate, a phosphate ester that is present at high concentrations in mammalian erythrocytes, and which shares with $Ins(1,4,5)P_3$ the possession of a vicinal pair of monoesterified phosphate groups, competitively inhibits $Ins(1,4,5)P_3$ 5-phosphatase with a K_i of approximately 0.35 mM (16). However, the $Ins(1,4,5)P_3$ 5-phosphatase and 2,3-diphosphoglycerate phosphatase activities of the erythrocyte appear to be catalysed by different enzymes: the former activity is inhibited by Ag^{2+} ions and is almost unaffected by low concentrations of Cu^{2+}, whereas the latter is activated by Ag^{2+} and inhibited by Cu^{2+} (see 16).

6. $Ins(1,4,5)P_3$ 5-phosphatase exists both in erythrocyte membranes and cytosol (16, C P Downes, unpublished data), with approximately half in each. To ensure that the apparently 'membrane-bound' activity is not a consequence of enzyme adsorbtion at the low ionic strengths normally used for isolation of erythrocyte membranes, we studied activity in a Triton-permeabilized sample of erythrocyte ghosts isolated under isoionic conditions (18): the membrane-associated activity remained essentially constant throughout the preparation of these ghosts (16).

7. It was assumed that the active centre of membrane-associated $Ins(1,4,5)P_3$ 5-phosphatase would be at the inner face of the plasma membrane, where it would have access to $Ins(1,4,5)P_3$ produced within stimulated cells. We checked this by using cytosol-free erythrocyte ghosts that had been isolated under isoionic conditions and then resealed in their original orientation (18,19): these ghosts are essentially impermeable, except to water and some ions (19). When they are progressively permeabilized by titration with saponin, there is a gradual activation of the inward-facing Ca^{2+}-pump ATPase, as a result of the enzyme gaining access to its substrate (data not shown; see, for example, ref 20). The sealed erythrocyte ghosts showed very little of the usual $Ins(1,4,5)P_3$ 5-phosphatase activity, but this activity was progressively unmasked by titration with saponin (Fig 1). Thus it appears that $Ins(1,4,5)P_3$, like the Ca^{2+}-pump ATPase, displays its active centre at the inner surface of the erythrocyte membrane.

8. The ability of the $Ins(1,4,5)P_3$ phosphatase of erythrocyte membranes also to specifically remove the 5-phosphate of $GroPIns(4,5)_2$, albeit at a relatively slow rate, has allowed this enzyme to be used as a tool with which to analyse the distribution of ^{32}P between the 4- and 5-monoesterified phosphates of $PtdIns(4,5)P_2$ (17).

9. It was recently discovered that red cell membranes also rapidly degrade $Ins(1,3,4,5)P_4$, removing the 5-phosphate and yielding $Ins(1,3,4)P_3$ (14). This raised two questions which have yet to be answered: (a) are $Ins(1,4,5)P_3$ and $Ins(1,3,4,5)P_4$ hydrolysed by the same phosphatase; and (b) if they are, then

which of these is the true cellular substrate of the erythrocyte membrane inositol polyphosphate 5-phosphatase

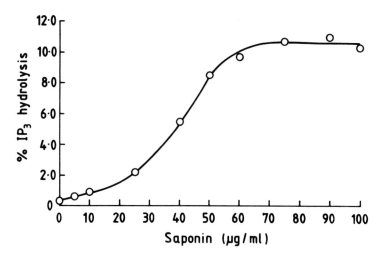

Fig. 1   Saponin activates the Ins(1,4,5)P₃ 5-phosphatase of resealed erythrocyte ghosts, indicating that the active site of this enzyme is at the cytoplasmic surface of the membrane.

As will be seen below, the membrane-bound $Ins(1,4,5)P_3$ 5-phosphatase of liver plasma membranes is very similar to that of the erythrocyte membrane. It is assumed that the role of the liver enzyme is to hydrolyse $Ins(1,4,5)P_3$ produced in response to activation of vasopressin, angiotensin or α_1-adrenergic receptors, and also of opiate receptors [which have recently been shown to activate $PtdIns(4,5)P_2$ hydrolysis in isolated rat hepatocytes: R P Leach, S B Shears and M A Titheradge, unpublished data]. At present, we know of no hormone receptors that employ this signalling mechanism and survive in mature human erythrocytes, so it seems likely that the $Ins(1,4,5)P_3$ 5-phosphatase of mature erythrocytes is a remnant of some hormonal signalling mechanism that was of importance during haemopoiesis.

THE PHOSPHATASES THAT DEPHOSPHORYLATE $Ins(1,4,5)P_3$ IN RAT LIVER

We expect that the intracellular concentrations of the inositol phosphates derived ultimately from $PtdIns(4,5)P_2$ hydrolysis will, under most normal physiological conditions, be very low, so we have concentrated on an examination of those enzyme activities that achieve hydrolysis of trace quantities of inositol phosphates in a medium of approximately physiological ionic strength and composition (21).

Initially we simply determined the first order rate constants of the dephosphorylation of ^{32}P-labelled $Ins(1,4,5)P_3$ (labelled in the 4- and 5-phosphates), $Ins(1,4)P_2$ (labelled in the 4-phosphate, and Ins1P in rat liver homogenates and in particulate and soluble fractions derived therefrom (Table 1 of ref 21). From these figures, we calculate that the half-life of $Ins(1,4,5)P_3$ in liver cells is likely to be of the order of 4 sec, with about three quarters of its degradation catalysed by a particle-bound 5-phosphatase and one-quarter by a cytosolic enzyme. The former, particle-bound, activity distributes in Percoll density gradients of liver homogenates in the same way as marker enzymes for

plasma membranes and Golgi membranes: it is clearly not associated with
mitochondria, lysosomes or endoplasmic reticulum (21). Further studies, in which
Golgi membranes were isolated relatively uncontaminated with plasma membranes,
have suggested that this activity is associated with plasma membranes rather than
Golgi membranes (22). Thus it seems that a major $Ins(1,4,5)P_3$ 5-phosphatase of
liver is associated with the plasma membrane, as in erythrocytes.

The cytosolic $Ins(1,4,5)P_3$ phosphatase was initially difficult to study further,
because of the presence of soluble $Ins(1,4)P_2$ phosphatase activity(ies) that
destroyed its product and thus liberated additional $^{32}P-P_i$. However, gel
filtration on Sephadex G-100 has allowed us to separate the soluble $Ins(1,4,5)P_3$
phosphatase (molecular weight approx. 75000) from $Ins(1,4)P_2$ phosphatase
activity(ies) (mol. wt. approx. 55000) (Fig 2). This separated $Ins(1,4,5)P_3$
phosphatase shares with the plasma membrane $Ins(1,4,5)P_3$ phosphatase an ability
specifically to remove the 5-phosphate and thus to destroy the Ca^{2+}-mobilizing
activity of $Ins(1,4,5)P_3$.

By analogy with cyclic AMP phosphodiesterase, it might be anticipated that
$Ins(1,4,5)P_3$ 5-phosphatase will be an enzyme subject to hormonal regulation. We
have therefore looked for stable alterations in activity in homogenates derived
from liver cells exposed to a variety of hormones, but without detecting any
changes (22). In addition, we found no changes in activity caused by variations
in $[Ca^{2+}]$ in the physiological intracellular range.

Our initial studies also showed rapid, and almost exclusively cytosolic,
degradation of $^{32}P-Ins(1,4)P_2$ to $^{32}P-P_i$ in liver homogenates (21).
Unexpectedly, however, the time-course of $Ins(1,4)P_2$ hydrolysis was biphasic:
rapid release of approximately one-half of the ^{32}P-labelled 4-phosphate as
$^{32}P-P_i$ was followed by an at least 10-fold slower liberation of the remainder.
This we interpreted as evidence for the presence of two pathways of $Ins(1,4)P_2$
metabolism: one is direct removal of the 4-phosphate, whilst the other involves
removal of the non-radioactive 1-phosphate followed by slow hydrolysis of the
resulting Ins4P (21). We have since confirmed this interpretation by directly
demonstrating the existence of separate $Ins(1,4)P_2$ 4-phosphatase and $Ins(1,4)P_2$
1-phosphatase activities. The cytosolic forms of both have molecular weights of
approximately 55000, but they differ in other respects. In particular: (a) the
4-phosphatase activity is all cytosolic, whereas approximately half of the
1-phosphatase activity is particle-bound (site undetermined); (b) the
4-phosphatase is inhibited by Li^+ at millimolar concentrations whilst the
1-phosphatase is not; and (c) less salt is required to elute the 4-phosphatase
than the 1-phosphatase from a DEAE-Sephacel column.

A third cytosolic activity that, by gel filtration, has a molecular weight of
approximately 55000 is Ins1P phosphatase (assayed either against ^{32}P-labelled
D-Ins1P or ^{14}C-labelled L-Ins1P). As is well known for other tissues, this
phosphatase is sensitive to millimolar concentrations of Li^+, so we wondered
whether the Ins1P phosphatase and $Ins(1,4)P_2$ 4-phosphatase activities of liver
cytosol might be catalysed by the same enzyme. However, this appears not to be
the case: although these soluble enzymes have the same apparent molecular
weight, elute together from DEAE-Sephacel and are both Li^+-sensitive, the Ins1P
phosphatase is less heat-labile than $Ins(1,4)P_2$ 4-phosphatase.

Ins4P phosphatase catalyses much the slowest reaction in the sequence leading
from $Ins(1,4,5)P_3$ to free inositol, and has so far been studied in less detail.
This activity is largely particle-bound (site undetermined, as yet) and is
insensitive to Li^+ at concentrations up to 50 mM.

The properties of the enzymes involved in degradation of $Ins(1,4,5)P_3$ to free Ins
in rat liver are summarised in Fig 2, in which the phosphate groups that are

^{32}P-labelled in inositol phosphates derived from labelled red cells are shown in bolder script than unlabelled phosphates. In particular, this scheme is designed to illustrate: (a) the rapidity of Ins(1,4,5)P$_3$ 5-phosphatase, the reaction that inactivates Ins(1,4,5)P$_3$, as compared with the subsequent steps that dephosphorylate Ins(1,4)P$_2$; and (b) the existence of two simultaneous routes of Ins(1,4)P$_2$ degradation, only one of which is detected by measurements of ^{32}P$_i$ release from ^{32}P-labelled Ins(1,4)P$_2$ and substantially inhibited by Li$^+$.

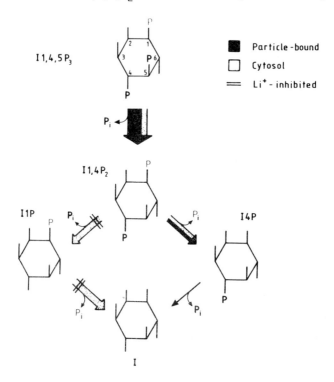

Fig. 2. A scheme that summarises the properties of the phosphatases involved in conversion of Ins(1,4,5)P$_3$ to free Ins in rat liver homogenates. The relative activities of the various enzymes are indicated approximately by the widths of the arrows.

DEGRADATION OF Ins(1,3,4)P$_3$ BY LIVER HOMOGENATES

Ins(1,3,4)P$_3$ is a phosphate ester that accumulates in stimulated liver and parotid gland more slowly than Ins(1,4,5)P$_3$ and also disappears more slowly on removal of a stimulus (12,13). At present, it is only available in trace quantities in ^{3}H-labelled form, and its metabolism therefore cannot yet be studied by the simple P$_i$ release assay that we have generally used for the other phosphatases. This problem has been circumvented by development of an isolation and enzyme assay technique that employs Mn^{2+} precipitation of InsP$_3$ from ammonium bicarbonate eluates from small Dowex-1 columns.

Briefly, Ins(1,3,4)P$_3$ was isolated as follows. Rat parotid fragments were labelled with ^{3}H-inositol and stimulated with carbamycholine for 30 min in the presence of 50 mM Li$^+$, to accumulate both Ins(1,4,5)P$_3$ and Ins(1,3,4)P$_3$ (11). An inositol trisphosphate fraction was isolated from the tissue by chromatography on small Dowex-1 columns, using 0.6 M ammonium bicarbonate as eluant (S B Shears, unpublished data). On addition of Mn^{2+}, the mixed $InsP_3^+$ of this fraction were co-precipitated with $MnCO_3$. The Mn^{2+} was removed with Dowex-50 (H^+ form), and the resulting solution of mixed $InsP_3$'s was destroyed by incubation with human erythrocyte ghosts, and the remaining $Ins(1,3,4)P_3$ was then recovered by a further cycle of Dowex chromatography, Mn^{2+} precipitation and deionization. HPLC was used to confirm that the resulting solution contained radiochemically pure ^3H-$Ins(1,3,4)P_3$. This was then incubated with rat liver homogenates, and the undegraded $Ins(1,3,4)P_3$ recovered by the same chromatography/precipitation procedure, acidified and its radioactivity determined: this allowed the rate of hydrolysis of $Ins(1,3,4)P_3$ to be measured.

The rate constant for $Ins(1,3,4)P_3$ hydrolysis thus determined (0.005 per sec for a 1 in 60 liver homogenate) suggests a half-life in liver cells of around 20 sec, as compared with the equivalent figure of approximately 4 sec for $Ins(1,4,5)P_3$ (21). This is in accord with the observed slower degradation of $Ins(1,3,4)P_3$ than $Ins(1,4,5)P_3$ in cells on removal of a stimulus (12). The site of initial attack on $Ins(1,3,4)P_3$ is yet to be determined.

EFFECTS OF INHIBITORY IONS, ESPECIALLY Li^+, ON INOSITOL PHOSPHATE PHOSPHATASES

Hallcher and Sherman's (23) demonstration that Li^+ inhibits Ins1P phosphatase led to the general adoption of Li^+ as a 'trapping agent' that facilitates the accumulation of inositol phosphates in stimulated tissues (24). The original expectation was that all inositol polyphosphates produced in such Li^+-treated cells would be partially dephosphorylated and would accumulate as Ins1P. However, it soon became apparent that $InsP_2$ (isomer undetermined) also accumulates in cells incubated with 1-10 mM Li^+, and that at even higher Li^+ concentrations there is a substantial accumulation of an $InsP_3$ (24, S Palmer and C J Kirk, unpublished) which turns out to be the 1,3,4-isomer (13).

Given this background, we have looked fairly carefully at the Li^+ sensitivity of the inositol phosphate phosphatases present in liver homogenates. As mentioned above, Ins1P phosphatase and $Ins(1,4)P_2$ 4-phosphatase are both sensitive to millimolar concentrations of Li^+, and this presumably explains why such Li^+ concentrations cause intracellular accumulation of both InsP and $InsP_2$: presumably the accumulation of $InsP_2$ is smaller because Ins1P breakdown can only occur by a single, Li^+-sensitive, reaction, whereas $Ins(1,4)P_2$ has the option of hydrolysis via the Li^+-insensitive Ins4P pathway (see Fig 2).

Unexpectedly, however, we have not obtained any evidence for inhibition of $Ins(1,4,5)P_3$ 5-phosphatase or $Ins(1,3,4)P_3$ phosphatase (site of attack undetermined) by 50 mM Li^+. A second possibility, yet to be tested, is that accumulated $Ins(1,4)P_2$ might act as a potent inhibitor of the phosphatase(s) that degrades $Ins(1,3,4)P_3$. Thus our enzyme studies do not yet offer an explanation for the selective accumulation of $Ins(1,3,4)P_3$ in stimulated cells incubated with high concentrations of Li^+.

A COMPARISON OF THESE RESULTS WITH THOSE OF OTHER GROUPS

Two other groups reported the presence of $Ins(1,4,5)P_3$ 5-phosphatase in rat liver plasma membranes at about the same time as us (26,27), though in neither case was the subcellular localization demonstrated with the same degree of precision. Seyfred and Wells (27), like us, used trace quantities of radiolabelled $Ins(1,4,5)P_3$, but without deriving appropriate first order rate constants from their data. Joseph and Williams (26) undertook a fuller kinetic analysis: the Km for $Ins(1,4,5)P_3$ of their plasma membrane 5-phosphatase was 30 µM, a value quite similar to that for the erythrocyte enzyme.

The most detailed study of a cytosolic Ins(1,4,5)P$_3$ 5-phosphatase thus far is that of Connolly et al. (28), who have isolated such an enzyme in pure form from human platelets, albeit in very low yield. They observed that about 84% of the total Ins(1,4,5)P$_3$ phosphatase activity is soluble in these cells, and this appears to exist as a monomeric protein of mol. wt. 38000. Since this is exactly half of the size we have determined for the liver cytosolic Ins(1,4,5)P$_3$ 5-phosphatase, we wonder whether the latter might be a dimeric enzyme. In most ways, the characteristics of the platelet cytosolic enzyme (activation by Mg^{2+}, for the 5-phosphate, Km for Ins(1,4,5)P$_3$, lack of inhibition by Li$^+$, etc) appear very similar to those of the previously known plasma membrane enzymes (see above). However, the platelet enzyme appears to differ by showing no activity against GroPIns(4,5)P$_2$. The only possibly physiological inhibitor that was identified in these studies was the Ca^{2+} ion, but with such a low affinity (K$_i$ = 70 μM) that it seems unlikely it is an important regulator of the platelet enzyme. Although Connolly et al. (27) reiterate the need for sequential removal of all three phosphates from Ins(1,4,5)P$_3$, in order constantly to replenish the free inositol needed for lipid resynthesis, they make the surprising claim that their platelet homogenates show no Ins(1,4)P$_2$ phosphatase activity. When taken together with the known slow rate of entry of inositol into human platelets, this could substantially limit the ability of these cells to resynthesise their inositol phospholipids after stimulation.

In another, much less detailed, study, Sasaguri and his colleagues (29) reported the presence of soluble Ins(1,4,5)P$_3$ phosphatase in porcine arterial smooth muscle. This enzyme, unlike those studied in our laboratory and by Connolly et al.(28), showed activation by [Ca^{2+}] within the normal intracellular range, a response that could readily serve as a feedback mechanism for attenuating the further mobilization of Ca^{2+} in response to Ins(1,4,5)P$_3$ formation in stimulated cells.

ACKNOWLEDGEMENTS

This work has been supported by grants and research studentships from MRC and SERC.

REFERENCES

1. Durell, J., Garland, J.T. and Friedel, R.O. (1969) Science 165, 862-866
2. Griffin, H.D. and Hawthorne, J.N. (1978) Biochem. J. 176, 541-552.
3. Allan, D. and Michell, R.H. (1978) Biochim. Biophys. Acta 508, 277-286.
4. Akhtar, R.A. and Abdel-Latif, A.A. (1980) Biochem. J. 527, 159-170.
5. Michell, R.H., Kirk, C.J., Jones, L.M., Downes, C.P. and Creba, J.A. (1981) Phil. Trans. Roy. Soc. Ser. B 296, 123-137.
6. Berridge, M.J. (1984) Biochem. J. 220, 345-360.
7. Berridge, M.J. and Irvine, R.F.I. (1984) Nature 312, 315-321.
8. Nishizuka, Y. (1984) Nature 308, 693-698.
9. Downes, C.P. and Michell, R.H. (1985) In Molecular Mechanisms of Transmembrane Signalling eds. P. Cohen and M.D. Houslay, pp 3-56. Elsevier, Amsterdam.
10. Berridge, M.J. (1983) Biochem. J. 212, 849-858.
11. Irvine, R.F., Letcher, A.J., Lander, D.J. and Downes, C.P. (1984) Biochem. J. 223, 237-243
12. Irvine, R.F., Anggard, E.A., Letcher, A.J. and Downes, C.P. (1985) Biochem. J. 229, 505-512
13. Burgess, G.M., McKinney, J.J., Irvine, R.F. and Putney, J.W. (1985) Biochem. J. 232, in press
14. Batty, I., Nahorski, S. and Irvine, R.F. (1985) Biochem. J., in press
15. Downes, C.P. and Michell, R.H. (1981) Biochem. J. 198, 133-140.

16. Downes, C.P., Mussat, M.C. and Michell, R.H. (1982) Biochem. J. 203, 169-177
17. Hawkins, P.T., Michell, R.H. and Kirk, C.J. (1984) Biochem. J. 218, 785-793.
18. Billah, M.M., Finean, J.B., Coleman, R. and Michell, R.H. (1976) Biochim. Biophys. Acta 433, 54-62.
19. Billah, M.M., Finean, J.B., Coleman, R. and Michell, R.H. (1977) Biochim. Biophys. Acta 465, 515-526.
20. Downes, C.P., Simmonds, S.H. and Michell, R.H. (1981) Cell Calcium 2, 473-482.
21. Storey, D.J., Shears, S.B., Kirk, C.J. and Michell, R.H. (1984) Nature 312, 374-376.
22. Shears, S.B., Storey, D.J., Michell, R.H. and Kirk, C.J. (1985) Biochem. Soc. Trans. 13, 944
23. Hallcher, L.M. and Sherman, W.R. (1980) J. Biol. Chem. 255, 10896-10901.
24. Berridge, M.J., Downes, C.P. and Hanley, M.R. (1982) Biochem. J. 206, 587-595.
25. Thomas, A.P., Marks, J.S., Coll, K.E. and Williamson, J.R. (1983) J. Biol. Chem. 258, 5716-5725.
26. Joseph, S.K. and Williams, R.J. (1985) FEBS Letters, 180, 150-154
27. Seyfred, M.A., Farrell, L.E. and Wells, W.W. (1984) J. Biol. Chem. 259, 13204-13208.
28. Connolly, T.M., Bross, T.E. and Majerus, P.W. (1985) J. Biol. Chem. 260, 7868-7874.
29. Sasaguri, T., Hirata, M. and Kuriyama, H. (1985) Biochem. J. 231, 497-503.

Résumé

L'inositol 1.4.5 triphosphate, second messager généré par les hormones et qui mobilise Ca^{2+} à partir d'une réserve intracellulaire est dégradé dans les erythrocytes humains et le foie de rat par une phosphomonoesterase Mg^{2+}-dépendante qui enlève le phosphate en position 5 et est présente sur la surface interne de la membrane plasmatique et dans le cytosol. L'inositol 1,3,4 triphosphate qui s'accumule également dans les tissus stimulés est dégradé dans le foie par une phosphatase qui est plusieurs fois moins active. L'inositol 1,4 biphosphate est dégradé, plus lentement que l'inositol 1,4,5 triphosphate par des phosphatases cytosoliques qui enlèvent les groupes phosphate 1 et 4. La phosphatase de l'inositol 1-phosphate est cytosolique , et beaucoup plus active que la phosphatase de l'inositol 4-phosphate qui est particulaire. Le Li^+ a des concentrations millimolaires, n'inhibe que l'inositol 1 phosphatase et l'inositol 1,4 biphosphate 4-phosphatase. Il semble donc que l'inositol 1,4,5 triphosphate 5 phosphatase catalyse une réaction rapide qui inactive l'inositol 1,4,5 triphosphate et que l'inositol qui en dérive peut revenir dans le pool d'inositol libre de la cellule par deux voies dont une seulement est bloquée par Li^+.

Hormones and cell regulation. Ed J. Nunez *et al.* Colloque INSERM/John Libbey Eurotext Ltd. © 1986. Vol. 139, pp. 81–93.

The role of inositol polyphosphates in the action of calcium-dependent hormones in liver

A.P. Thomas, S.K. Joseph and J.R. Williamson

University of Pennsylvania School of Medicine, Department of Biochemistry and Biophysics, Philadelphia, Pennsylvania 19104 and Hahnemann University School of Medicine, Department of Pathology, 235 N. 15th Street, Philadelphia, Pennsylvania 19102, USA.

Liver cells utilize an elevation of cytosolic free Ca^{2+} as a key coupling step in mediating the intracellular effects of a range of different hormones including α_1-adrenergic agents and vasopressin. Hormone-induced increases of cytosolic free Ca^{2+} are accompanied by the breakdown of phosphatidylinositol 4,5-bisphosphate to yield diacylglycerol and inositol 1,4,5-trisphosphate with time courses which are consistent with a role for these compounds in the Ca^{2+} mobilization pathways. By using saponin-permeabilized hepatocytes it has been shown that inositol 1,4,5-trisphosphate releases Ca^{2+} from an intracellular Ca^{2+} pool at concentrations similar to the inositol trisphosphate concentrations measured in intact hormone-stimulated liver cells. This inositol 1,4,5-trisphosphate-sensitive Ca^{2+} pool is a non-mitochondrial, ATP-dependent vesicular pool which is probably a subfraction of the endoplasmic reticulum. The Ca^{2+} release pathway is not dependent on the ATP-hydrolysing component of the Ca^{2+} pump which fills the pool. In general the characteristics of the Ca^{2+} release induced by inositol 1,4,5-trisphosphate are suggestive of a ligand-activated channel mechanism which is electrogenic and consequently requires the presence of a compensating permeant cation. Although the intracellular inositol 1,4,5-trisphosphate-sensitive Ca^{2+} pool is sufficient to account for the initial rise of cytosolic free Ca^{2+} after hormone treatment, it is apparent that Ca^{2+}-dependent hormones also act to alter Ca^{2+} fluxes at the hepatocyte plasma membrane in a manner which allows prolonged maintenance of an elevated cytosolic free Ca^{2+} concentration. Whether there is a mechanistic relationship between this latter effect on Ca^{2+} fluxes and the changes of inositol lipid metabolites has yet to be determined.

INTRODUCTION

A wide variety of hormones in many different cell types employ an elevation of cytosolic free Ca^{2+} as a key step in their intracellular stimulus-response coupling mechanisms. Michell first pointed out that many of these systems have another effect in common, namely the activation of inositol lipid metabolism (1). Over the past five years it has become apparent that hormone-induced Ca^{2+} mobilization is generally correlated with increased breakdown of phosphoinositidies, particularly phosphatidylinositol 4,5-bisphosphate (2-5). Most recently the pathway of phosphatidylinositol 4,5-bisphosphate breakdown has been shown to involve phosphodiesteric cleavage of the inositol phosphate head group through a phospholipase C activity to yield inositol (1,4,5)-trisphosphate (6-8). This latter compound is able to release Ca^{2+} from intracellular Ca^{2+} storage sites and it is now generally believed to be the second messenger for hormone-induced Ca^{2+} release in mammalian tissues (9-13). In addition to inositol polyphosphates, phospholipase C action also yields diacylglycerol and this compound probably also has a second messenger function through its ability to activate protein kinase C (12-14).

In the liver several agonists act to increase cytosolic free Ca^{2+} and these include α_1-adrenergic agents, vasopressin, and angiotensin II (8,13,15-19). An elevation of cytosolic free Ca^{2+} probably has many metabolic effects in the liver but one key effect is the stimulation of phosphorylase kinase which phosphorylates and activates glycogen phosphorylase resulting in enhanced glycogenolysis. Isolated hepatocytes have proved to be a very useful model for studying the relationship between hormone-induced inositol-lipid breakdown and the elevation of cytosolic free Ca^{2+}. In the following pages some of the properties of this system in the liver will be discussed.

Hormone-sensitive Ca^{2+} Pools

Treatment of hepatocytes with a Ca^{2+}-dependent hormone such as vasopressin results in a rapid (<10 s) increase of cytosolic free Ca^{2+} as measured using the fluorescent Ca^{2+} indicator Quin 2, and this is closely followed by an increase in phosphorylase a activity (8,13,15-19). The initial rise of cytosolic free Ca^{2+} and phosphorylase a activity is unaffected in the absence of extracellular Ca^{2+} and this is generally taken to indicate that the hormone-sensitive Ca^{2+} pool is of intracellular origin (13,17-19). Early results tended to suggest that this intracellular Ca^{2+} pool was associated with the mitochondria (15-17). However, recent fractionation studies (20) together with experiments using permeabilized hepatocytes (10) have implicated an intracellular vesicular Ca^{2+} pool which is distinct from the mitochondria.

Although the initial rise of cytosolic free Ca^{2+} in response to hormone treatment is independent of extracellular Ca^{2+}, physiological levels of extracellular Ca^{2+} are required for prolonged elevation of cytosolic free Ca^{2+} and activation of phosphorylase (13,19). This implies that in addition to causing a release of Ca^{2+} from an internal Ca^{2+} storage site, hormones must also alter Ca^{2+} fluxes at the plasma membrane of the hepatocyte. One means by which this may be achieved is through an inhibition of the ATP-dependent Ca^{2+} pump. Evidence for such a hormone-induced inhibition of the Ca^{2+} extrusion pump has been obtained from both intact cells (21) and isolated hepatic plasma membranes (22,23). However, studies with $^{45}Ca^{2+}$ have also suggested that Ca-dependent hormones may act to enhance the inward flux of Ca^{2+} across the plasma membrane in liver (24,25).

Direct evidence for a hormone-induced activation of Ca^{2+} influx across the plasma membrane comes from experiments using Quin 2-loaded hepatocytes similar to that shown in Fig. 1. Isolated hepatocytes were loaded with Quin 2 and suspended in a calcium-containing medium. Shortly before hormone addition EGTA was added to chelate extracellular Ca^{2+}. Reduction of the extracellular free Ca^{2+} concentration to micromolar levels caused a small and fairly slow decline in cytosolic free Ca^{2+} as measured using Quin 2. When vasopressin was added subsequent to the EGTA there was a very rapid increase in cytosolic free Ca^{2+} which was identical to the initial Ca^{2+} mobilization observed with cells in physiological Ca^{2+} medium (not shown, see Ref. 19). These results are in agreement with the data discussed above indicating that hormones initially rely on an intracellular Ca^{2+} pool for elevation of cytosolic free Ca^{2+}. However, with cells suspended in Ca^{2+}-depleted medium the cytosolic free Ca^{2+} concentration did not remain elevated but rapidly declined towards its basal level so that it was indistinguishable from the control cells not treated with hormone by 2 min after addition of vasopressin. Restoration of the extracellular free Ca^{2+} concentration to a physiological level after the transient of hormone-induced intracellular Ca^{2+} mobilization caused a return of cytosolic free Ca^{2+} to a partially elevated state. This presumably reflects flux through a Ca^{2+} influx pathway at the hepatic plasma membrane which is clearly much less active in cells not treated with hormone (lower trace of Fig. 1).

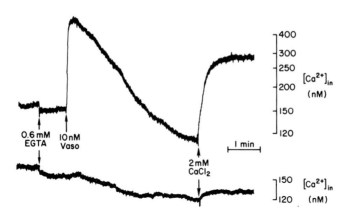

Fig. 1. The role of intracellular and extracellular Ca^{2+} in hormone-induced increases of cytosolic free Ca^{2+}. Isolated rat liver cells were loaded with Quin 2 and suspended in Hanks medium containing 0.6 mM $CaCl_2$ maintained at 37°C in the cuvette of a fluorometer (8). At the points indicated 0.6 mM EGTA was added followed by 10 nM vasopressin (upper trace only) and the resulting changes of Quin 2-Ca fluorescence were followed. After completion of the vasopressin-induced Ca^{2+} transient 2 mM $CaCl_2$ was added to both samples. Calibration of the fluorescence signals in terms of cytosolic free Ca^{2+} was carried out as described previously (8).

Taken together the data outlined above suggest that hormones mobilize Ca^{2+} into the cytosol of liver cells by two distinct mechanisms. Initially Ca^{2+} is rapidly released from an intracellular Ca^{2+} storage pool but this pool quickly becomes depleted of Ca^{2+} (13,19). In order to maintain a partially elevated steady state level of cytosolic free Ca^{2+} there is probably an inhibition of the plasma membrane ATP-dependent Ca^{2+} extrusion pump (21-23) and a concomitant activation of an as yet unidentified Ca^{2+} influx pathway (19,24,25). Despite these alterations of plasma membrane Ca^{2+} transport processes hormone treatment always results in a net loss of cell Ca^{2+} whether extracellular Ca^{2+} is present or absent (17,21,26). This net movement of Ca^{2+} out of the cell may be taken as an indication of the predominance of the internal Ca^{2+} mobilizing system over the plasma membrane changes occurring during the early phase of hormonal stimulation.

Kinetic Relationships Between Inositol Lipid Metabolism and Ca^{2+} Mobilization

Before any change of inositol lipid metabolism can be proposed to be causal in hormone-induced Ca^{2+} mobilization it is necessary to show that the components involved are altered within a time frame consistent with the time course of cytosolic free Ca^{2+} increases. During early studies the primary focus of research was on the parent inositol lipid, phosphatidylinositol (1, 27-30). Although this lipid makes up about 95% of the inositol lipid pool in liver (30)

83

it shows only small and rather slow breakdown after hormone treatment (3,29,30). Michell and coworkers first showed that two minor derivatives of phosphatidylinositol, phosphatidylinositol 4-phosphate and phosphatidylinositol 4,5-bisphosphate breakdown much more rapidly in response to hormones (2). From careful measurements of polyphosphoinositide breakdown rates and dose response studies we were able to conclude that phosphatidylinositol 4,5-bisphosphate breaks down in response to vasopressin-treatment of hepatocytes with a time course which is consistent with a role for this compound in Ca^{2+} mobilization (3). Fig. 2 compares the time courses of cytosolic free Ca^{2+} increase, phosphorylase activation and phosphatidylinositol 4,5-bisphosphate breakdown in vasopressin-stimulated hepatocytes.

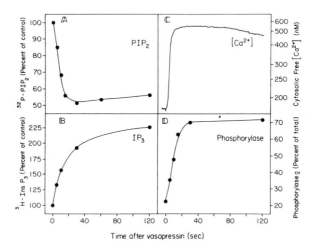

Fig. 2. Time courses of phosphatidylinositol 4,5-bisphosphate breakdown and Ca^{2+} mobilization after vasopressin addition to hepatocytes. Isolated hepatocytes were preincubated for 60-90 min with ^{32}Pi (panel A) or myo-[2-^{3}H]-inositol (panel B) or without isotopes (panels C and D) before addition of 20 nM vasopressin. In A the breakdown of phosphatidylinositol 4,5-bisphosphate (PIP$_2$) was followed by quantitation of ^{32}P-PIP$_2$ separated using thin layer chromatography (3). In B ^{3}H-inositol trisphosphate [IP$_3$] was separated using anion exchange chromatography and quantitated by liquid scintillation counting (8). Changes of cytosolic free Ca^{2+} (panel C) were followed using the fluorescence Ca^{2+} indicator Quin 2 (8,18) and phosphorylase a activity (panel D) was determined in parallel incubations as described previously (3).

Phosphatidylinositol 4,5-bisphosphate appears to breakdown through the action of a hormone-activated phospholipase C as shown by the accumulation of inositol trisphosphate subsequent to hormonal stimulation in a variety of tissues (6,7,12). The vasopressin-induced breakdown of phosphatidylinositol 4,5-bisphosphate in liver is also accompanied by inositol trisphosphate generation (8) and as shown in Fig. 2B this compound increases significantly during the first 5-10 s when hormone-induced Ca^{2+} mobilization and phosphorylase activation are maximal. It has recently been found that much of the inositol trisphosphate generated in hormone-stimulated mammalian cells is not the predicted inositol 1,4,5-trisphosphate but is inositol 1,3,4-trisphosphate, a compound with no known parent lipid (31,32). Studies in which time courses for the generation of these two isomers have been compared indicate that for the first 10 s or so of stimulation only inositol 1,4,5-trisphosphate is generated but at later times this product reaches a plateau and the predominant isomer generated becomes inositol 1,3,4-trisphosphate (32,33). Thus, measurements of total inositol

84

trisphosphate changes occurring over the first few seconds after hormone treatment probably do reflect changes of the active inositol 1,4,5-trisphosphate, but at later times these measurements are complicated by the presence of the second isomer whose function is unknown. Interestingly, Li^+, a known inhibitor of inositol monophosphatase (12, 34), is able to cause substantial increases in hormone-stimulated peak levels of total inositol trisphosphate without altering any of the Ca^{2+}-related process in liver (8). It is probable that Li^+ specifically causes accumulation of inositol 1,3,4-trisphosphate (33) and if this compound is not active in mobilizing Ca^{2+} then an explanation for our previously reported lack of effect of Li^+ on cytosolic free Ca^{2+} (8) may be at hand.

In addition to inositol mono-, bis-, and tris-phosphates, even higher phosphorylated forms of inositol have recently been found in mammalian cells. Batty et al (35) have identified inositol 1,3,4,5-tetrakisphosphate in brain slices and this compound appears to increase in parallel with inositol 1,4,5-trisphosphate subsequent to carbachol stimulation. Clearly inositol 1,3,4,5-tetrakisphosphate could well be the precursor of inositol 1,3,4-trisphosphate since the latter compound is unlikely to be generated by isomerization of inositol 1,4,5-trisphosphate. Very recently yet further complexity has been added to the family of inositol phosphates with the discovery that inositol penta- and hexaphosphates also occur in mammalian cells (36). The source of these higher inositol polyphosphates remains to be elucidated but recent work has shown that mammalian tissues contain an inositol polyphosphate kinase capable of phosphorylating inositol 1,4,5-trisphosphate to generate more highly phosphorylated forms (C. Hanson and J. R. Williamson, unpublished observations). The potential role of these compounds as second messengers opens the way to a diversification of the basic inositol lipid/diacylglycerol transduction system well beyond that which has been characterized so far.

Inositol 1,4,5-trisphosphate as a Ca^{2+}-Mobilizing Second Messenger

Measurements of the kinetics of formation and breakdown of inositol polyphosphates showed that these compounds could potentially be second messengers for Ca^{2+} mobilization (6,8,12,13,32,33), but it was necessary to demonstrate direct effects on Ca^{2+} fluxes. This was first achieved by Streb et al (9) who showed that inositol 1,4,5-trisphosphate could release Ca^{2+} from an intracellular pool in "leaky" pancreatic acinar cells. Subsequent to this we were able to show that inositol 1,4,5-trisphosphate releases Ca^{2+} from saponin-permeabilized hepatocytes (10) and similar effects have now been reported using a variety of different cell and tissue types (See 12 for Refs.).

The data of Fig. 3 demonstrate the type of experiment which can be carried out to investigate inositol 1,4,5-trisphosphate effects on intracellular Ca^{2+} homeostasis in permeabilized cells. Isolated hepatocytes were permeabilized by using saponin which selectively disrupts the plasma membrane by removing cholesterol. The permeabilization medium was KCl-based buffer (see legend of Fig. 3) containing 75 μM Quin 2 and $CaCl_2$ to give a free Ca^{2+} concentration of about 250 nM. The Quin 2 was used to buffer the free Ca^{2+} concentration at levels similar to the cytosolic free Ca^{2+} of unstimulated cells and at the same time the fluorescent properties of Quin 2 (10) permitted measurement of changes of the total amount of Ca^{2+} in the medium. Because the cytosolic volume is effectively expanded by several orders of magnitude in permeabilized cells, the free Ca^{2+} changes are much reduced compared to what would occur in the cytoplasm of intact cells. However, the absolute amounts of Ca^{2+} taken up and released in this system allow direct quantitation of the various Ca^{2+} pool sizes of intracellular organelles.

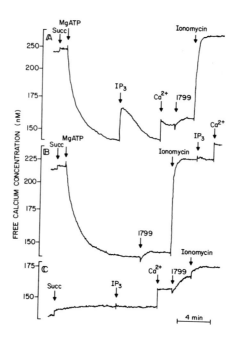

Fig. 3. Inositol 1,4,5-trisphosphate causes Ca^{2+} release from an ATP-dependent Ca^{2+} storage pool in permeabilized hepatocytes. Hepatocytes (5 mg dry weight/ml) were permeabilized by incubation for 5 min with 15 µg/mg saponin in medium composed of 110 mM KCl, 10 mM NaCl, 1 mM KH_2PO_4, 5 mM $KHCO_3$, 20 mM HEPES, 20 mM creatine phosphate, 10 U/ml creatine kinase, 0.3 mM $MgCl_2$, 75 µM Quin 2, 3 µg/ml oligomycin, 0.5 µg/ml rotenone at pH 7.2 and 37°C. Subsequently, $CaCl_2$ was added to give the indicated starting free Ca^{2+} concentrations and changes of the medium Quin 2-Ca concentration were followed fluorometrically (8,10). Other additions were as indicated: Succ, 5 mM succinate; MgATP, 3 mM MgATP; IP_3, 0.5 µM inositol 1,4,5-trisphosphate; Ca^{2+}, 2 nmol/ml $CaCl_2$; 1799, mitochondrial uncoupler 1799; ionomycin, calcium ionophore ionomycin.

In the experiment of Fig. 3A succinate was added first as a substrate for mitochondrial energization but although mitochondria accumulated Ca^{2+} very effectively at higher free Ca^{2+} concentrations, no Ca^{2+} uptake occurred at 250 nM Ca^{2+}. Subsequent addition of MgATP with oligomycin present to prevent mitochondrial ATP synthesis or hydrolysis induced a rapid sequestration of Ca^{2+} from the medium until a new steady state was achieved. This steady state reflects the point at which the ATP-dependent Ca^{2+} pool is effectively saturated since further Ca^{2+} additions did not result in significant Ca^{2+} uptake. Addition of 0.5 µM inositol 1,4,5-trisphosphate after ATP-dependent Ca^{2+} uptake resulted in a very rapid release of Ca^{2+} into the medium which was complete within about 5 s. The total amount of Ca^{2+} which could be released by inositol 1,4,5-trisphosphate was about 0.5 nmol/mg cell dry weight which is approximately 30% of the total ATP-dependent intracellular Ca^{2+} pool. The Ca^{2+} release caused by inositol 1,4,5-trisphosphate was transient and the Ca^{2+} was subsequently resequestered. Using ^{32}P-labeled inositol 1,4,5-trisphosphate we were able to show that the Ca^{2+} resequestration coincided with dephosphorylation of the added inositol 1,4,5-trisphosphate by cellular phosphatases (10). Further additions of inositol 1,4,5-trisphosphate caused additional cycles of Ca^{2+} release and resequestration (data not shown). In Fig. 3A $CaCl_2$ was added for a calibration subsequent to the inositol 1,4,5-trisphosphate and then the mitochondrial uncoupler 1799 was added to release mitochondrial Ca^{2+} followed by ionomycin to release Ca^{2+} from other vesicular storage pools. Figs. 3B and 3C show that inositol 1,4,5-trisphosphate is neither able to release Ca^{2+} into the medium after addition of 1799 and ionomycin to deplete vesicular Ca^{2+} pools nor is it able to release Ca^{2+} in the absence of ATP-dependent Ca^{2+} uptake.

The data described above provide evidence that inositol 1,4,5-trisphosphate is able to release Ca^{2+} from a non-mitochondrial, ATP-dependent Ca^{2+} pool in isolated hepatocytes. The size of this Ca^{2+} pool is sufficient to account for the hormone-sensitive Ca^{2+} pool measured in intact hepatocytes (21) and the amount of inositol 1,4,5-trisphosphate giving maximal release (0.5 µM; Ref.10) is very similar to the measured increase of inositol trisphosphate which correlates with a maximal rate of Ca^{2+} mobilization in vasopressin-stimulated hepatocytes (0.6 µM; Ref. 8).

One possible mechanism for the Ca^{2+} release system is that inositol 1,4,5-trisphosphate is able to reverse the ATP-dependent Ca^{2+} pump which is responsible for filling the relevant Ca^{2+} pool. However, the experiment of Fig. 4 tends to rule out this possibility. The protocol for this experiment is similar to that in Fig. 3. In the upper panel of Fig. 4 the time course of Ca^{2+} release over the first 40 s subsequent to inositol 1,4,5-trisphosphate addition is shown on an expanded time scale. Then, after complete resequestration of the released Ca^{2+} and addition of a $CaCl_2$ standard, glucose and hexokinase were added to remove ATP and hence allow reversal of the ATP-dependent Ca^{2+} pump. In the lower panel of Fig. 4 1 mM vanadate was added after completion of the ATP-dependent Ca^{2+} uptake.

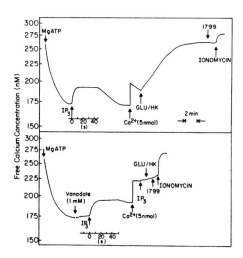

Fig. 4. Effect of vanadate on Ca^{2+} uptake and release by the inositol 1,4,5-trisphosphate-sensitive Ca^{2+} pool in permeabilized hepatocytes. Experimental conditions were essentially as described for Fig. 3 with the extra additions of 1 mM orthovanadate (Vanadate) and 10 mM glucose + 10 U/ml hexokinase (GLU/HK). The time base is expanded over the initial period of Ca^{2+} release induced by inositol 1,4,5-trisphosphate.

This amount of vanadate was sufficient to completely block ATP-induced Ca^{2+} sequestration (not shown), but it had no effect on the Ca^{2+} release induced by inositol 1,4,5-trisphosphate. However, in contrast to the situation without vanadate (upper panel) there was no reaccumulation of the released Ca^{2+} and removal of ATP did not result in Ca^{2+} release since the ATP-dependent Ca^{2+} pump was presumably inhibited in both the forward and reverse directions. These results tend to rule out a reversal of the ATP-dependent Ca^{2+} pump as being the mechanism of Ca^{2+} release by inositol 1,4,5-trisphosphate, although it is still possible that ATP hydrolysis becomes dissociated from the Ca^{2+} transport activity of the pump. Another interesting observation from Fig. 4 is that inositol 1,4,5-trisphosphate did not cause a second Ca^{2+} release when vanadate was present to prevent reaccumulation of the Ca^{2+} released by the first addition even though sufficient vesicular Ca^{2+} was still available. This provides further evidence that the pool of Ca^{2+} released by inositol 1,4,5-trisphosphate is a subfraction of the total ATP-dependent Ca^{2+} pool in permeabilized hepatocytes.

By analogy with the Ca^{2+} release systems in excitable cells the most likely mechanism for Ca^{2+} release by inositol 1,4,5-trisphosphate is through some form of ion specific channel. Some evidence for this suggestion comes from the finding that the response to inositol 1,4,5-trisphosphate is kinetically very insensitive to lowering the incubation temperature (S. K. Joseph and J. R. Williamson, unpublished observations). Furthermore, the Ca^{2+} release is probably electrogenic in nature since a permeant cation is apparently necessary as a counter ion in order to observe maximal Ca^{2+} release with inositol 1,4,5-trisphosphate (37).

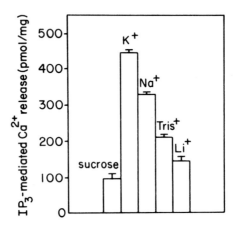

Fig. 5. Effect of different media on $^{45}Ca^{2+}$ release from preloaded permeabilized hepatocytes treated with inositol 1,4,5-trisphosphate. Permeabilized hepatocytes were loaded with $^{45}Ca^{2+}$ in the presence of ATP in medium buffered with 20 mM Tris/HEPES and containing 240 mM sucrose as osmotic support. Subsequently the $^{45}Ca^{2+}$ loaded cells were diluted into media in the presence or absence of 1 μM inositol 1,4,5-trisphosphate containing 20 mM Tris/HEPES, 0.5 mM EGTA, and either 240 mM sucrose or 120 mM of the gluconate salts of the indicated ion. After 10 s the cells were separated from the medium by rapid filtration using glass fibre filters and after washing the amount of $^{45}Ca^{2+}$ remaining in association with the cells was determined. The results are expressed in terms of the increment of $^{45}Ca^{2+}$ lost to the medium due to the presence of inositol 1,4,5-trisphosphate.

In the experiment shown in Fig. 5 the ATP-dependent Ca^{2+} pools of permeabilized hepatocytes were loaded with $^{45}Ca^{2+}$ in sucrose-based medium and then the cells were diluted into different media with or without inositol 1,4,5-trisphosphate to examine the effect of various cations on the Ca^{2+} release process. In all cases the anionic species was gluconate which is a non-permeant ion. Fig. 5 shows the amount of Ca^{2+} release which was caused specifically by the presence of inositol 1,4,5-trisphosphate. The K^+ gave the greatest release of Ca^{2+}, while sucrose or the non-permeant cation tris were strongly inhibitory. When Na^+ was substituted for K^+ the release of Ca^{2+} was only partially inhibited while Li^+ with its much larger hydrated radius was much more inhibitory. These data suggest the presence of a cation channel which favours K^+ to allow charge compensation for the electrogenic movement of Ca^{2+} in response to inositol 1,4,5-trisphosphate. Thus, inositol 1,4,5-trisphosphate appears to release Ca^{2+} from an intracellular ATP-dependent Ca^{2+} pool which has channel-like properties in the liver cell. Purification and reconstitution of the component(s) involved in this process is clearly an important next step which may be facilitated using the recently described photoaffinity label of inositol 1,4,5-trisphosphate (38).

CONCLUSIONS

Fig. 6 shows a scheme describing the Ca^{2+} fluxes occurring in hormone-stimulated liver cells combined with the known inositol lipid changes involved in regulating these Ca^{2+} fluxes. Agonist binding to receptors at the hepatocyte plasma membrane results in activation of a phospholipase C [PL-C] (39) probably through the mediation of a guanine nucleotide binding protein [G-protein] (40-44). This leads to degradation of phosphatidylinositol 4,5-bisphosphate [PIP_2] (2,3) to form diacylglycerol [DG] (3,45) and inositol 1,4,5-trisphosphate [IP_3] (8,33). Diacylglycerol activates protein kinase C [PK-C] leading to various protein phosphorylation events whose relationship to Ca^{2+} homeostasis in the liver cell is currently unknown (13). Inositol 1,4,5-trisphosphate on the other hand has been clearly shown to release Ca^{2+} from an intracellular, ATP-dependent Ca^{2+} pool which is a subfraction of the endoplasmic reticulum (10). The resulting elevation of cytosolic free Ca^{2+} causes activation of cytosolic enzymes such as phosphorylase kinase (15-19) as well as causing increases of Ca^{2+} entry into the mitochondria where other Ca^{2+}-sensitive enzymes are located (46,47). In the long term intracellular Ca^{2+} pools are insufficient for the maintenance of elevated cytosolic free Ca^{2+} concentraitons and alterations of Ca^{2+} cycling at the plasma membrane must play the key role in prolonged activation (13,19).

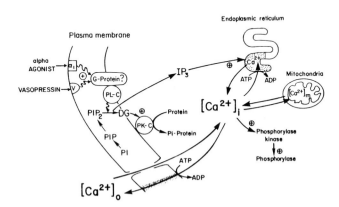

Fig. 6. Scheme for the hormone-induced increase of cytosolic free Ca^{2+} in liver.

REFERENCES

1. Michell, R. H. (1975) Inositol phospholipids and cell surface receptor function. Biochim. Biophys. Acta 415, 81-147.
2. Creba, J. A., Downes, C. P., Hawkins, P. T., Brewster, G., Michell, R. H. and Kirk, C. J. (1983) Rapid breakdown of phosphatidylinositol 4-phosphate and phosphatidylinositol 4,5-bisphosphate in rat hepatocytes stimulated by vasopressin and other Ca^{2+}-mobilizing hormones. Biochem. J. 212, 733-747.
3. Thomas, A. P., Marks, J. S., Coll, K. E. and Williamson, J. R. (1983) Quantitation and early kinetics of inositol lipid changes induced by vasopressin in isolated and cultured hepatocytes. J. Biol. Chem. 258, 5716-5725.
4. Weiss, S. J., McKinney, J. S., and Putney, J. W. (1982) Receptor mediated net breakdown of phosphatidylinositol 4,5-bisphosphate in parotid acinar cells. Biochem. J. 206, 555-560.
5. Billah, M. M. and Lapetina, E. G. (1982) Rapid decrease of phosphatidylinositol 4,5-bisphosphate in thrombin-stimulated platelets. J. Biol. Chem. 257, 12705-12708.
6. Berridge, M. J., Dawson, R. M. C., Downes, C. P., Heslop, J. P. and Irvine, R. F. (1983) Changes in the levels of inositol phosphates after agonist-dependent hydrolysis of membrane phosphoinositides. Biochem. J. 212, 473-482.
7. Agranoff, B. W., Murthy, P. and Sequin, E. B. (1983) Thrombin-induced phosphodiesteratic cleavage of phosphatidylinositol 4,5-bisphosphate in human platelets. J. Biol. Chem. 258, 2076-2078.
8. Thomas, A. P., Alexander, J. and Williamson, J. R. (1984) Relationship between inositol polyphosphate production in the increase of cytosolic free Ca^{2+} induced by vasopressin in the isolated hepatocyte. J. Biol. Chem. 259, 5574-5584.
9. Streb, H., Irvine, R. F., Berridge, M. J. and Schulz, I. (1983) Release of Ca^{2+} from a nonmitochondrial intracellular store in pancreatic acinar cells by inositol 1,4,5-trisphosphate. Nature (London) 306, 67-69.
10. Joseph, S. K., Thomas, A. P., Williams, R. J., Irvine, R. F., and Williamson, J. R. (1984) Myo-inositol 1,4,5-trisphosphate: a second messenger for the hormonal mobilization of intracellular Ca^{2+} in liver. J. Biol. Chem. 259, 3077-3081.
11. Burgess, G. M., Godfrey, P. P., McKinney, J. S., Berridge, M. J., Irvine, R. F., and Putney, J. W. (1984) The second messenger linking receptor activation to internal Ca^{2+} release in liver. Nature (London) 309, 63-66.
12. Berridge, M. J. (1984) Inositol trisphosphate and diacylglycerol as second messengers. Biochem. J. 220, 345-360.
13. Williamson, J. R., Cooper, R. H., Joseph, S. K., and Thomas, A. P. (1985) Inositol trisphosphate and diacylglycerol as intracellular second messengers in liver. Am. J. Physiol. 248, C203-C216.
14. Nishizuka, Y. (1984) The role of protein kinase C in cell surface signal transduction and tumor promotion. Nature (London) 308, 693-698.
15. Williamson, J. R., Cooper, R. H., and Hoek, J. B. (1981) Role of calcium in the hormonal regulation of liver metabolism. Biochim. Biophys. Acta 639, 243-259.
16. Exton, J. H. (1981) Molecular mechanisms involved in α-adrenergic responses. Mol. Cell. Endocrinol. 23, 233-264.
17. Blackmore, P. F., Hughes, B. P., Shuman, E. A., and Exton, J. H. (1982) α-Adrenergic activation of phosphorylase in liver cells involves mobilization of intracellular calcium without influx of extracellular calcium. J. Biol. Chem. 257, 190-197.

90

18. Charest, R., Blackmore, P. F., Berthon, B., and Exton, J. H. (1983) Changes in free cytosolic Ca^{2+} in hepatocytes following α_1-adrenergic stimulation. J. Biol. Chem. 258, 8769-8773.
19. Joseph, S. K., Coll, K. E., Thomas, A. P., Rubin, R. and Williamson, J. R. (1985) The role of extracellular Ca^{2+} in the response of the hepatocyte to Ca^{2+}-dependent hormones. J. Biol. Chem. 260, 12508-12515.
20. Shears, S. B. and Kirk, C. J. (1984) Determination of mitochondrial calcium content in hepatocytes by a rapid cellular fractionation technique. Biochem. J. 219, 383-389.
21. Joseph, S. K. and Williamson, J. R. (1983) The origin, quantitation, and kinetics of intracellular calcium mobilization by vasopressin and phenylephrine in hepatocytes. J. Biol. Chem. 258, 10425-10432.
22. Lin, S.-H., Wallace, M. A., and Fain, J. N. (1983) Regulation of Ca^{2+}-Mg^{2+}-ATPase activity in hepatocyte plasma membranes by vasopressin and phenylephrine. Endocrinol. 113, 2268-2275.
23. Prpic, V., Green, K. C., Blackmore, P. F. and Exton, J. H. (1984) Vasopressin, angiotensin II, and α_1-adrenergic induced inhibition of Ca^{2+} transport by rat liver plasma membrane vesicles. J. Biol. Chem. 259, 1382-1385.
24. Mauger, J.-P., Poggioli, J., Guesdon, F., and Claret, M. (1984) Noradrenaline, vasopressin, and angiotensin II increases Ca^{2+} influx by opening a common pool of Ca^{2+} channels in isolated rat liver cells. Biochem. J. 221, 121-127.
25. Reinhart, P. H., Taylor, W. M., and Bygrave, F. L. (1984) The action of α-adrenergic agonists on plasma membrane calcium fluxes in perfused rat liver. Biochem. J. 220, 43-50.
26. Reinhart, P. H., Taylor, W. M., and Bygrave, F. L. (1982) Calcium ion fluxes induced by the action of α-adrenergic agonists in perfused rat liver. Biochem. J. 208, 619-630.
27. Lapetina, E. G. and Michell, R. H. (1973) Phosphatidylinositol metabolism in cells receiving extracellular stimulation. FEBS Lett. 31, 1-10.
28. Kirk, C. J., Verrinder, T. R. and Hems, D. A. (1977) Rapid stimulation by vasopressin and adrenaline of inorganic phosphate incorporation into phosphatidylinositol in isolated hepatocytes. FEBS Lett. 83, 267-271.
29. Kirk, C. J., Michell, R. H., and Hems, D. A. (1981) Phosphatidylinositol breakdown in rat hepatocytes stimulated by vasopressin. Biochem. J. 194, 155-165.
30. Prpic, V., Blackmore, P. F., and Exton, J. H. (1982) Phosphatidylinositol breakdown induced by vasopressin and epinephrine in hepatocytes is calcium dependent. J. Biol. Chem. 257, 11323-11331.
31. Irvine, R. F., Letcher, A. J., Lander, D. J. and Downes, C. P. (1984) Inositol trisphosphate in carbachol-stimulated rat parotid glands. Biochem. J. 223, 237-243.
32. Irvine, R. F., Anggard, E. E., Letcher, A. J., and Downes, C. P. (1985) Metabolism of inositol 1,4,5-trisphosphate in rat parotid glands. Biochem. J. 229, 505-511.
33. Burgess, G. M., McKinney, J. S., Irvine, R. F. and Putney, J. W. (1985) Inositol 1,4,5-trisphosphate and inositol 1,3,4-trisphosphate formation in Ca^{2+} mobilizing hormone-activated cells. Biochem. J. 232, 237-248.
34. Hallcher, L. M., and Sherman, W. R. (1980) The effects of lithium ion and other agents on the activity of myo-inositol-1-phosphate from bovine brain. J. Biol. Chem. 255, 10896-10901.
35. Batty, I. R., Nahorski, S. R., and Irvine, R. F. (1985) Rapid formation of inositol 1,3,4,5-tetrakisphosphate following muscarinic receptor stimulation of rat cerebral cortical slices. Biochem. J. 232, 211-215.

36. Heslop, J. P., Irvine, R. F., Tashjian, A. H., and Berridge, M. J. (1985) Inositol tetrakis- and pentakis-phosphates in GH_4 cells. J. Exp. Biol. 119, in press.
37. Muallem, S., Schoeffield, M., Pandol, S. and Sachs, G. (1985) Inositol trisphosphate modification of ion transport in rough endoplasmic reticulum. Proc. Natl. Acad. Sci. USA 82, 4433-4437.
38. Koga, T. (1985) Irreversible inhibition of Ca^{2+} release in saponin-treated macrophages by the photoaffinity derivative of inositol 1, 4,5-trisphosphate. Nature (London) 317, 723-725.
39. Seyfred, M. A. and Wells, W. W. (1984) Subcellular site and mechanisms of vasopressin stimulated hydrolysis of phosphoinositides in rat hepatocytes. J. Biol. Chem. 259, 7616-7672.
40. Gomperts, B. D. (1983) Involvement of guanine nucleotide binding protein in the gating of Ca^{2+} by receptors. Nature (London) 306, 64-66.
41. Haslam, R. J. and Davidson, M. M. L. (1984) Receptor induced diacylglycerol formation in permeabilized platelets: possible role for a GTP binding protein. J. Receptor Res. 4, 605-629.
42. Cockcroft, S. and Gomperts, B. D. (1985) Role of guanine nucleotide binding protein in the activation of polyphosphoinositide phosphodiesterase. Nature (London) 314, 534-536.
43. Wallace, M. A. and Fain, J. N. (1985) Guanosine 5'-0-thiotriphosphate stimulates phospholipase C activity in plasma membranes of rat hepatocytes. J. Biol. Chem. 260, 9527-9530.
44. Uhing, R. J., Jiang, H., Prpic, V. and Exton, J. H. (1985) Regulation of a liver plasma membrane phosphoinositide phosphodiesterase by guanine nucleotides and calcium. FEBS Lett. 188, 317-320.
45. Bocckino, S. B., Blackmore, P. F. and Exton, J. H. (1985) Stimulation of 1,2-diacylglycerol accumulation in hepatocytes by vasopressin, epinephrine, and angiotensin II. J. Biol. Chem. 260, 14201-14207.
46. Denton, R. M. and McCormack, J. G. (1980) The role of the calcium transport cycle in heart and other mammalian mitochondria. FEBS Lett. 119, 1-8.
47. McCormack, J. G. (1985) Studies on the activation of rat liver pyruvate dehydrogenase and 2-oxoglutarate dehydrogenase by adrenaline and glucagon. Biochem. J. 231, 597-608.

Résumé

Les cellules hépatiques utilisent une augmentation de Ca^{2+} cytosolique libre comme une étape clé de couplage des effets intracellulaires d'une série d'hormones différentes telles que les agents α_1 adrenergiques et la vasopressine. L'augmentation du Ca^{2+} libre cytosolique induite par les hormones est accompagnée de la dégradation du phosphatidylinositol 4,5 biphosphate avec production de diacylglycerol et d'inositol 1,4,5 triphosphate selon des cinétiques qui sont compatibles avec un rôle de ces composés dans les voies mobilisatrices de Ca^{2+}. En utilisant des hépatocytes perméabilisés par la saponine il a été montré que l'inositol 1,4,5 triphosphate libère Ca^{2+} à partir d'un pool intracellulaire de Ca^{2+} et cela à des concentrations similaires à celles d'inositol triphosphate qui sont mesurées dans des cellules hépatiques intactes stimulées par l'hormone. Ce pool Ca^{2+} sensible au 1,4,5 inositol triphosphate est un pool vésiculaire non mitochondrial, ATP dépendant, probablement une sous fraction du réticulum endoplasmique. La voie de libération de Ca^{2+} n'est pas dépendante du constituant de la pompe Ca^{2+} qui hydrolyse l'ATP et qui remplit le pool. En général les caractéristiques de libération du Ca^{2+} induite par l'inositol 1,4,5 triphosphate évoquent un mécanisme de canal activé par le ligand qui est électrogénique et exige par conséquent la présence d'un cation de permeation compensateur. Bien que le pool Ca^{2+} sensible à l'inositol triphosphate intracellulaire soit suffisant pour rendre compte de l'augmentation initiale au Ca^{2+} cytosolique libre après traitement hormonal il apparaît que les hormones dépendant de Ca^{2+} agissent également en altérant les flux de Ca^{2+} au niveau de la membrane plasmique de l'hépatocyte d'une manière permettant le maintien prolongé d'une augmentation élevée de la concentration de Ca^{2+} cytosolique libre. Il n'a pas encore été déterminé s'il existe une relation mécanistique entre ce dernier effet sur les flux de Ca^{2+} et les changements en métabolites des lipides à inositol.

Peptide hormone action
Action des hormones peptidiques

Hormones and cell regulation. Ed J. Nunez *et al.* Colloque INSERM/John Libbey Eurotext Ltd. © 1986. Vol. 139, pp. 97–109.

Hormonal regulation of testicular steroidogenesis via different transducing systems

F.F.G. Rommerts, R. Molenaar, A.P.N. Themmen and H.J. van der Molen

Department of Biochemistry (Division of Chemical Endocrinology), Erasmus University Rotterdam, PO Box 1738, 3000 DR Rotterdam, The Netherlands

SUMMARY

Testicular steroidogenesis is determined by the number and activity of the Leydig cells in the testis and these parameters are regulated by gonadotrophins. Proliferation and differentiation of Leydig cells in immature rats depends on LH and FSH. However, Leydig cell repopulation in mature rats after selective destruction of existing Leydig cells with ethylene dimethyl sulfonate (EDS) requires LH but can take place without FSH. This indicates that FSH may have a function as a differentiation hormone during development but not in the adult testis. In addition to LH other locally produced factors can regulate steroid production in Leydig cells. Different transducing systems for transmitting signals of hormones or locally produced factors have been identified by comparing the effects of hormones or other agonists on steroid production as well as on phosphorylation and synthesis of specific proteins. Trophic and acute stimulatory effects of LH may involve stimulation of protein kinase A and C and Ca^{2+} fluxes. An LHRH agonist appears not to utilize these transducing systems, but may act via stimulation of phospholipid metabolism in the cell membrane.

INTRODUCTION

In the testis Leydig cells are the source of androgens. The gonadotrophins luteinizing hormon (LH) and follicle–stimulating hormone (FSH) regulate mitosis and maturation of Leydig cells but the cellular and molecular aspects of these trophic effects are still largely unknown. The steroidogenic activity of Leydig cells depends on long term (trophic) as well as on short term stimulatory effects of hormones. LH plays an important role in both processes but there is increasing evidence that locally produced factors may also be important for normal cell function. Investigations of the sequence of events after stimulation of Leydig cells with artificial or naturally occurring regulators may give further insight in the intracellular mechanisms for acute and trophic regulation of Leydig cell function.

The following aspects of regulation of Leydig cell function will be discussed:
1. Regulation of proliferation and differentiation
2. (Intra)cellular transducing systems
3. Effects of locally produced regulators

Details of methods for isolation and incubation of Leydig cells as well as analytical methods have been described elsewhere (1-4) or are given in the tables and legends to figures.

REGULATION OF PROLIFERATION AND DIFFERENTIATION

During the postnatal growth of the mammalian testis the primitive mesenchymal cells in the interstitial tissue play an important role as precursor cells for the development of the adult population of Leydig cells. However the knowledge of the exact origin of Leydig cells and the regulation of cell divisions and maturation is limited (5).

It is widely accepted that LH is the hormone that exerts a major stimulatory influence on the development of Leydig cells but it has been shown that FSH has also clear stimulatory effects (6-8). These effects of FSH are probably mediated via secretion products of Sertoli cells since these cells are the main target cells for FSH (9). Testicular macrophages, which may be influenced by FSH, can also stimulate testosterone production by Leydig cells (10).

It is obvious that for most of the studies on the effects of FSH immature rats have been used since these animals offer the best possibilities to study the maturation of Leydig cells. In adult rats Leydig cells form a stable population of cells and under normal conditions Leydig cell renewal does not take place. Only when high doses of hCG are administered to normal adult rats over a period of several weeks an increase in the number of Leydig cells is observed; however, nothing was reported about the mechanism behind this Leydig cell hyperplasia (11). These studies in immature rats have been complicated by the fact that during testicular development in addition to Leydig cells other cells such as Sertoli cells also differentiate (12). Maturation of Leydig cells may therefore depend largely on the development of Sertoli cells or other testicular cells.

In order to exclude such interactions between developing systems, it might be advantageous to study Leydig cell maturation in fully differentiated testes. This appears to be possible since it was shown that after administration of ethylene dimethyl sulfonate (EDS) to mature male rats the existing population of Leydig cells was selectively destroyed, and that in approximately 35 days repopulation occurred (4,13). Results of these experiments will be discussed below.

After administration of EDS to mature rats, Leydig cell degeneration was observed in tissue sections within 24 h and the testosterone concentration in plasma of these animals was suppressed to approximately 35% of control values "Fig. 1". At 72 h after EDS Leydig cells could not be detected microscopically and plasma testosterone levels had dropped to less than 5% of the normal value. Damaging effects of EDS on germinal cells in the seminiferous tubules or macrophages in the interstitial tissue were not observed. Results of similar experiments in immature 22 days old rats indicate that Leydig cells in these animals cannot be destroyed by administration of EDS. Inhibiting effects of EDS on in vitro steroid production by isolated Leydig cells from mature rats could also be demonstrated, but there were no indications for Leydig cell degeneration during a 24 h incubation period (14). Compounds which are homologous to EDS, such as butylene dimethyl sulfonate (busulfan) or ethylmethyl sulfonate at the same dose as EDS had no effects on Leydig cells, neither in vivo nor in vitro, while these compounds specifically injure spermatogonia in vivo (15).

The molecular mechanisms which form the basis for the specific effects of the different methyl sulfonates are unknown although some information has been collected. The compounds are known as alkylating agents and it has been shown that EDS inhibits within 3 h the synthesis of a few specific proteins in Leydig cells

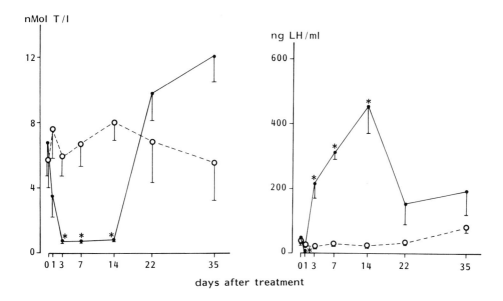

Fig. 1 Temporal changes in concentrations of serum testosterone (left panel) and
LH concentration (right panel) following administration of EDS (solid
line) or vehicle (broken line). Means ± S.E.M., n>5
* significantly different from day 0.

(14). A better insight in the mechanisms which determine the specificity of the
action of EDS ("EDS receptors"?) may be useful for the development of other methyl
sulfonates since these compounds are known to be potent anti-cancer drugs.

For a period of approximately 2 weeks after destruction of Leydig cells by EDS,
peripheral testosterone levels remained at castration levels whereas gonadotrophin
levels were very high. After approximately 14 days testosterone levels increased
and gonadotrophin levels decreased. At this time small groups of Leydig cells
could be identified in the testis. During the following weeks complete restoration
of the Leydig cell population and steroid production took place "Fig. 1".
The hormonal regulation of this process of cell repopulation was studied in
hypophysectomized, Leydig cell depleted rats which were injected daily with saline
hCG (as a substitute for LH) or FSH. The results in Table 1 show that 35 days
after destruction of the Leydig cells only hCG treatment stimulated the generation
of a new population of Leydig cells, while FSH had no effects. In rats with
testosterone implants which suppressed LH below the detection limit of the assay
but maintained FSH levels, Leydig cell repopulation did also not occur. It was
concluded from these observatins that there are precursor cells for Leydig cells
in the testis which are resistant to EDS and dependent on LH for their
multiplication and differentiation to proper Leydig cells. In contrast to Leydig
cell differentiation in immature rats during normal sexual maturation, FSH is not
required for this development in mature rats. It may be possible that in mature

Table 1. Effect of daily injections of hCG and FSH on steroidogenic activities. Measurements were performed in Leydig cells from hypophysectomized rats 35 days following EDS administration.

treatment	plasma testosterone (nmol/1)	pregnenolone production in vitro pmol/3x10^5 nucleated cells/h		
		basal	LH stimulated	22R-OH-cholesterol induced
vehicle	<0.9(7)	0.9 ± 0.4(6)	6.9 ± 9.0(6)	36.3 ± 40.4(5)
hCG 15 i.u.	36.4 ± 24.6*(3)	104.3 ± 57.5*(3)	308.8 ± 171.4*(3)	1491.9 ± 83.2*(3)
FSH 5 µg	<0.9(3)	0.6 ± 0.1(3)	0.8 ± 0.1(3)	4.1 ± 1.2(3)

Means ± S.D. Number of cell preparations in parentheses.
* significantly different from control (p<0.005).

rats Sertoli cells can support the development of Leydig cells independent of FSH. A similar situation exists for the support of spermatogenesis by Sertoli cells which is in immature rats dependent of FSH and testosterone and which can be maintained in mature rats by testosterone alone. Both observations can be understood if FSH has a major function as a differentiation hormone. Although adult Sertoli cells may have become less dependent of FSH, this does not imply that they are less active. To the contrary, evidence for the importance of locally produced factors for regulation of trophic and acute regulation of Leydig cells is growing (6-10).

INTRACELLULAR TRANSDUCING SYSTEMS

LH stimulates the differentiation of Leydig cells and increases the rate of steroid production. Although the production of androgens appears to be the physiologically most important function of Leydig cells, there are indications that these cells also synthesize prostaglandins and opiates (16,17). Furthermore, it has been shown that, in addition to LH, other compounds may regulate steroid production e.g. epidermal growth factor (18), LHRH (19) and vasopressin (20). Leydig cells may therefore be considered as multifunctional cells regulated by a variety of hormones with presumably different modes of action.
It is generally accepted that in transmission of the LH effect on steroid production cAMP plays an important role, since stimulation of the mitochondrial cholesterol side-chain cleavage enzyme by LH is accompanied by increased levels of cAMP, activation of protein kinase A and increased protein phosphorylation of at least six proteins (21,22,23). However, several discrepancies have been observed between hormone-induced dose response curves for cAMP and for steroid production (24,25). This has been frequently explained by assuming that hormone-induced changes in cAMP levels occur in specific intracellular pools of cAMP. However cAMP can pass the cell membrane and can accumulate in the extracellular space. It is therefore difficult to envisage isolated intracellular pools. An alternative explanation is that LH activates other messenger systems in addition to adenylate cyclase. Involvement of other transducing systems may also explain why isoproterenol can stimulate cAMP production more than 10-fold in freshly isolated Leydig cells without any change in steroid production whereas LH can stimulate steroid production more than 10-fold without any detectable change in cAMP (24).

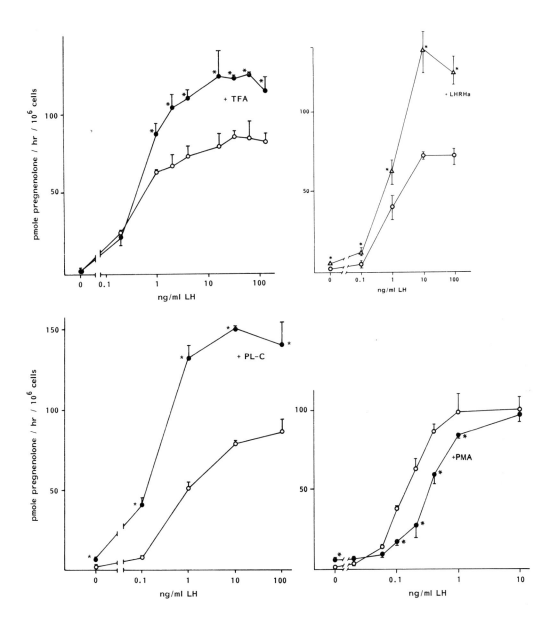

Fig. 2. Effect of 100 μM 9(tetrahydro-2-furyl)adenine (TFA), 10 nM LHRH agonist HOE 766 (LHRHa), 1 U/ml phospholipase C (PL-C) and 100 ng/ml 4β-phorbol-12-myristate-13-acetate (PMA) on LH stimulated pregnenolone production by Leydig cells from immature rats. Leydig cells were incubated for three hours with different concentrations of LH with or without additions. Values shown are means ± S.D. (4) of two experiments.

* significantly different (P<0.01) from corresponding productions with LH alone.

The observation that the ED_{50} of hCG stimulated testosterone production by Leydig cells is 60 times lower than the ED_{50} of choleratoxin, while the dose response curves for stimulation of cAMP production by hCG and choleratoxin are similar (25), also indicates that LH may employ alternative pathways to stimulate steroid production.

We have investigated if cAMP plays an obligatory role in the regulation of steroid production by studying the effect of inhibition of adenylate cyclase. In complementary studies compounds which are supposed to interfere with cAMP-independent pathways have been tested.

Incubations of LH-stimulated Leydig cells in the presence of 100 µM 9(tetrahydro-2-furyl)adenine (TFA) (a P site agent) leads to inhibition of adenylate cyclase. This could be inferred from a 22% decrease in the intracellular levels of cAMP which was independent of LH concentration between 5 and 100 ng LH/ml (26).

This inhibition of cyclase did not coincide with an inhibition of steroid production but surprisingly a stimulation of steroid production was observed which was largest at high doses of LH "Fig. 2".

Similar results on cAMP and steroid production were obtained with 2'5'-dideoxyadenosine (DDA), another inhibitor of adenylate cyclase (26). At present it cannot be explained how inhibition of the cyclase leads to stimulation of steroid production. These results clearly do not support the obligatory role of cAMP for regulation of steroid production, but they further illustrate that other transducing systems may function in Leydig cells.

Intracellular calcium has been shown to be also involved in the mechanism of stimulation of steroidogenesis by LH (27), but less experiments on the role of calcium have been carried out. Possible reasons for this are that specific changes in intracellular calcium levels and calcium dependent pathways are difficult to measure, especially when they interact with cAMP dependent pathways.

Experiments with an LHRH agonist have shown that steroid production in Leydig cells could be stimulated without changes in cAMP levels and some suggestions for a role of calcium in this process have been made (28,29). It is unlikely that LHRH is produced within the testis and can act as a paracrine hormone, but it cannot be excluded that a high molecular weight LHRH-like molecule is produced (31). The relative importance of calcium in the action of the LHRH agonist and LH on steroid production was determined by utilizing diltiazem as a calcium channel blocker. It was found that the action of LH on steroid production was partly inhibited when diltiazem was present "Fig. 3". LHRH agonist stimulated steroid production was not inhibited. Non-specific effects of diltiazem on the cholesterol side chain cleavage activity in the presence of 25-hydroxycholesterol were not observed "Fig. 3". These results indicate that calcium fluxes may play a more important role in the LH rather than in the LHRH agonist action. This observations supports the notion that LH action involves different transducing systems. No further insight was however obtained in the intracellular signals triggered by LHRH agonist. Since the effects of the LHRH agonist on steroid production are relatively small it may be possible that the LHRH agonist acts partly via the transducing system employed by LH. Therefore, combination experiments with LH and the LHRH agonist were carried out in order to test whether LH and the LHRH agonist act independently. The results in Figure 2 show that the LHRH agonist could further stimulate the maximally LH-stimulated steroid production. This is an indication that the LHRH agonist stimulates cellular activities which are not under the influence of LH.

The possibility that phospholipid metabolism or activation of protein kinase C could play a major role in the action of the LHRH agonist was studied by comparing the effects of phospholipase C and phorbol esters on steroid production "Fig. 2". Phospholipase C added to intact cells could mimic the synergistic effects of the LHRH agonist suggesting that the LHRH agonist may also stimulate phospholipid metabolism. This confirms studies in which effects of the LHRH agonist on phosphatidylinositol turnover have been described (32,33). Specific activation of protein kinase C by 4β-phorbol-12 myristate-13-acetate (PMA) caused a small

102

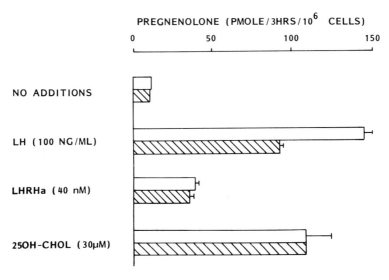

PREGNENOLONE (PMOLE/3HRS/10^6 CELLS)

Fig. 3. Effects of the calcium-channel blocker diltiazem on pregnenolone production by immature rat Leydig cells in the presence of LH, LHRHa or 25-hydroxycholesterol. Cells were incubated for 3 h with the indicated hormone concentrations in the absence (open bars) or presence of 100 µM diltiazem (hatched bars). Values given are means ± S.D. (n=4) of two experiments.
* p<0.01 vs corresponding incubation without diltiazem.

stimulation of steroid production when LH was absent, but in the presence of non-saturating doses of LH a small inhibition was observed. PMA exerted a constant but small stimulatory effect when steroid production was stimulated with increasing doses of dibutyryl-cAMP (dcAMP) (data not shown). 4β-Phorbol-13 monoacetate (PA) which does not activate protein kinase C had no effects on steroid production (data not shown). It can be inferred from the results of these experiments with PMA in combination with LH or dcAMP, that PMA specifically inhibits the adenylate cyclase and that there are no effects on the post cAMP-dependent steps that control steroid production. This was recently confirmed by a direct demonstration of the inhibitory effects of PMA on cAMP production (34). It was previously shown in this chapter that inhibition of adenylate cyclase with TFA and DDA leads to stimulation of steroid production. The effects of TFA and DDA on steroid production are thus in contrast with those of PMA. It is known that TFA and DDA act on the catalytic component of the adenylate cyclase (35) and that PMA presumably acts at the G-protein which mediates the interaction between LH receptors and adenylate cyclase (34). It can therefore be concluded that inhibition of adenylate cyclase alone cannot explain the effects of PMA on steroid production and the results support the notion that the G-protein may interact with different proteins in the plasma membrane, among which proteins that play a role in regulation of calcium transport (36).

There is abundant evidence that the primary effects of hormones on the cell membrane determine the degree of phosphorylation of specific proteins which play a role in an integrated network of regulatory pathways that control cell function (37). The divergence in the effects of PMA, TFA, LHRH agonist and phospholipase C on steroid production was therefore further investigated by analyzing the pattern of protein phosphorylation after cells had been incubated with the indicated compounds and radioactively labelled phosphate. The results of these experiments

are shown in Figure 4. LH and PMA both stimulated the phosphorylation of 17K and
33K proteins, whereas the LHRH agonist and phospholipase C had no detectable
effect on protein phosphorylation. The similarities between the effects of PMA and
LH on protein phosphorylation are striking since PMA and LH predominantly activate
different kinases and have completely different effects on steroid production. The
observation that PMA and LH cause increased phosphorylation of the same proteins
may indicate that protein kinase A and C have common properties or that they
interact. These interactions may also occur inside the cell membrane before or
during the activation step and the possible involvement of the G-protein has
already been mentioned. Protein kinase A and C may also independently
phosphorylate a common protein at different sites. This was shown to be the case
in erythrocytes (38). At present it is unknown which of these alternatives occurs
predominantly in Leydig cells. The results obtained with PMA show that a
significant increase in the phosphorylation of the 17K and 33K proteins does not
coincide with an increased steroid production as was observed with LH. Thus the
phosphorylation of these proteins does not appear to be an essential step in the
regulation of steroid production, although a permissive role of these proteins
cannot be excluded. Other phosphoproteins which have been detected in tumour cells
(2) have not yet been demonstrated in Leydig cells of immature rats neither have
specific functions of these phosphoproteins for steroid production been
demonstrated (39). It may even be possible that these strongly phosphorylated
proteins mainly represent structural elements of the cell, such as the
cytoskeleton, which play a major role in the trophic effects of hormones rather
than in the acute stimulatory actions. This notion is supported by the observation
that the LHRH agonist and phospholipase C have no detectable effect on general
protein phosphorylation while these agents can amplify LH-stimulated steroid

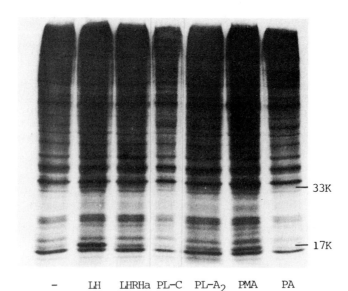

       −   LH   LHRHa PL-C PL-A$_2$ PMA   PA

Fig. 4. Protein phosphorylation pattern of immature rat Leydig cells after
incubation of the cells for 3 h with radioactively labelled phosphate and
hormones. ^{32}P-labelled phosphoproteins were isolated from cells incubated without
further additions or with LH (100 ng/ml), LHRH agonist (40 nM), phospholipase C
(PL-C; 1 U/ml), phospholipase A$_2$ (5 U/ml), PMA (100 ng/ml) or PA (100 ng/ml).
Proteins were separated on SDS-PAGE. The autoradiograph of the dried gel is shown.

production. Investigations on phosphorylation of specific proteins which are directly involved in the relation of cholesterol side-chain cleavage activity might be more meaningful, but it has not yet been possible to identify these proteins; LH-dependent phosphoproteins have not been detected in mitochondria of Leydig cells (2). However, two proteins have been proposed for regulation of adrenal steroid production at the mitochondrial level. Sterol carrier protein (SCP$_2$) is supposed to play a role in the supply of cholesterol for the cholesterol side-chain cleavage enzyme located in the inner mitochondrial membrane (40) and a rapidly synthesized activator peptide of molecular weight 2K has been suggested to stimulate the mitochondrial enzyme activity (41). In future studies on regulation of steroid production in Leydig cells possible hormone effects on these specific proteins have to be investigated.

The previous results have illustrated that the mechanism of action of the different hormones or probes can be classified only partly by investigations of patterns of protein phosphorylation. The intracellular pathways of LH, PMA, LHRH agonist and phospholipase C action were therefore further characterized by investigating the pattern of newly synthesized proteins as a fingerprint of long

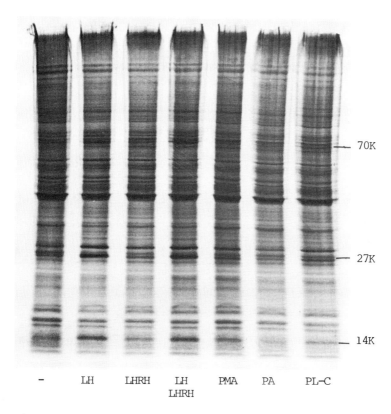

      — 70K

      — 27K

      — 14K

     —   LH   LHRH   LH   PMA   PA   PL-C
                           LHRH

Fig. 5. Protein synthesis pattern of immature rat Leydig cells after incubation of the cells for 3 h with radioactively labelled ^{35}S-methione. Radioactively labelled proteins were isolated from the cells and separated on SDS polyacrylamide gels, the autoradiograph of the dried gel is shown. The following additions were made: LH (100 ng/ml), LHRHa (40 nM), LH together with LHRHA, PMA (100 ng/ml), PA (100 ng/ml), phospholipase C (PL-C; 1 U/ml).

term (trophic) effects. The synthesis of these proteins appears not to be relevant for the acute effects on steroid production since hormonal effects on protein synthesis were only shown after incubation periods of several hours whereas stimulation of steroid production can be obtained within several minutes. On the other hand the hormone-induced changes in protein synthesis may be considered as a long term functional consequence of alterations in protein phosphorylation.
The results presented in figure 5 show that all four compounds tested were able to exert stimulatory actions on synthesis of specific proteins, although the degree of stimulation varied. The effects of the 4 stimulatory agents were mainly expressed in the synthesis of only 3 proteins. The effects of LH and PMA were stronger than those of phospholipase C and the LHRH agonist. LH and PMA both stimulated the synthesis of 14, 27 and 70K protein whereas the LHRH agonist only stimulated the 27K protein.
The functions of these newly synthesized proteins are not known. The similarities between the pattern of LH- and PMA-induced proteins on one hand and that of LHRH agonist and phospholipase C on the other hand reinforce the hypothesis that LHRH agonist and phospholipase C employ transducing pathways that are different from those of LH and PMA.
The effects of the 4 investigated stimulators on steroid production, protein phosphorylation and protein synthesis lead to the conclusion that LH action may involve activation of protein kinase A, protein kinase C and calcium transport through the membrane. Because of analogies between the effects of phospholipase C and LHRH agonist it is proposed that LHRH may act through activation of phospholipid metabolism in the cell membrane. It remains to be demonstrated how these different activities finally regulate cholesterol side-chain cleavage activity inside the mitochondria.

EFFECTS OF LOCALLY PRODUCED REGULATORS

Different transducing systems present in Leydig cells may play a role in the regulation of cell growth and differentiation and of steroid production.
It has been discussed earlier that Sertoli cells or testicular macrophages can influence Leydig cell function. Recently it was shown that interstitial fluid from

Table 2. Effect of serum (S) or testicular fluid (TF) on steroid production by Leydig cells incubated in the presence of 1% foetal calf serum (FCS) and LHRH or 22R-hydroxycholesterol

added to incubation medium	pmol pregnenolone/0.5×10^6 cells/3 h			ratio $\frac{b}{a}$
	foetal calf serum (1%)	S (40%) (a)	TF (40%) (b)	
control	2.9 ± 0.2	7.2 ± 0.3	18.7 ± 0.6*	2.6
LHRH (50 ng/ml)	11.5 ± 0.4	15.9 ± 0.2	46 ± 3*	2.9
LH (100 ng/ml)	80 ± 5	72 ± 2	326 ± 22*	4.5
22R-hydroxy- cholesterol (80 µM)	712 ± 29	783 ± 37	2389 ± 47*	3.1

Means ± S.D. of quadruplicate incubations from 2 different experiments.
*P<0.01 when compared to serum

rat testes could enhance the LH-stimulated testosterone production by isolated Leydig cells and this effect could not be suppressed by antiserum to LH or LHRH antagonists (42). These observations indicate that in the fluid which surrounds Leydig cells specific intratesticular regulators are present which may employ one of the available transducing systems for regulation of Leydig cell functions. We have therefore attempted to study effects of compounds in interstitial fluid on pregnenolone production, protein phosphorylation and protein synthesis.

Addition of a medium containing 40% testicular fluid to Leydig cells resulted in an approximately 6-fold increased steroid production "Table 2". Addition of the same amount of serum gave only a two-fold increase. Leydig cells maximally stimulated with LHRHa or LH (28) showed a 3-fold higher steroid production when testicular fluid was present "Table 2". Serum had no effect under these conditions. The effect of testicular fluid was also apparent when cells were incubated with 22R-hydroxycholesterol to measure the maximal endogenous cholesterol side-chain cleavage activity not limited by the supply of cholesterol. Addition of 22R-OHcholesterol stimulated steroid production more than 200-fold. Under these conditions addition of testicular fluid resulted in a further increase of the steroid production. Thus, addition of testicular fluid causes a significantly higher steroid production than addition of serum, irrespective of the activity of pregnenolone production which can vary more than 300-fold depending on the additions to the incubation medium. This indicates that factors in testicular fluid can act as amplifiers of steroid production.
The effects of testicular fluid on 22R-hydroxycholesterol stimulated steroid production point to a specific effect of intratesticular modulators. None of the compounds previously tested (LH, LHRH, PMA, phospholipase C) exerted similar effects. Moreover, since testicular fluid can further increase steroid production already maximally stimulated by LHRH or LH, it appears that the active principles in the testicular fluid utilize stimulatory pathways which differ from those employed by LHRH and LH. The active principle in testicular fluid was heat sensitive, could not be removed by charcoal treatment and was retained behind a filter with a cut off point of a molecular weight of 25 kD. This is in agreement with earlier reported results (42). Since rat testicular fluid can only be obtained in small quantities (50 µl per testis), we have tried to find alternative sources. Testicular lymph from boars may be useful since 20 ml portions can easily be obtained after cannulation of testicular lymph vessels. Characterizations of this testicular fluid has recently been started.

The results presented in this paper show that intratesticular factors in addition to LH are involved in the regulation of steroidogenesis and that different transducing systems play a role in these regulatory processes.

REFERENCES

1. Rommerts FFG, Molenaar R, van der Molen HJ (1985) Meth in Enzymol 107:275-288
2. Bakker GH, Hoogerbrugge JW, Rommerts FFG, van der Molen HJ (1981) Biochem J 198:339-346
3. Bakker GH, Hoogerbrugge JW, Rommerts FFG, van der Molen HJ (1982) Biochem J 204:809-815
4. Molenaar R, de Rooy DG, Rommerts FFG, Reuvers RJ, van der Molen HJ (1985) Biol Reprod., accepted for publication
5. Christensen AK (1975) In: Hamilton DW, Greep RO (eds) Handbook of Physiology. Section 7. Vol. 5. Williams and Wilkins, Baltimore, pp 57-94
6. Odell WD, Swerdloff RS (1976) Recent Prog Horm Res 32:245-288
7. van Beurden WHO, Roodnat B, de Jong FH, Mulder E, van der Molen HJ (1976) Steroids 28:847-866
8. Kerr JB, Sharpe RM (1985) Endocrinology 116:2592-2604

9. Means AR, Fakundig JL, Huckins C, Tindall DJ, Vitale R (1976) Recent Prog Horm Res 32:477-527
10. Yee JB, Hutson JC (1985) Endocrinology 116:2682-2685
11. Christensen AK, Peacock KC (1980) Biol Reprod 22:383-391
12. Ritzen EM, Syed V (1985) In: Saez JM, Forest MG, Dazord A, Bertrand J (eds) Recent Progress in Cellular Endocrinology of the Testis. Inserm Paris, pp 141-155
13. Molenaar R, de Rooy DG, Rommerts FFG, van der Molen HJ (1985) Endocrinology: submitted
14. Rommerts FFG, Grootenhuis AJ, Hoogerbrugge JW, van der Molen HJ (1985) Molec Cell Endocr 42:105-111
15. de Rooy DG, Kramer MF (1968) Z Zeelforsch Mikrosk Anat 92:400-406
16. Haour F, Kouznetzova B, Dvay F, Saez JM (1979) Life Sci 24:2151-2158
17. Chen CLC, Mather JP, Morris PL, Bardin CW (1984) Proc Natl Acad Sci 81:5672-5675
18. Melner MH, Lutin WA, Puett D (1982) Life Sci 30:1981-1986
19. Sharpe RM, Cooper I (1982) Molec Cell Endocr 27:199-211
20. Adashi EY, Hsueh AJW (1981) Endocrinology 109:1793-1798
21. Cooke BA, Lindh LM, Janszen FHA (1976) Biochem J 160:439-446
22. Podesta E, Dufau ML, Solano AR, Catt KJ (1978) J Biol Chem 253:8994-9001
23. Bakker GH, Hoogerbrugge JW, Rommerts FFG, van der Molen HJ (1983) Molec Cell Endocr 33:243-253
24. Cooke BA, Golding M, Dix CJ, Hunter MG (1982) Molec Cell Endocr 27:221-231
25. Dufau ML, Horner KA, Hayashi K, Tsuruhura T, Conn PM, Catt KJ (1981) J Biol Chem 253:3721-3729
26. Themmen APN, Hoogerbrugge JW, Rommerts FFG, van der Molen HJ (1985) Biochem Biophys Res Comm 128:1164-1172
27. Hall PF, Osawa S, Mrotek J (1981) Endocrinology 109:1677-1682
28. Rommerts FFG, Molenaar R, Themmen APN, van der Molen (1984) In: McKerns KW (ed) Hormonal Control of the Hypothalamo-pituitary Gonadal Axis. Plenum Publ. Corp. pp 423-436
29. Sullivan MHF, Cooke BA (1984) Molec Cell Endocr 34:17-22
30. Hedger MP, Robertson DM, Browne CA, de Kretser DM (1985) Molec Cell Endocr 42:163-174
31. Bhasin S, Swerdloff RS (1984) Biochem Biophys Res Comm 122:1071-1075
32. Lowitt S, Farese RV, Sabir AM, Root AW (1982) Endocrinology 11:1415-1417
33. Molcho J, Zakut H, Naor Z (1984) Endocrinology 114:1048-1050
34. Mukhopadhyay AK, Schumacher M (1985) FEBS Lett 187:56-60
35. Florio VA, Ross EM (1983) Molec Pharmacol 24:195-202
36. Gomperts BD (1983) Nature 306:64-66
37. Cohen P (1982) Nature 296:613-620
38. Horne WC, Leto TL, Marchesi VT (1985) J Biol Chem 260:9073-9077
39. Rommerts FFG, Bakker GH, van der Molen HJ (1983) J Steroid Biochem 19:367-373
40. Chanderban R, Noland BJ, Scallen TJ, Vahouny GV (1982) J Biol Chem 257:8928-8934
41. Pederson RC, Brownie AC (1983) Proc Natl Acad Sci 80:1882-1886
42. Sharpe RM, Cooper I (1984) Molec Cell Endocr 37:159-168

Résumé

La stéroïdogénèse testiculaire est déterminée par le nombre et l'activité des cellules de Leydig dans les testicules et ces paramètres sont régulés par les gonadotropines. La prolifération et la différenciation des cellules de Leydig chez le rat immature dépendent de la LH et la FSH. Cependant, la repopulation en cellules de Leydig chez le rat mature après destruction sélective des cellules de Leydig existantes par l'éthylène dimethylsulfonate (EDS) exige LH mais peut intervenir sans FSH. Ceci indique que FSH peut avoir une fonction de différenciation hormonale pendant le développement mais non dans le testicule adulte. Outre la LH d'autres facteurs produits localement peuvent réguler la production de stéroïdes dans les cellules de Leydig. Différents systèmes de transduction responsables de la transmission des signaux hormonaux ou de facteurs produits localement ont été identifiés en comparant les effets des hormones ou autres agonistes sur la production de stéroïdes ainsi que sur la phosphorylation et la synthèse de protéines spécifiques. Les effets trophiques et aigus de LH peuvent impliquer la stimulation des protéines kinases A et C et les flux de Ca^{2+}. Un agoniste, LHRH, apparaît ne pas utiliser ces systèmes de transduction mais peut agir via la stimulation du métabolisme phospholipidique dans la membrane cellulaire.

Hormones and cell regulation. Ed J. Nunez *et al.* Colloque INSERM/John Libbey Eurotext Ltd. © 1986. Vol. 139, pp. 111–128.

Luteinizing hormone receptor-adenylate cyclase regulation in Leydig cells: desensitization, receptor internalization and recycling and roles of cyclic AMP and Ca^{2+}

B.A. Cooke, C.J. Dix, A.D. Habberfield and M.H.F. Sullivan

Department of Biochemistry, Royal Free Hospital School of Medicine, (University of London), Rowland Hill Street, London NW3 2PF, UK

The mechanisms of desensitization, receptor internalization and recycling have been investigated in testis Leydig cells and tumour Leydig cells. The factors controlling the production of cyclic AMP and Ca^{2+} and their possible roles in these processes have been assessed. It was found using the fluorescent intracellular Ca^{2+} chelator quin-2, that both LH and cyclic AMP analogues stimulate increases in intracellular Ca^{2+} ($[Ca^{2+}]_i$) in the testis Leydig cells indicating that cyclic AMP is a modulator of $[Ca^{2+}]_i$ in this cell type. This increased $[Ca^{2+}]_i$ was derived mainly from the extracellular medium. Experiments carried out with the tumour cells and stimulated by LH, forskolin and cholera toxin demonstrated that the initial homologous LH-induced desensitization is rapid (within minutes) and independent of cyclic AMP, whereas the heterologous desensitization is cyclic AMP dependent and requires 4–8 hours <u>in vitro</u>. Adenosine was found to potentiate LH-stimulated cyclic AMP production and prevent LH-induced desensitization. A phorbol ester was shown to mimic and be more potent than LH in its effects on desensitization. However, in contrast to LH-desensitized cells, the plasma membranes from phorbol ester treated cells did not have a decreased response to LH and LH plus GppNHp.

In the testis Leydig cells the binding of LH to its receptor is followed by rapid internalization of the hormone–receptor complex, dissociation of the hormone–receptor complex in endocytic vesicles, metabolism of the hormone and return of the receptors to the cell surface. The evidence for the roles of Ca^{2+} and cyclic AMP in desensitization and internalization is discussed.

Key words: Leydig cell, Ca^{2+}, cyclic AMP, desensitization, tumour, internalization, recycling, monensin, quin-2, LH-receptor.

INTRODUCTION

The initial action of luteinizing hormone (LH) or choriogonadotropin (hCG) on the testis Leydig cell is to stimulate the formation of the male androgen testosterone, a response which is thought to be mediated through the production of cyclic AMP (1,2). However, apart from this stimulatory action, LH and hCG also cause a refractoriness or desensitization of that same steroidogenic response (3). This hormone-induced loss of LH responsiveness may involve a loss of plasma membrane receptors for LH (4,5,6,7) an uncoupling of the LH receptor from the adenylate cyclase (8,9,10) an increase in the metabolism of cyclic AMP due to increased phosphodiesterase activity (11) and a decrease in the activities of some of the enzymes in the steroidogenic pathways from pregnenolone to testosterone (12).

In other systems there is recent evidence which suggests that Ca^{2+}-mediated processes may be involved in desensitization and internalization of receptors. In particular activation of protein kinase C (via increased phosphoinositide metabolism) leading to phosphorylation of receptors and/or GTP binding protein may be taking place (see review, (13)).

In this review we wish to summarise our recent studies on a) the control of intracellular Ca^{2+} and cyclic AMP b) the effects of modulators of desensitization and internalization c) the possible roles of Ca^{2+} and cyclic AMP in these processes. These studies have been carried out <u>in vitro</u> using cells from the normal rat testis and from a Leydig cell tumour.

(I) REGULATION OF CYCLIC AMP AND $[Ca^{2+}]_i$

Using the recently developed fluorescent intracellular Ca^{2+} chelator quin-2 (14) we have measured the intracellular calcium concentrations $([Ca^{2+}]_i)$ in purified rat Leydig cells (15). The effects of LH and cyclic AMP analogues on $[Ca^{2+}]_i$ have been determined and compared with cyclic AMP and testosterone production.

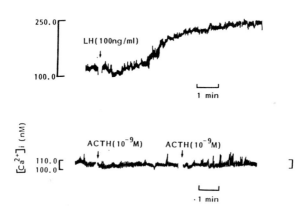

Fig. 1   Effects of LH and ACTH on $[Ca^{2+}]_1$ in rat testis Leydig cells.

Typical fluorescence traces are shown which were obtained after loading the cells with quin-2 and then LH (100ng/ml) or ACTH (10^{-9}M) was added as indicated (data from (15)).

A typical fluorescence trace for the effect of LH on $[Ca^{2+}]_i$ is shown in shown in Fig 1. It was found that there was no increase above basal (which was 89.4±16.6nM Ca^{2+}, mean± S.D. n=25) during the first 1-2 min which was followed by a linear increase in $[Ca^{2+}]_i$ to reach a maximum after 6-8 min. There was a similar lag time for all the compounds investigated that increased $[Ca^{2+}]_i$, the only exception being the Ca^{2+} ionophore ionomycin which produced a very rapid increase in $[Ca^{2+}]_i$.

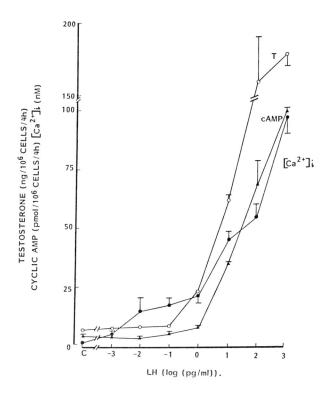

Fig. 2 Effects of LH on $[Ca^{2+}]_i$, cyclic AMP and testosterone production in rat testis Leydig cells

$[Ca^{2+}]_i$ (●) represents the increase above basal $[Ca^{2+}]_i$ (89.4±16.6nM, mean S.D., n=25). Cyclic AMP (▲) and testosterone (O) production in the Leydig cells was determined after 4h in the presence of various concentrations of LH. All data are means ±S.E.M. (n=3) and are representative of 2 experiments. Data from (15).

In Fig. 2 the results of the effects of different amounts of LH on $[Ca^{2+}]_i$, cyclic AMP and testosterone levels are given. Significant increases in $[Ca^{2+}]_i$ and testosterone production were detectable with .01 pg/ml of LH and in cyclic AMP with 0.1 pg/ml LH. Maximum steroidogenesis was reached with 100pg/ml LH whereas both $[Ca^{2+}]_i$ and cyclic AMP levels continued to increase in parallel with amounts of LH up to 1ng/ml (Fig 2). Even with LH concentrations up to 1000ng/ml maximum $[Ca^{2+}]_i$ and cyclic AMP levels were not reached.

The parallel increase in LH-stimulated $[Ca^{2+}]_i$ and cyclic AMP levels suggested that they might be inter-related and therefore the effects of cyclic AMP and cyclic AMP analogues on $[Ca^{2+}]_i$ were investigated. It was found that dibutyryl cyclic AMP increased $[Ca^{2+}]_i$; the increases obtained to give maximum testosterone production were similar to those required with LH (78-80pM and 60-90nM respectively). 8Br-cyclicAMP also gave similar increases in $[Ca^{2+}]_i$; with .01, 0.5 and 5.0mM the $[Ca^{2+}]_i$ increases in were 29.7 ± 8.3; 107.9 ±6.3 and 265.3±13.3nM respectively (means ± range from 2 separate experiments). None of the other compounds tested (butyrate, AMP, ADP, adenosine and ATP) gave a detectable increase in $[Ca^{2+}]_i$. Essentially similar results were obtained with these compounds with respect to their effects on testosterone production.

The effects of lowering the extracellular $[Ca^{2+}]$ on LH stimulated cyclic AMP was investigated. Only small effects of depleting the $[Ca^{2+}]_e$ from 2.5mM to 1.1μM were found; with the highest amounts of LH added the cyclic AMP production was approximately 25% lower.

The effects of lowering the extracellular $[Ca^{2+}]$ on $[Ca^{2+}]_i$ were investigated. Basal $[Ca^{2+}]_i$ was decreased to 61.3±5.7nM (means±S.E.M., n=20) although basal testosterone production was not affected. With 100pg/ml LH a substantial increase in $[Ca^{2+}]_i$ was still obtained with 1.1μM $[Ca^{2+}]_e$ and was similar to that obtained with 2.5mM (50nM). No further increase occurred with 100ng/ml LH with the low Ca^{2+} medium. These LH-increased $[Ca^{2+}]_i$ were still sufficient to give near maximum steroidogenesis because testosterone production was only decreased by 35% in the low Ca^{2+} medium.

These results demonstrate that LH and cyclic AMP stimulate increases in intracellular Ca^{2+} concentrations in rat Leydig cells and that the increases in $[Ca^{2+}]_i$ are mainly derived from the extracellular medium. Whether there is also a mobilization of intracellular stores via the calcium mobilizer inositol 1,4,5-trisphosphate (IP_3) as has been shown in other cells (16) remains to be determined. LH has been shown to stimulate phosphoinositide metabolism in rat Leydig cells (17,18). However the time required to stimulate Leydig cell $[Ca^{2+}]_i$ by LH and cyclic AMP analogues (2-3 min) is much slower than in other tissues (within seconds) where IP_3 has been shown to be involved. This possibly indicates another mechanism. If cyclic AMP does directly modulate $[Ca^{2+}]_i$ it may be activating plasma membrane Ca^{2+}-channel by cyclic AMP dependent protein kinase mediated phosphorylation. This has been suggested for heart cells (19).

(II) DESENSITIZATION OF ADENYLATE CYCLASE IN TUMOUR LEYDIG CELLS

Because of the limitations of cell numbers with the normal Leydig cell, studies were initially carried out with the tumour cells which have been shown to have many of the characteristics of normal Leydig cells (20). In this system LH produces a rapid (within 16-60 min) desensitization of subsequent stimulation with LH; however the response to cholera toxin in intact cells and to guanine nucleotides and NaF in plasma membranes from LH-desensitized cells is unaltered (10). In the same studies it was shown that exposure of the cells to dibutyryl cyclic AMP had no effect on subsequent responses of intact cells or plasma membranes. These findings led to the conclusions that LH induces a lesion at or proximal to the GTP-binding protein (N_s-protein) and that this lesion is specific to the LH response and is not mediated by cyclic AMP (10). In order to characterise further this desensitization the stimulatory effects of the diterpene forskolin (21,22) on cyclic AMP

114

production was compared with those of LH (23).

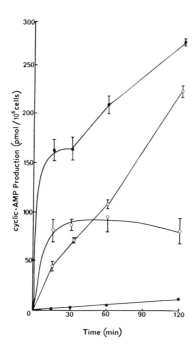

Fig. 3   Effect of forskolin, LH or forskolin + LH on the kinetics of cyclic AMP production in Leydig tumour cells

The cells were incubated at $32^{\circ}C$ in media containing isobutylmethylxanthine (0.5mM) for various times in the presence of LH (O; 1μg/ml), forskolin (□;100μM), LH (1μg/ml) + forskolin (■;100μM) or in medium alone (●) and the cyclic AMP contents were determined.   Results are means ± S.D. (represented by the bars) for duplicate determinations on triplicate samples.   Data from (23).

The rate of cyclic AMP production stimulated by LH was non-linear whereas forskolin-stimulated cyclic AMP production was linear over the time period studied (2h) (Fig.3). The kinetics of the forskolin + LH-stimulated cyclic AMP production showed the characteristics of the kinetics of these stimulants alone.   Although the initial rate of cyclic AMP production was faster than that with LH or forskolin alone, once the LH-stimulated component was desensitized the rate of production was similar to that with forskolin alone.

In contrast with the two-component kinetics seen with LH + forskolin-stimulated cyclic AMP production, a maximal concentration of cholera toxin (5μg/ml) and forskolin (100μM) produced linear kinetics.   After an initial characteristic lag phase, the kinetics of cholera-toxin-stimulated cyclic AMP production were linear for up to 2h. The combination of forskolin and cholera toxin also stimulated cyclic AMP production with linear kinetics.   Whereas the initial rate of cyclic AMP production (0-15 min) stimulated by LH + forskolin showed a clear potentiation at all time points studied, the cholera-toxin + forskolin-stimulated cyclic AMP production only showed additive effects (23).

In order to investigate the effect of time on desensitization, cells were incubated for up to 8h with either LH (1.0μg/ml) or forskolin (100μM), and the subsequent responses to various stimuli were examined. LH produced a rapid desensitization to a second challenge with LH, the cells showing complete desensitization by 2h. However, the rate of desensitization of the forskolin response was slower and only became significantly different from the control response by 4h, and reached the degree of desensitization seen with LH by 8h. The pattern of response to LH + forskolin in these LH-desensitized cells was a combination of the patterns with LH and with forskolin alone, with an initial rapid rate of desensitization followed by a slower decline to a degree of desensitization similar to that of the response to either LH or forskolin by 8h.

Forskolin has been reported to stimulate cyclic AMP production in a variety of cell types (22,24). In common with these findings, forskolin stimulated cyclic AMP production in a dose-dependent maner in isolated rat Leydig tumour cells. The stimulation of cyclic AMP production by forskolin was rapid and did not have a lag phase, as reported (25) for S49 lymphoma cells. Forskolin acted synergistically with LH, but only had an additive effect on cholera-toxin-stimulated cyclic AMP production. This is again in common with the findings of other studies (26,27,28). It is apparent from the kinetics of LH + forskolin-stimulated cyclic AMP production that the synergism is lost once the LH response is desensitized. Since the LH-induced desensitization involves a lesion between the LH receptor and the G-protein (10), the kinetics of LH + forskolin-stimulated cyclic AMP production suggests that the coupling between the LH receptor and the G-protein needs to be functional for synergism between LH and forskolin. Similar conclusions were reached for the actions of isoprenaline and forskolin on genetic variants of S49 lymphoma cells (29).

These studies (23) have characterized more clearly the desensitization of adenylate cyclase in the tumour Leydig cells i.e. the initial homologous LH-induced desensitization is rapid (within 15 min) and independent of cyclic AMP whereas the heterologous desensitization is cyclic AMP dependent and requires 4-8 hr.

(III) EFFECT OF ADENOSINE

Adenosine has beeen reported to interact with the adenylate cyclase system in various cell types and of relevance to our studies was the demonstration that an analogue of adenosine (N^6-(phenylisopropyl)adenosine) inhibits glucagon-induced uncoupling of adenylate cyclase in hepatocytes (30). The action of adenosine on LH-stimulated cyclic AMP production and LH-induced desensitization in the tumour cells was therefore investigated (31).

Adenosine caused a dose-dependent potentiation of LH-stimulated cyclic AMP production in rat Leydig tumour cells over a concentration range of 0.1-10μM. Over the same concentration range there was no detectable increase in cyclic AMP production when adenosine alone was added. The kinetics of LH-stimulated cyclic AMP production (Fig. 4) are characteristic of a desensitizing dose of LH (1 μg/ml) and show a rapid loss of cyclic AMP production with time. However, in the presence of adenosine (10μM) the kinetics of LH-stimulated cyclic AMP production were linear over the same time period (Fig. 4).

116

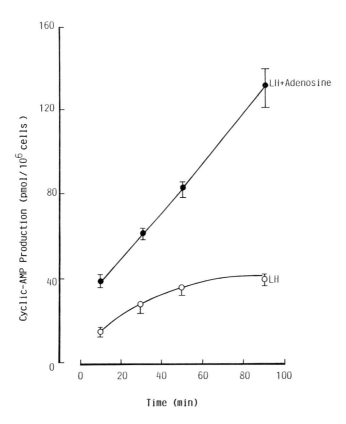

Fig. 4   Effect of LH or LH + adenosine on the kinetics of cyclic AMP
production in Leydig tumour cells.

The cells were incubated at $32^{O}C$ in media for various times in the presence
of LH (1.0 μg/ml) (O) or LH (1μ g/ml) + adenosine (10μM) (●), and the cyclic AMP
contents were determined.   Results are means ±S.D. (represented by the bars)
for duplicate determinations on triplicate samples (data from (31)).

Preincubation of the rat Leydig tumour cells with LH caused a typical
desensitization, as assessed by a second challenge with LH (Fig. 5). When
these desensitized cells were re-challenged with adenosine alone, a
potentiation of both basal and LH-stimulated cyclic AMP production was seen
(Fig. 5), but the desensitization was still evident in the presence
of adenosine. However, when adenosine was included along with the LH during
the preincubation, desensitization was completely inhibited (Fig. 5).

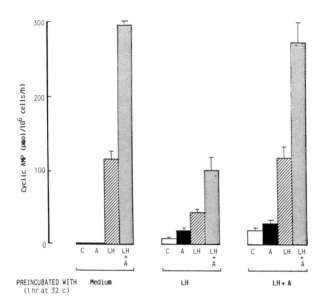

Fig. 5 Effect of pretreatment with LH or LH + adenosine on the subsequent
cyclic AMP production in Leydig tumour cells.

The cells were incubated at $32^{o}C$ for 1h either alone or with LH (1.0µg /ml)
or with LH (1.0µg/ml) + adenosine (10µM). The cells were then washed (three
times) and incubated for a further 1h at $32^{o}C$ with medium alone (C) or with
medium containing adenosine (A) (10µM), LH (1.0µg/ml) or LH (1.0µg/ml) +
adenosine (10µM). The cyclic AMP was then determined.   Results are means±
S.D. for duplicate determinations on triplicate samples. (Data from (33)).

The mechanisms of these adenosine effects are different from those reported
for luteal cells (32), where it was found that adenosine was selectively
transported and utilized for conversion into ATP and subsequently for the LH-
sensitive adenylate cyclase. In the rat Leydig tumour cells this does not
seem to be the case, since dipyridamole, a specific nucleoside-transport
inhibitor, had no inhibitory effect on the potentiating action of adenosine
over a concentration range which inhibited up to 90% of the uptake of
[^{3}H]adenosine by the rat Leydig tumour cells. These results suggest that
adenosine was acting at the cell surface. In support of this, 2-
deoxyadenosine, which has a greater specificity for the P-site receptor (33)
which is situated on the cytoplasmic face of the plasma membrane, was without
effect on both basal and LH-stimulated cyclic AMP production.

118

N^6-(Phenylisopropyl)adenosine, which has a higher specificity for the externally facing R-type receptors (34) showed the same potentiating effects as adenosine. The action of adenosine through an R-type receptor is further supported by the inhibition of the potentiating effects by isobutylmethylxanthine, which acts as a competitive antagonist at the R-type adenosine receptor (35) and thereby blocks its actions. The other phosphodiesterase inhibitor used, Ro-10-1724, which does not compete with adenosine, had no effect.

The inhibition of LH-induced desensitization by adenosine is consistent with the linear kinetics of cyclic AMP production seen in the presence of LH plus adenosine. Since LH-induced desensitization is due to an uncoupling of the LH receptor from the N-protein, then the action of adenosine through its receptor must also be acting to inhibit at this same site. This is supported by the finding that adenosine does not potentiate either the forskolin- or the cholera-toxin-stimulated cyclic AMP production, and therefore has its potentiating effects through an action at or proximal to the N-protein.

The mechanism by which adenosine acts in rat Leydig tumour cells is far from clear. R-type receptors have been classified as inhibitory or stimulatory on the adenylate cyclase system (35). However, in rat Leydig tumour cells there is no evidence that adenosine alone either stimulates or inhibits cyclic AMP production. Adenosine acts through a R-type receptor to prevent both glucagon-induced desensitization and the blockage of insulin-induced activation of the cyclic AMP phosphodiesterase by glucagon in hepatocytes (30). It was suggested that this action of adenosine was through a novel mechanism and not through adenylate cyclase, and tentatively suggested a possible involvement of Ca^{2+} mobilization. Analogues of adenosine which act to inhibit adenylate cyclase through R-type receptor also inhibit a membrane-bound low-K_m cyclic AMP phosphodiesterase through a GTP-dependent mechanism (36), therefore a role for guanine nucleotide regulatory protein in the regulation of this phosphodiesterase was postulated.

It is clear that the major action of adenosine in rat Leydig tumour cells on preventing LH-induced desensitization does not involve stimulation or inhibition of adenylate cyclase or the inhibition of phosphodiesterase.

IV EFFECTS OF PHORBOL ESTERS

The ability of tumour promoting phorbol esters to cause desensitization of adenylate cyclase has been reported in several systems. TPA (12-0-tetradecanoylphorbol 13-acetate) causes desensitization of β-adrenergic agonist stimulated adenylate cyclase in rat glioma C6 cells (37,38). Two groups have reported TPA induced desensitization in avian erythrocytes via phosphorylation of the β-adrenergic receptor (39,40). We have investigated the effect of TPA on LH-stimulated cyclic AMP production in the Leydig tumour cells (41).

It was found that preincubation of the cells for 1 hour with TPA caused a very marked dose and time-dependent inhibition of subsequent LH stimulated cyclic AMP. The effect of TPA was also very rapid; within 5 min it had almost reached a maximum, whereas the effect of LH was less rapid and was still decreasing after 40min incubation. Unlike the effect on LH desensitization, it was found that addition of adenosine to the preincubation medium was unable to reverse the effect of TPA. The latter was also found to decrease cholera toxin and forskolin stimulation of cyclic AMP. For example in cells pretreated with 10ng/ml of TPA for 2 hours the LH, cholera toxin and forskolin stimulated cyclic AMP levels were decreased by 53.3, 26.2 and 24.1% in the subsequent 2 hour incubation. The effects of TPA were found not to be due to a reduction in the number of binding sites for ^{125}I-hCG. The Leydig cells bound 8.4±0.72, 2.64±0.09 and 7.38±1.29 fmoles ^{125}I-hCG/10^6 cells after preincubation for 2 hours with cells in medium, LH (1 µg/ml) and TPA (500ng/ml) respectively.

It was found that plasma membranes prepared from TPA treated cells were not different from control cells with respect to the effects of added GppNHp, LH and NaF. This is in marked contrast to the effects of pretreating with LH; the plasma membranes showed a typical decrease in response to LH and LH plus GppNHp.

It is concluded that although TPA mimics and is more potent than LH in its effects on desensitization in intact cells there are differences between the two compounds. It cannot be concluded from the evidence obtained that they are working through the same mechanism. Similarly it has been reported that phorbol esters and β-adrenergic agonists mediate desensitization of adenylate cyclase in rat Glioma C6 cells by distinct mechanisms. It has been reported that TPA-induced desensitization was similar to hCG-induced desensitization in a murine Leydig Tumour Cell line by every criteria tested (43).

INTERNALIZATION AND RECYCLING OF THE LH-RECEPTOR

Desensitization has been proposed to be the first step in the clustering and internalization of the hormone-receptor complex, most notably in the β - adrenergic system (44,45,46). Initial studies in the rat testis using in vivo experiments with hCG (37) refer to an occupancy stage where there is little or no net loss of LH receptors in Leydig cells and which lasts up to 24 hours. This is then followed by a prolonged loss of receptors which is maximal after 2 days. Thus down regulation of the LH receptor and early desensitization of the Leydig cell did not appear to be connected (10,47). However it was assumed that there is little movement or turnover of LH receptors occuring in the Leydig cell, because no net loss in their number during short term studies could be measured, and this does not take into account the existence of an endocytic pathway leading to recycling of receptors. We have investigated the dynamics of LH receptor binding, internalization and recycling in purified rat testis Leydig cells (48).

120

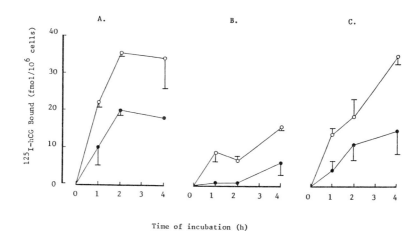

Fig. 6   Effect of monensin and cycloheximide on surface binding and internalization of ^{125}I-hCG in rat testis Leydig cells

The cells were incubated for 30 min in the absence of inhibitors (A) or presence (B) of monensin (25μM) or (C) of cycloheximide (2.5μg/ml) at 34°C. ^{125}I-hCG was then added to all the cells which were further incubated for the times shown.   Surface bound (O) hormone was determined by assaying the acid wash of the cells and the internalized hormone-receptor complexes (●) by assaying the residual cells dissolved in NaOH.   All the points are corrected for non-specific binding and are the means±S.E.M. for triplicate determinations for each time point. (Data from (48)).

Internalization of ^{125}I-hCG by rat testis Leydig cells at 34°C was found to be a rapid process (Fig. 6A).   In the presence of 50ng/ml ^{125}I-hCG, internalization occurred at a rate of 0.4 fmol/min.   Over the same period of time the surface binding increased at a slower rate. Both the surface binding and internalization reached a plateau after 2 hours.

At 4°C there was no significant increase in internalized levels of hormone even up to 20 hours of exposure to ^{125}I-hCG. Surface binding however, increased to a level of 6.75 fmol/10^6 cells after 4 hours and remained at that level for up to 20 hours.

The effect of preincubation for 2h in the presence of ^{125}I-hCG, followed by a further 4 hour incubation after removal of the hormone from the medium at 4°C, 21°C and 34°C was investigated.   At 4°C the surface binding remained at about 11 fmol/10^6 cells and there was no detectable increase in ^{125}I-labelled release. At 21°C over the 2 hour pre-incubation, internalized levels rose to 25 fmol/10^6 cells and remained at that level for the subsequent 4 hour incubation without hormone. Cell surface-bound levels dropped to insignificant levels after 1 hour but then increased to 8 fmol/10^6 cells at 4 hours. At the same time there was an increase in released radioactivity, which rose to the equivalent of 6 fmol/10^6 cells of ^{125}I-hCG. At 34°C after a 2 hour pre-incubation, the level of internalized hormone reached 35 fmol/10^6 cells (Fig 7), compared to surface bound levels of 10 fmol. Over the 4 hour incubation, internalized levels fell to 7.5 fmol/10^6 cells, a drop of 27.5 fmol whilst acid-soluble release rose to an ^{125}I-hCG equivalent of 20.7 fmol/10^6 cells. The surface-bound levels increased by 7 fmol/10^6 cells.

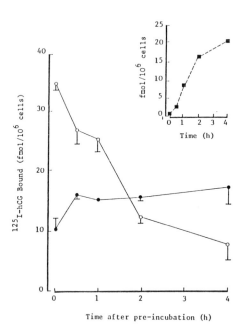

Fig. 7 Surface binding internalization and degradation of ^{125}I-hCG in rat testis Leydig cells.

The cells were pre-incubated at 34°C for 3 hours, ^{125}-IhCG was added after the first hour. The cells were washed with medium and then incubated for the times shown at 34°C. Determination of surface bound (●) and internalized (O) hormone were as described in Fig. 6. The total radioactivity released from the cells (■) was determined in the incubation medium. All points are corrected for non-specific binding and are the means ± S.E.M. for triplicate determinations for each time point. (Data from (48)).

The effect of monensin (which prevents acidification of endocytic vesicles) and cycloheximide (a protein synthesis inhibitor) on the internalization and surface binding of ^{125}I-hCG was investigated at 34°C. Incubation with ^{125}I-hCG in the presence of 25μM monensin caused a large reduction in total binding of hormone to the cells compared to control (Fig. 6B). In control cells total binding reached 55 fmol/10^6 cells after a 2 hour incubation compared to 7 fmol/10^6 cells in monensin-treated cells. Surface binding over this time was reduced to undetectable levels by monensin. There was no effect of monensin on binding of ^{125}I-hCG to the cells under control binding conditions at 4°C.

Treatment with cycloheximide (Fig. 6C) did not remove all the surface binding as monensin had done but did reduce total binding over a 2 hour incubation to 29 fmol/10^6 cells or 53% of control cell binding. The level of binding was reduced equally for both the surface bound and internalized components. At 4 hours of incubation however, there was no difference in the levels of binding between control and cycloheximide-treated cells.

The results obtained demonstrate that binding of LH (hCG) to its receptor in isolated rat testis Leydig cells leads to rapid internalization of the hormone-receptor complex. This appears to be followed by dissociation of the hormone-receptor complex in endocytic vesicles and the return of the receptors to the cell surface. During these short term incubations there is no net loss of receptors, i.e. down regulation does not occur. However, contrary to previous results which indicated that the hormone-receptor complex was stable until after prolonged exposure to the hormone (3) it is now apparent that the rat testis Leydig cell contains a highly dynamic internalization-recycling system.

That internalization of the ^{125}I-hCG receptor complex is a dynamic process, was demonstrated by incubation of the testis Leydig cells at 4°C. At this temperature the binding sites were saturated by 4 hours after exposure to the hormone compared with 2 hours at 34°C. The level of binding however was much lower at 4°C and remained surface-bound up to 20 hours after exposure to the hormone. The energy dependence of LH receptor internalization was shown by incubation of testis Leydig cells with ^{125}I-hCG in the presence of NaN$_3$ at 34°C. This compound is known to deplete the cells of ATP and was found to inhibit the internalization of the LH receptors.

The large level of internalization which occurs in the first hour after exposure to hormone, compared to the delay which was found in release of degraded hormone (48), suggests that at least some of the cell-associated hormone measured after incubation at 34°C is not receptor bound. This hormone may well remain in the endocytic vesicle and lysosomes of the cell as part of its 'processing', prior to the release of the degradation products. This could have important implications in the determination of receptor affinity and number, using Scatchard analysis of whole cells. It is also interesting to note that at 34°C after 2 hours, where the level of binding appears to reach a plateau, must represent a point where internalization into the cell and release of degraded hormone from the cell have reached a steady state, rather than saturation of the binding sites.

It was calculated that the Leydig cell internalizes 240 receptors per minute, we would therefore expect that in view of the limited number of receptors available, surface binding would be lost in the first 16 minutes after exposure to the hormone. To account for the continued increase in surface binding that is normally seen and the increase in internalized levels, there must be either new synthesis and/or recycling of the receptors to the cell surface. Monensin, a carboxylic ionophore (49) has been used by many investigators to examine the mechanism of receptor return. It has been used to interrupt endosomal acidification (50) which is a necessary pre-requisite for hormone receptor dissociation (51). Incubation of the testis Leydig cells with monensin reduced total binding of ^{125}I-hCG by 65% after 2 hours of incubation as compared to control cells, with the largest relative decrease being in surface bound levels. This evidence coupled with the lack of effect of cycloheximide after 4 hours indicated that recycling of LH receptors is taking place.

Finally, studies were carried out to determine if receptor internalization occurs at physiological levels of LH. The normal testis Leydig cells were preincubated with 1pg/ml to 100ng/ml LH for 2 and 8 hours in the presence of monensin to inhibit recycling. The cells were then cooled to 4°C, acid washed to remove surface-bound hormone and then incubated at 4°C with ^{125}I-hCG to determine the numbers of plasma membrane receptors. It was found that with all amounts of LH and at both incubation times there was a dose related loss of surface receptors (52).

In summary the data show that following binding of hormone to the LH receptor in the testis Leydig cell, the hormone-receptor complex is rapidly internalized into endosomal vesicles. These pass through a temperature sensitive endocytic pathway and seem to be capable of acidifying their contents. This appears to be a necessary step prior to the recycling of the LH receptor to the cell surface. Following dissociation and segregation of the H-R complex, the hormone is degraded (which probably occurs in the lysosomes) and is eventually released in a highly degraded form from the cell.

CONCLUSIONS

This study has been carried out with normal testis Leydig cells and tumour Leydig cells. It is difficult to interchange results obtained from both cell types because completely identical studies have not yet been carried out. However, current studies e.g. with adenosine and phorbol esters in the normal testis Leydig cell indicate they have similar effects to those obtained in the tumour cells. Both cell types are rapidly desensitized after exposure to LH and display many other common properties (20).

In the normal testis Leydig cell LH increases both cyclic AMP and $[Ca^{2+}]_i$. This positive effect on both these intracellular messenger and their requirement in steroidogenesis is unusual; in most systems either one or the other is increased and they usually have opposite effects (53,54). Dibutyryl cyclic AMP has however, been shown to stimulate Ca^{2+} influx in pancreatic B cells (55). The formation of cyclic AMP is clearly established to occur via the GTP binding protein N_s and activation of adenylate cyclase (Fig. 8). The increase in $[Ca^{2+}]_i$ in other systems is thought to occur via PI metabolism leading to increased formation of the Ca^{2+} mobilizer IP_3 (Fig. 8). However, in the Leydig cell, cyclic AMP appears to be a stimulator of intracellular Ca^{2+} mobilization. This may occur in addition to IP_3 formation.

124

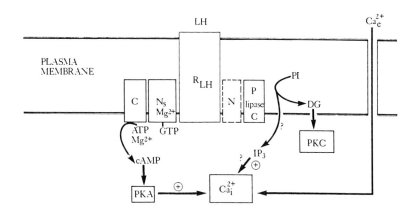

Fig. 8. Regulation of cyclic AMP and Ca^{2+} in Leydig cells.

R_{LH}, LH receptor; N_s, GTP binding protein; N, hypothetical GTP binding
protein; P lipase C, phospholipase C; PI, phosphatidylinositol; DG,
diacylglycerol; PKC, protein kinase C; IP_3, inositol 1,4,5-trisphosphate;
Ca_e^{2+}, extracellular Ca^{2+}; Ca_i^{2+}, intracellular calcium; C, adenylate
cyclase; PKA, cyclic AMP dependent protein kinase.

The initial LH-induced homologous desensitization in the Leydig cell is
independent of cyclic AMP, so what is the evidence for Ca^{2+} being involved
in this process? A high influx of Ca^{2+} into normal Leydig cells caused by
the addition of the Ca^{2+} ionophore A23187 does markedly decrease LH-
stimulated cyclic AMP production (56). Although part of this effect is
probably via activation of a cyclic AMP-phosphodiesterase, it was only
partially inhibited in the presence of the phosphodiesterase inhibitor methyl
isobutylxanthine, indicating that uncoupling had taken place and that Ca^{2+}
is involved in desensitization. LH has also been shown to increase
phosphoinositide metabolism in Leydig cells (17,18) which in other systems
leads to diacylglycerol release and activation of protein kinase C. Phorbol
esters cause desensitization of Leydig cell adenylate cyclase and these
compounds activate the Ca^{2+} dependent prottein kinase C. It is possible
therefore that LH activates both adenylate cyclase and phospholipase C via GTP
regulatory proteins (Fig. 8) leading to protein kinase C phosphorylation of,
initially the LH receptor (homologous desensitization) and, subsequently of
the N_s protein (heterologous desensitization). This has been proposed for
the β-adrenergic desensitization of various adenylate cyclase systems (13).

It is not known if uncoupling of the LH receptor from N_s is a pre-requisite
for internalization of the LH receptor. In the normal Leydig cell activation
and desensitization of the adenylate cyclase and internalization all occur
over the same dose range of added LH. The amounts used range from sub-
physiological to pharmacological. The time scales are also very similar. It
would seem therefore that the Leydig cell <u>in vitro</u> is a highly dynamic system
which responds to physiological amounts of LH resulting in cyclic AMP

production, uncoupling of the receptor from the N_s protein, internalization of the hormone-receptor complex, degradation of the hormone, and recycling of the receptor back to the plasma membrane.

ACKNOWLEDGEMENTS

We are grateful to the Medical Research Council, The Wellcome Trust and The Peter Samuel Royal Free Fund for financial support.

REFERENCES

1. Cooke, B.A., Dix, C.J., Maggee-Brown, R., Janszen, F.H.A. and van der Molen, H.J. (1981) Adv. Cycl. Nucl. Res. Vol. 14 (eds. Dumont, J.E., Greengard, P., and Robison, G.A.) 593-609 (Raven, New York).
2. Dufau, M.L. & Catt, K.J. (1978) Vitam. Horm. (N.Y.) 36, 461-592.
3. Catt, K.J., Harwood, J.P., Agiulera, G.R. & Dufau, M.L. (1979) Nature (London) 280, 109-116.
4. Hsueh, A.J.W., Dufau, M.L. & Catt, K.J. (1977) Proc. Natl. Acad. Sci. U.S.A. 74, 592-595..
5. Sharpe, R.M. (1977) Biochem. Biophys. Res. Commun 75, 711-717.
6. Purvis, K., Torjesen, P.A., Haug, E. & Hansson, V. (1977) Mol. Cell Endocrinol. 8, 73-80.
7. Haour, F. & Saez, J.M. (1977) Mol. Cell. Endocrinol. 7, 17-24.
8. Saez, J.M., Haour, F. & Cathiard, A.M. (1978) Biochem. Biophys. Res. Commun. 81, 552-558.
9. Jahnsen, T., Purvis, K., Torjesesn, P.A. & Hansson,, V. (1981) Arch. Androl. 6, 155-162.
10. Dix, C.J., Schumacher, M. & Cooke, B.A. (1982) Biochem. J. 202, 739-745.
11. Purvis, K. & Hansson, V. (1978) Arch. Androl. 2, 89-91.
12. Tsuruhara, T., Dufau, M.L., Cigorraga, S. & Catt, K.J. (1977) J. Biol. Chem. 252, 9002-9009.
13. Sibley, D.R. & Lefkowitz, R.J. (1985) Nature 317, 124-129.
14. Tsien, R.Y., Pozzan, T. and Rink, T.J. (1982) J. Cell. Biol. 94, 325-334.
15. Sullivan, M.H.F. & Cooke, B.A. submitted for publication.
16. Berridge, M.J., (1984) Biochem. J. 220, 345-360.
17. Lowitt, S., Farese, R.V., Sabir, M.A. & Root, A.W. (1982) Endocrinology 111, 1415-1417.
18. Molcho, J., Zakut, H. & Naor, Z. (1984) Endocrinology, 114, 1048-1050.
19. Cachelin, A.B., de Peyer, J.E., Kokubun, S. and Reuter, H. (1983) Nature 304, 462-464.
20. Cooke, B.A., Janszen, F.A., van Driel, M.J.A. & van der Molen, H.J. (1979) Biochim. Biophys. Acta 583, 320-331.
21. Metzger, H. & Lindner, E. (1981) IRCS Med. Sci. 9, 99.
22. Seamon, K.B. & Daly, J.W. (1981) J. Biol. Chem. 256, 9799-9806.
23. Dix, C.J., Habberfield, A.D. & Cooke, B.A. (1984)Biochem.J. 220, 803-809.
24. Siegl, A.M., Daly, J.W. & Smith, J.B. (1982) Mol. Pharmacol. 21, 680-687.
25. Clark, R.B., Goka, T.J., Green, D.A., Barber, R. & Butcher, R.W. (1982) Mol. Pharmacol. 22, 609-613.
26. Fradkin, J.E., Cook, G.H. Kilhoffer, M.C. & Wolff, J. (1982) Endocrinology 111, 849-856.
27. Forte, L.R. (1983) Arch. Biochem. Biophys, 225, 898-905.
28. van Sande, J., Cochaux, P., Mockel, J. & Dumont, J.E. (1983) Mol. Cell. Endocrinol. 29, 109-119.

126

29. Darfler, F.J., Mahan, L.C., Koachman, A.M. & Insel, P.A. (1982) J. Biol. Chem. 257, 11901-11907.
30. Wallace, A.V., Heyworth, C.M. & Housley, M.D. (1984) Biochem. J. 222, 177-182.
31. Dix, C.J., Habberfield, A.D. & Cooke, B.A. (1985) Biochem. J. 230, 211-216.
32. Behrman, H.R., Ohkawa, R., Preston, S.L. & MacDonald, G.J. (1983) Endocrinology 113, 1132-1140.
33. Londos, C., Cooper, D.M.F. & Wolff, J. (1980) Proc. Natl. Acad. Sci. U.S.A. 77, 2551-2554.
34. Sattin, A. & Rall, T.W. (1970) Mol. Pharmacol. 6, 13-23.
35. Fain, J.M. & Malbonn, C.C. (1979) Mol. Cell. Biochem. 25, 143-169.
36. de Mazancourt, P. & Giudicelli, Y. (1984) FEBS Lett. 173, 385-388.
37. Mallorga, P., Tallman, J.F., Henneberry, R.C., Hirata, F., Strittmatter, W.T., and Axelrod, J. (1980) Proc. Natl. Acad. Sci. U.S.A. 77, 1341-1345.
38. Brostrom, M.A., Brostrom, C.O., Brotman, L.A., Lee, C.S., Wolff, D.J. and Geller, H.M. (1982) J. Biol. Chem. 257, 6758-6765.
39. Sibley, D.R., Nambi, P., Peter, J.R. and Lefkowitz, R.J. (1984) Biochem. Biophys. Res. Commun. 121, 973-979.
40. Kelleher, D.J., Pesin, J.E., Ruoho, A.E., and Johnson, G.L. (1984) Proc. Natl. Acad. Sci. U.S.A. 81, 4316-4320.
41. Dix, C.J., Habberfield, A.D. & Cooke, B.A. (unpublished results).
42. Kassis, S., Zaremba, T., Patel, J. & Fishman, P.H. (1985) J. Biol. Chem. 260, 8911-8917.
43. Rebois, R.F. & Patel, J. (1985) J. Biol. Chem. 260, 8026-8031.
44. Strulovicci, B., Stadel, J.M. and Lefkowitz, R.J. (1983) J. Biol. Chem. 258, 6410-6414.
45. Toews, M.L. and Perkins, J.P. (1984) J. Biol. Chem. 259, 2227-2235.
46. Stadel, J.M., Strulovici, B., Nambi, P., Lavin, T.N., Briggs, M.M., Caron, M.G., and Lefkowitz, R.J. (1983) J. Biol. Chem. 258, 3032-3038.
47. Wu, F.C.W., Zhang, G.Y., Williams, B.C. and de Kretser, D.M. (1985) Mol. Cell. Endocrinol. 40, 45-46.
48. Habberfield, A.D., Dix, C.J., Cooke, B.A. (1986) Biochem. J. (in press).
49. Pressman, B.C. (1976) Ann. Rev. of Biochem. 45, 501-530.
50. Yamashiro, D.J., Fluss, S.R. and Maxfield, F.R. (1983) J. Cell Biol. 97, 929-934.
51. Wolkoff, A.W., Klausner, R.D., Ashwell, G. and Harford, J. (1984) J. Cell. Biol. 98, 375-381.
52. Habberfield, A.D., Dix, C.J. & Cooke, B.A. (unpublished results).
53. Tsuruta, K., Grieve, C.W., Cote, T.E., Eskay, R.L. and Kabian, J.W. (1982) Endocrinology 110, 1133-1140.
54. Feinsteinn, M.B., Egan, J.J., Sha'afi, R.I. and White, J. (1983) Biochem. Biophys. Res. Commun. 113, 598-604.
55. Henquin, J-C and Meissner, H.P. (1983) Biochem. Biophys. Res. Comm. 12, 614-620.
56. Sullivan, M.H.F. & Cooke, B.A. (1984) Mol. Cell. Endocrinol. 34, 17-22.

127

Résumé

Le mécanisme de la désensibilisation, de l'internalisation des récepteurs et de leur recyclage a été étudié dans des cellules de Leydig testiculaires et tumorales. Les facteurs contrôlant la production de cAMP et de Ca^{2+} et leur rôle possible dans ces processus ont été établis. En utilisant le chelateur intracellulaire fluorescent de Ca^{2+}, $quin^{-2}$, qu'aussi bien LH que des analogues de cAMP stimulent l'augmentation de Ca^{2+} intracellulaire (Ca^{2+}) dans les cellules de Leydig du testicule ce qui indique que le cAMP est un modulateur de (Ca^{2+}) dans ce type cellulaire. Ce Ca^{2+} augmenté dérive principalement du milieu extracellulaire. Des expériences effectuées avec les cellules tumorales stimulées par LH, la forskoline et la cholera toxine démontrent que la désensibilisation initiale homologue induite par LH est rapide (minutes) et indépendante du cAMP alors que la désensibilisation hétérologue est cAMP dépendante et demande 4-8 heures in vitro. L'adenosine potentialise la production de cAMP stimulée par la LH et empêche la désensibilisation induite par LH. Un ester de phorbol mimetise et est plus actif que LH dans ses effets sur la désensibilisation. Cependant, contrairement aux cellules désensibilisées par LH, les membranes plasmiques provenant des cellules traitées par l'ester de phorbol ne présentaient pas de diminution de réponse à la LH ou à la LH plus GppNHp.

Dans les cellules de Leydig testiculaires la liaison de la LH à son récepteur est suivie d'une internalisation rapide du complexe hormone-récepteur, de la dissociation du complexe hormone-récepteur dans les vésicules d'endocytose, du métabolisme de l'hormone et du retour des récepteurs à la surface de la cellule. Les preuves concernant le rôle de Ca^{2+} et du cAMP dans le desensibilisation et l'internalisation sont discutées.

Protein phosphorylation and sulfation

Phosphorylation et sulfatation des protéines

Hormones and cell regulation. Ed J. Nunez *et al.* Colloque INSERM/John Libbey Eurotext Ltd. ©1986. Vol. 139, pp. 131–140.

Role of protein kinase C in initiation and termination of cell functions

Tullio Pozzan[1], Francesco Di Virgilio[1], Susan Treves[2], Daria Milani[2], Lucia M. Vicentini[2], Daniel P. Lew[3], Barry Jacopetta[4] and Jan Louis Carpentier[4]

[1]*Institute of General Pathology, CNR Unit for the Study of the Physiology of Mitochondria, University of Padova, Padova, Italy,* [2]*Dept. of Pharmacology, University of Milano, Milano, Italy,* [3]*Infectious Diseases Unit, Dept. of Internal Medicine, University of Geneva, Geneva 4, Switzerland* and [4]*Dept. of Histology, University Medical Center, University of Geneva, Geneva 4, Switzerland*

INTRODUCTION

Protein kinase C was discovered more than 8 years ago (Takai et al., 1979), but only recently has its pivotal role in the regulation of cell activation been appreciated (Nishizuka, 1984) .Most of the information avaiable on the role of protein kinase C is based on the use of pharmacological activators, such as the tumor promoters phorbol esters and synthetic diacylglycerols.Compelling evidence from a number of laboratories indicates that protein kinase C is a key enzyme in the cascade of events triggered by receptors linked to polyphosphoinositide breakdown and Ca2+ mobilization (Berridge and Irvine, 1984).The extensive literature (see for a recent review Nishizuka, 1984) on phorbol esters is essentially devoted to unraveling the stimulatory role of protein kinase C ;in fact in a number of cell types active phorbl esters ,or dyacilglycerols , mimick completely or in part ,the stimulatory action of natural agonists.In the past few months however a second critical role for protein kinase C in the termination,rather than initiation of cell stimulation, has been discovered (Vicentini et al., 1985; Drummond, 1985; Sagi-Eisenberg et al., 1985) .In this report we will review some of our recent data in this field.Since the stimulatory action of protein kinase C is universally recognized we will mainly focus on the feedback regulatory role of this enzyme.

MATERIALS AND METHODS

Human neutrophils were obtained from healthy donors and purified according to standard procedures (Pozzan et al., 1983). PC12 cells ,RINm5F and HL60 cell lines were cultured as described (Pozzan et al., 1984; Wollheim and Pozzan, 1984) in RPMI 1640 medium supplemented with 10% foetal calf serum .
Cytosolic free Ca2+,[Ca2+]i ,membrane potential,superoxide anion production, secretion, and electron microscopy autoradiography were performed as described previously(Pozzan et al., 1983; Pozzan et al., 1984; Wollheim and Pozzan, 1984; Carpentier et al., 1978).
Radioactive material was obtained from Amersham, quin2/AM from CalBiochem or Sigma,media and sera from Flow Lab .All other materials were analytical or highest avaiable grade. Unless otherwise specified all experiments were performed at 37 C.

RESULTS AND DISCUSSION

Protein Kinase C was so named by Nishizuka and coworkers to indicate a novel protein kinase whose activity depended on Ca2+ but not calmodulin.The concept of its Ca2+ dependency has survived convincing evidence that in the presence of phorbol esters this enzyme becomes practically insensitive to Ca2+ concentration in the range where all true Ca2+ dependent enzymes are known to be regulated.While in their original paper Castagna et al.(1982) concluded that in vitro Protein kinase C required both Ca2+ and phorbol ester for activity ,more recently,under similar conditions,Nidel et al.(1983) reached a very different conclusion.In fact up to 70-80 % of its maximal phosphorylating activity can be obtained in the presence of negatively charged phospholipids and phorbol miristate acetate ,PMA,in the absence of Ca2+ in the medium and in the presence of mmolar concentrations of EGTA.Although the actual free Ca2+ concentration was not determined by Nidel et al.(1983),a rough calculation (assuming contaminating Ca2+ in the micromolar range) suggests that in those experiments Ca2+ was below 10^{-7} M.No known [Ca2+] activated enzyme is known to be operative under those conditions.A similar conclusion for purified protein kinase C was reached recently by Castagna and coworkers(1985). In permeabilized human platelets Knight et al.(1984) concluded that protein kinase C dependent phosphorylation was maximally activated at [Ca2+]i of 10^{-8} M. Fig 1 shows that PMA induced protein phosphorylation was indistinguishable whether this drug was added to cells at resting [Ca2+]i or whether [Ca2+]i was reduced 20 times below resting level ,i.e from 100 to 5 nmolar.In intact neutrophils reduction of [Ca2+]i to such low levels could be achieved by loading the Ca2+ indicator and chelator quin2 in Ca2+ free medium ,in the presence of EGTA. Under these conditions the intracellular generation of quin2 cannot be compensated by the influx of Ca2+ from the extracellular medium to maintain Ca2+ homeostasis.It is quite obvious that these manipulations of [Ca2+]i are artificial,since in intact cells [Ca2+]i will never be decreased to such low levels.However this experiment confirms the conclusion of Niedel et al.(1983) and Castagna and coworkers (1985) that ,when activated by PMA ,protein kinase C can be considered a Ca2+ independent enzyme,even in intact cells.

By this approach it has also been possible to tackle a very important biological question, i.e. whether protein kinase C activated functions are themselves Ca2+ dependent or independent processes.

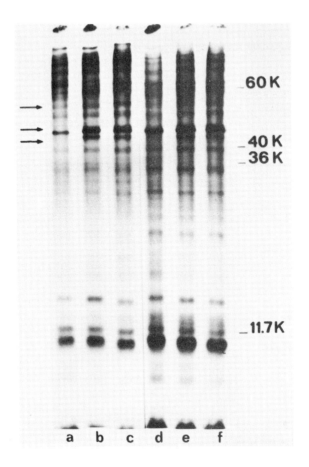

Fig.1. 32P autoradiography of SDS polyacrylamide gels of human neutrophils stimulated with PMA at resting or low [Ca2+]i.Human neutrophils were loaded with quin2 in Ca2+ containing(lines A,B,C) and Ca2+ free media + EGTA (lines D,E,F,). The [Ca2+]i was 100nM and 10 nM respectively.The cells were also incubated with ^{32}P for two hours. Cells loaded in Ca2+ free were stimulated in Ca2+ free and cells loaded in Ca^{2+}medium were stimulated in Ca2+ medium. After loading the cells were washed,resuspended in ^{32}P free medium and stimulated for 15 sec (lines B and E)and 30 sec (lines C and F) with 100 nM PMA. Lines A and D represent the unstimulated controls.

This is not a trivial question since among other functions activated by phorbol esters there is secretion which involves membrane fusion and cytoskeletal rearrangements,i.e. processes which have always been assumed to depend on [Ca2+]i elevations.Table 1 shows that in human neutrophils stimulation of superoxide anion generation and secretion of secondary granule markers are indistinguishable whether PMA is added at resting (100nM)or at low [Ca2+]i (5nM). Interestingly however the dose of PMA necessary to activate the same extent of secretion or superoxide anion generation decreases as [Ca2+]i increases from 5 to 100 to 800 nM (Di Virgilio et al., 1984). This suggests that an elevation of [Ca2+]i sensitizes the enzyme to submaximal doses of the activator(PMA or diacylglycerol).

TABLE 1

Effect of [Ca2+]i on secretion and superoxide anion production stimulated by PMA in neutrophils.

	Vit.B12 binding protein release %total content
Resting [Ca2+]i (100 nM)	37%
Low [Ca2+]i (10 nM)	36%

	Superoxide anion generation % unloaded controls
Resting [Ca2+]i (100 nM)	100%
Low [Ca2+]i (10 nM)	98%

Loading conditions as in Fig 1.The cells were stimulated for 5 minutes with 30nM PMA for assaying Vit.B12 binding protein release.Total enzyme content was measured in a parallel batch of cells treated with Triton X100.For Vit.B12 binding protein release the cells were pretreated with 5 ug/ml of cytochalasin B.Superoxide anion production was monitored continuously by measuring the rate of oxigen consumption.The results are expressed as % of the maximal rate of oxigen consumption measured in the same batch of neutrophls not loaded with quin2.

In this respect our interpretation of the role of [Ca2+]i rises in protein kinase C dependent activation observed with physiological agonists in intact cells is opposite to that of Nishizuka and coworkers. We think that Ca2+ sensitizes protein kinase C to suboptimal doses of diacylglycerol,while Nishizuka proposes that diacylglycerol sensitizes the enzyme to small rises of [Ca2+]i.However ,though there is a fondamental difference in interpreting the role of Ca2+ ,the synergism between the two signals ,Ca2+ and diacylglycerol, remains the basis for the explanation of cell activation via receptors linked to polyphosphoinositide breakdown.
While PMA has become more and more popular among biochemists and cell biologist a number of evidence has accumulated indicating that this compound has a variety of effects that are not easily predicted on the basis of the simple scheme outlined below: agonist-receptor interaction

polyphosphoinositide breakdown   inositoltrisphosphate and diacyglycerol formation   [Ca2+]i rise and protein kinase C activation   cell function.In fact according to this scheme PMA ,bypassing the receptor step,should mimick in part or completely the action of agonists;however it has been observed that PMA is not only able to stimulate but also to inhibit some cellular functions.For example we initially reported (Tsien et al., 1982) that,in mouse lymphocytes ,PMA pretreatment could prevent and reverse the rise in [Ca2+]i induced by the mitogenic lectin Concanavalin A.We interpreted that result as an indication that PMA could stimulate the plasma membrane Ca2+ pump .This conclusion was supported by the finding that PMA was capable of slightly decreasing the resting [Ca2+]i of these cells (Tsien et al., 1982). We recently offered a similar interpretation in order to explain that ,in human neutrophils,PMA slightly decreases the resting level and severely curtailes the elevation of [Ca2+]i induced by the chemotactic peptide f-Met-Leu-Phe (Lagast et al., 1984). In support of this interpretation we also showed that inside out plasma membrane vescicles isolated from neutrophils pretreated with PMA exhibited a significantly higher Ca2+ pumping activity (Lagast et al., 1984). More recent evidence from our and other laboratories suggests that the effect of PMA is not quite so simple.In fact ,as shown in Table. 2,PMA pretreatment of PC12 cells,a line of neurosecretory cells originally derived from a rat pheocromocytoma ,did not affect resting [Ca2+]i but largely prevented the rises of [Ca2+]i and the production of inositolphosphates induced by carbachol.Carbachol ,acting via the muscarinic acetylcholine receptor,activates polyphosphoinositide breakdown ,inositoltrisphosphate production and Ca2+ mobilization (Vicentini et al., 1985).

TABLE 2

Effect of PMA pretreatment on inositol phosphate production
and [Ca2+]i rise induced by carbachol in PC12 cells

	Inositol phosphate production	[Ca2+]i rises
	% unstimulated controls	
No addition	185%	200%
PMA 100nM	153%	120%

PC12 cells were incubated for 24 hours with ^{3}H-myoinositol washed and loaded with quin2 as described (Vicentini et al., 1985). The cells were stimulated with 500 µM charbacol with or without preincubation with PMA ,30 nM,for 5 minutes.For inositol phosphate determination,30 seconds after stimulation the samples were quenched with TCA and processed as described (Vicentini et al., 1985). [Ca2+]i was monitored in parallel with quin2.

Thus in this cell line the effect of PMA seems to be primarily on the generation of the messenger leading to [Ca2+]i rises,rather than on the Ca2+ extruding mechanism .Confirming that this is in fact the case,we demonstrated that when PC12 cell are preloaded with 45Ca and resuspended in 45Ca free medium they loose the tracer at the same speed irrespective of whether they were treated or not with PMA (Di Virgilio et al., 1985). An inhibition of inositoltrisphosphate generation could in large part explain the effect of PMA on the rises of [Ca2+]i induced by f-Met-Leu-Phe in neutrophils,but in this cell type, as well as in lymphocytes, the decrease in basal [Ca2+]i could only be explained if the Ca2+ extruding mechanism had been activated by PMA pretreatment.It is quite possible that different cell types possess different targets for protein kinase C .In particular,we have no evidence for a direct phosphorylation of the Ca2+ pump by PMA in neutrophils.One explanation for the increase in Ca2+ extrusion elicited by phorbol esters is that this is due to the insertion of new Ca2+ pump molecules in the plasma membrane .In fact ,in neutrophils, PMA induces a large secretion and it is not unlikely that preformed Ca2+ pumps are present in the membrane of the secretory vescicles.If this is the case during fusion new Ca2+ pump units will get inserted in the plasma membrane .This interpretation is consistent with the observation that in inside out plasma membrane vescicles PMA pretreatment increases the Vmax and not on the Km for Ca2+ of the Ca2+ pump (Lagast et al., 1984). In cells which would not possess such a reservoir of Ca2+ pumps the effect of PMA on Ca2+ extrusion will not be observed.

The inhibitory effect of PMA ,and thus probably of protein kinase C , on those processes which eventually lead to activation of this enzyme are another example of feedback inhibition ,a process which is very commonly observed in enzymatic cascade reactions.More surprising ,on the other hand,is the observation that PMA can inhibit events which are not directly linked to protein Kinase C activation.In Table 3 it is shown that PMA pretreatment of PC12 cells or RINm5F cells ,a cell line derived from a rat insulin secreting tumor,largely inhibits the rise of [Ca2+]i induced by membrane depolarization.Both these cell lines possess voltage gated Ca2+ channels ,whose opening probability is increased several folds when the internal negative membrane potential is collapsed. Depolarization is not elicited by receptor triggering nor is there any evidence that the rises of [Ca2+]i induced by opening of voltage gated Ca2+ channels activate protein kinase C.It could be argued that this inhibition by PMA is not due to inhibition of the channel,but rather to activation of Ca2+ extruding mechanisms. However:

a)The increase of [Ca2+]i due to voltage gated Ca2+ channels opening is an extremely rapid event,u sec range ,and one would expect that an activation of Ca2+ extrusion or sequestration would not affect the initial rise,but would make it more transient.

b)Rises of [Ca2+]i similar to those obtained with membrane potential depolarization but obtained with Ca2+ ionophores are not affected by PMA pretreatment (Pozzan et al., 1984).

c)As stated above no evidence for increase 45Ca extrusion is observed in PMA treated cells.

d)Unidirectional 45Ca influx as well as [Ca2+]i rises are inhibited by PMA pretreatment(Di Virgilio et al., 1985). Activation of Ca2+ extrusion or sequestration would not have affected unidirectional 45Ca influx.

e) Very recently it has been reported that in root ganglia PMA reduces the Ca2+ currents as measured with the patch clamping technique (Rane and Dunlap, 1985).

TABLE 3

	[Ca2+]i rise % controls
PC12 cells	50
RINm5F cells	40

The cells were loaded with quin2 as described (Pozzan et al., 1984; Wollheim and Pozzan, 1984) and then stimulated with 50 mM KCl with or without pretreatment with 30 nM PMA.Depolarization with KCl resulted in a rapid rise of [Ca2+]i (from 85 to 260 nM and from 110 to 480 nM in PC12 and RINm5F cells respectively)wich very slowly declined back towards resting level.In the presence of PMA the maximum rise of [Ca2+]i after KCl depolarization was to 160 nM and to 190 nM in PC12 and RINm5F cells respectively.

Further evidence suggests that this effect of PMA is directly or indirectly due to protein kinase C activation:
a) The effect of PMA is observed in a concentration range (10^{-9}M-10^{-7}M) considered specific for protein kinase C.
b) The effect of PMA is rapid ,maximal around 5 minutes of preincubation, but not immediate, as expected for a process triggered by protein phosphorylation.
c) It is observed ,although reduced ,with exogenously added diacylglycerol.
What might the physiological role of this unexpected finding be ? There are at least two experimental observations that might be explained by an effect of protein kinase C on voltage gated Ca2+ channels:
1)In neurons it has been shown that agonists which activate polyphosphoinositide breakdown often result in the reduction of Ca2+ current elicited by membrane depolarization (Belluzzi et al., 1985).
2) In bovine adrenal medulla cells it has been shown that acetylcholine causes catecholamine secretion mainly via activation of the nicotinic receptor which depolarizes the cells .Surprisingly in these cells muscarinic stimulation is inhibitory (Swilem et al., 1983) ,while in permeabilized cells PMA synergizes with Ca2+ (Knight and Baker, 1983) .The simplest interpretation of this result is that muscarinic stimulation,which activates polyphosphoinositide breakdown ,and thus protein kinase C, will reduce the amplitude or the duration of the Ca2+ rises induced by voltage gated Ca2+ channel opening.
Table 4 shows another effect of PMA which might be important in terminating the signal generated by ligand receptor interaction.In the human promyelocitic cell line HL60 PMA induces the internalization of insulin and transferrin receptor complexes.This effect of PMA is identical whether Ca2+ is at resting(100nM) ,below resting (10nM) or above resting (1μM) levels,again supporting the notion that protein kinase C can be activated by phorbol esters independently of [Ca2+]i.It is worth pointing out that neither transferrin nor insulin are ligands coupled to protein kinase C activation, indicating that this enzyme can affect receptor down regulation independently of a classical feedback regulation.

TABLE 4

Effect of PMA and [Ca2+]i on receptor down regulation.

[Ca2+]i

	Low		Resting		High	
	T	I	T	I	T	I
Controls	50	45	55	55	40	40
PMA,30 nM	72	68	75	70	65	68

HL60 cells were loaded with quin2 in Ca2+(resting and high [Ca2+]i) and Ca2+ free medium (low [Ca2+]i).Low [Ca2+]i cells were washed and resuspended in Ca2+ free medium + EGTA,incubated with ^{125}transferrin (T) or insulin (I)with or without PMA.After 15 minutes the cells were fixed and processed for quantitative electron microscopy autoradiography (Carpentier et al., 1978). High [Ca2+]i cells were washed and resuspended in Ca2+ medium,treated with 1 μM ionomycin together with insulin and transferrin and then processed as above. Resting [Ca2+]i cells were washed and resuspended in Ca2+ medium and processed as above without further manipulation.The results are expressed as % transferrin or insulin internalized after 15 minutes of incubation.

The use of the HL60 cell line,which possess both transferrin and insulin receptors has allowed us to obtain another important piece of information on this effect of PMA.In fact PMA induces the internalization of both occupied and unoccupied transferrin receptors, while it only induces internalization of occupied insulin receptors.It might be suggested that this difference is due to the fact that insulin receptors become concentrated in coated pits only after binding to the ligand,while transferrin receptors are known to be physiologically clustered in coated pits independently of ligand binding.It might also be speculated that the stimulation of receptor down regulation by protein kinase C might be an important mechanism by which receptors linked to polyphosphoinositide breakdown regulate cell activation:It is now well established that a number of ligands known to cause protein kinase C activation,not only down regulate their own receptors ,but other receptors as well (Wrann et al.,1980;Brown et al.,1984).While the down regulation of autologous receptors is a general phenomenon ,cross desensitization is mainly or exclusively observed with receptors linked to polyphosphoinositide breakdown .We suggest that both autologous and cross down regulation can be mediated by protein kinase C activation.In the case of autologous down regulation the same enzymatic pathway initiates the cascade of events leading both to expression of cellular responses and the feedback inhibitory pathway which prevents excessive stimulation of the cells .By down regulating the receptors for other ligands protein kinase C could cause the cell engaged in a specific pathway to become refractory to other stimuli.This refractoriness might be particulary useful since the specific program initiated by the first stimulus might be altered by the activation of another receptor dependent pathway.In this respect it is interesting to note that a nutrient,such as transferrin,is internalized as well as a hormone,such as insulin,in response to PMA .However the increased internalization of these two ligands will have different effects on the pathways dependent on transferrin and insulin respectively.In the case of transferrin the ligand releases the nutrient, Fe2+, after internalization and the ligand-receptor complex recycles back to the plasma membrane.Insulin,on the other hand,is degraded together with its receptor in the lysosomal compartment.Thus the net result of down regulating these two receptors leads in the case of transferrin to a net increase of nutrient uptake,while, in the case of insulin,internalization of the ligand-receptor complex will terminate the signal provided by the hormone.The differential effect of protein kinase C on occupied and unoccupied receptors provides another useful mechanism for finely regulating this process.

ACKNOWLEDGMENTS

This work was in part supported by grants from the Italian Ministery of Public Education (40 ,60 %) to T.P.and F.Di V., from grants from the Italian CNR ,Special project "Oncology" and "Oncologia e rischio Tossicologico" to T.P. and by grants from the Swiss National Science Foundation N.3.460.83 and 3.990.084 .D.P. Lew is a recipient of a Max Cloetta career developement award. S.T. was supported by a grant from the "Banca Popolare di Padova" and D.M. from a grant from FIDIA Lab. ,Abano ,Italy.During part of this work T.P. was also supported by a grant from the ZYMA Foundation, Nyon Switzerland.The authors are indebted to Dr. C.B. Wollheim,J. Meldolesi and L.Orci for helpful discussions and for permission to quote unpublished results

REFERENCES

Belluzzi, O., Sacchi, O. and Wanke, E. (1985) J. Physiol. 358, 109-129.
Berridge M.J. and Irvine, R.F. (1984) Nature 312, 315-321.
Brown, K.D., Baly, J., Irvine, R.F., Heslop, J. and Berridge, M.J. (1984) Biochem. Biophys. Res. Commun. 123, 377-384.
Carpentier, J.L., Gorden, P., Amherdt, M., Van Obberghen, E., Kahn, C.R. and Orci, L. (1978) J. Clin Invest 61, 1057-1070.
Castagna, M., Takai, Y., Kaibuchi, K., Sano, K., Kikkawa, U. and Nishizuka, Y. (1982) J. Biol. Chem. 257, 7847-7851.
Cognez, P., Couturier, A. and Castagna, M. (1985): 10th European Symposium on Hormones and Cell Regulation. Mont St. Odile (Abstract).
Di Virgilio, F., Lew, D.P. and Pozzan, T. (1984) Nature 310, 691-693.
Di Virgilio, F., Pozzan, T., Wollheim, C.B., Vicentini, L.M. and Meldolesi, J. (1985) J. Biol. Chem., in press.
Drummond, A.H. (1985) Nature 315, 752-755

Knight, D.E. and Baker, P.F. (1983) FEBS Lett. 160, 98-100.

Knight, D.E. and Scrutton, M.C. (1984) Nature 309, 66-68.

Lagast, J., Pozzan, T., Waldvogel, F.A. and Lew, D.P. (1984) J. Clin. Inv. 73, 878-885.Niedel, J.E., Kuhn, L.J. and Wanderbark, G.R. (1983) Proc. Natl. Acad. Sci. U.S.A. 80, 36-40.

Nishizuka, Y. (1984) Nature 308, 693-698.

Pozzan, T., Lew, D.P., Wollheim, C.B. and Tsien, R.Y. (1983) Science 221, 1413-1415.

Pozzan, T., Gatti, G., Dozio, N., Vicentini, L.M. and Meldolesi, J. (1984) J. Cell. Biol. 99, 628-638.

Rane, S.G. and Dunlap, K. (1985) Neuroscience, Abs. N.218.9.

Sagi-Eisenberg, R., Lieman, H. and Pecht, I. (1985) Nature 8, 59-60.

Swilem, A.F., Hawthorne, J.N. and Azila, N. (1983) Biochem. Pharmacol. 32, 3873-3874.

Takai, Y., Kishimoto, A., Kikkawa, A., Mori, T. and Nishizuka, Y. (1979) J. Biol. Chem. 254, 3692-3695.

Tsien, R.Y., Pozzan, T. and Rink, T.J. (1982) Nature 295, 68-71.

Vicentini, L.M., Ambrosini, A., Di Virgilio, F., Pozzan, T. and Meldolesi, J. (1985) J. Cell Biol. 100, 1330-

Vicentini, L.M., Di Virgilio, F., Ambrosini, A., Pozzan, T. and Meldolesi, J. (1985) Biochem. Biophys. Res. Commun. 127, 310-317.

Wollheim, C.B. and Pozzan, T. (1984) J. Biol. Chem. 259, 2262-2267.

Wrann, M., Fox, C.F. and Ross, R. (1980) Science 210, 1363-1365.

Résumé

Dans ce rapport on examinera le rôle de la protéine kinase C lors de l'initiation et la terminaison de l'activation cellulaire. On verra que lors de l'activation de la protéine kinase C par les esters de phorbol, l'on obtient une stimulation de la phosphorylation des protéines et des fonctions cellulaires indépendamment de la concentration intracellulaire de Ca^{2+}. On insistera particulièrement sur le rôle inhibiteur de la protéine kinase C à la fois comme régulateur de retro-contrôle de sa propre activation que de son rôle général dans le processus d'activation cellulaire.

Hormones and cell regulation. Ed J. Nunez *et al.* Colloque INSERM/John Libbey Eurotext Ltd. © 1986. Vol. 139, pp. 141–157.

Calcium (calmodulin)-dependent protein kinase II*

Thomas R. Soderling, Charles M. Schworer, M. Elizabeth Payne,
Mary Frances Jett, Devra K. Porter, Janet L. Atkinson and Neil M. Richtand

*Howard Hughes Medical Institute and Department of Molecular Physiology and Biophysics,
Vanderbilt Medical School, Nashville, TN 37232, USA.*

ABSTRACT

Ca^{++}(CaM)-dependent protein kinase II is widely distributed in many tissues
and species and exists in several isozyme forms differing in apparent M_r and
with subunits in the range of 50 to 60 kDa. Biochemical studies with purified
kinase and substrates have demonstrated that the Ca^{++}(CaM)-dependent protein
kinase II has a rather broad substrate specificity. Proteins phosphorylated in
vitro include brain synapsin, tyrosine hydroxylase, tryptophan hydroxylase,
phenylalanine hydroxylase, glycogen synthase, pyruvate kinase, tubulin,
microtubule-associated protein 2 (MAP-2), ribosomal protein S6, and cardiac
phospholamban. Although many of these same proteins are phosphorylated in vivo
in response to calcium mobilizing conditions, definitive proof for the catalytic
involvement of the Ca^{++}(CaM)-dependent protein kinase II remains to be
established in most instances. Studies using synthetic peptides indicate that
the Ca^{++}(CaM)-dependent protein kinase II and the cAMP-dependent protein
kinase share certain recognition determinants in their substrates but also have
some distinct differences. Therefore, one would predict that these two kinases
would have some overlap in substrate specificities as well as having some unique
and different substrates.

(Key words: Protein phosphorylation, protein kinase, calmodulin, calcium)

INTRODUCTION

The role of covalent protein phosphorylation and dephosphorylation in the
regulation of diverse cellular functions has become well established over the
past twenty years (for reviews see 1,2). Work by Krebs and Fischer (3) in the
1950s-1960s on the cascade enzyme system catalyzing skeletal muscle glycogen
metabolism was foundational in the field of protein phosphorylation. These
studies led to the identification first of phosphorylase kinase (4), a
calcium-dependent protein kinase, and subsequently to the discovery of the
cAMP-dependent protein kinase (5). The identification of a protein kinase whose
activity was regulated by cAMP acted as a catalyst in the development of this

---

*This article is dedicated to Professor Luis Leloir on the occasion of his
eightieth birthday.

area because of the studies by Sutherland and colleagues establishing cAMP as an intracellular second messenger in numerous physiological systems (6). In the 1970s many proteins were identified as in vitro substrates of the cAMP-dependent protein kinase. Subsequently, many of these proteins have been established as in vivo substrates of this kinase using the criteria of Krebs (7).

During the course of studies of protein phosphorylation, numerous other protein kinases have been identified. For example, the enzyme glycogen synthase, which has seven or more serine residues phosphorylated in vitro and in vivo, is a substrate for many protein kinases (8,9). In contrast to the cAMP-dependent protein kinase which catalyzes phosphorylation of many proteins, most other protein kinases are thought to be more limited in their protein substrate specificities. Generally, specific protein kinases are named according to their substrate (e.g., phosphorylase kinase), whereas general protein kinases are designated according to a regulatory molecule if known (e.g., cAMP-dependent protein kinase). These kinases catalyze phosphorylation of serine and/or threonine residues. More recently, protein kinases which phosphorylate tyrosine residues have been characterized (10).

As stated above, one of the first protein kinases to be characterized was skeletal muscle phosphorylase kinase, a calcium-dependent protein kinase (4). The mechanism by which calcium activates phosphorylase kinase was provided by the discovery that calmodulin (CaM), the ubiquitous calcium binding protein, is a constitutive subunit of the kinase (11). This contrasts with other physiological systems in which elevated intracellular calcium first binds to free CaM which subsequently binds to the regulated enzyme or protein (12). Table 1 lists several calcium-stimulated protein kinases and a phosphoprotein phosphatase which have been characterized. All of them except protein kinase C involve CaM as the mediator of the calcium effect. Protein kinase C is regulated by a combination of the phospholipid phosphatidylserine plus diacylglycerol and calcium ion (13). Of the calcium-dependent protein kinases listed in Table 1, phosphorylase kinase and myosin light chain kinase are thought to be specific for their designated substrates in vivo. Protein kinase C and Ca^{++}(CaM)-dependent protein kinase II [1], on the other hand, are thought to have a rather general substrate specificity. This chapter will be devoted to a discussion of the properties of the Ca^{++}(CaM)-dependent protein kinase II.

The involvement of our laboratory with calmodulin-dependent protein kinase II evolved from our studies on the phosphorylation of glycogen synthase. Several laboratories had shown that hormones which act through elevations of intracellular calcium ion (such as vasopressin and angiotensin II) produce not only activation of phosphorylase but also an inactivation of glycogen synthase (14,15). Although it was known that phosphorylase kinase could catalyze the phosphorylation and partial inactivation of glycogen synthase in vitro (16-18), we questioned whether this reaction occurred in the presence of high intracellular concentrations of phosphorylase in vivo. Therefore, we began to search for another calcium-dependent protein kinase which could phosphorylate glycogen synthase. Such an enzyme was identified and purified from rabbit liver (19,20).

---

[1]The Ca^{++}(CaM)-dependent protein kinase II has also been referred to as CaM-dependent glycogen synthase kinase, CaM-dependent synapsin I kinase II, or CaM-dependent protein kinase II by various investigators.

This calmodulin-dependent glycogen synthase kinase was differentiated from the two other calmodulin-dependent protein kinases known at that time, namely, phosphorylase kinase and myosin light chain kinase, on the basis of several physical and catalytic properties (19-22). The name was chosen because initial studies did not identify any protein substrates other than glycogen synthase (19). However, as the properties of this kinase became better characterized, it became apparent that it might be very similar to other calmodulin-dependent protein kinases being described at that time in the literature. In particular, the laboratory of Paul Greengard was working on a calmodulin-dependent protein kinase that phosphorylates the brain protein synapsin (23,24). Several other laboratories were also characterizing Ca^{++}(CaM)-dependent protein kinases which could phosphorylate tryptophan hydroxylase and tyrosine hydroxylase (25,26), smooth muscle myosin light chain (27) and tubulin (28,29). When we tested synapsin and tyrosine hydroxylase as substrates for our liver

Table 1. Properties of purified calcium-dependent protein kinases and phosphatase

KINASES:	DISTINGUISHING PROPERTIES	REFERENCES
Phosphorylase kinase	$(\alpha\beta\gamma\delta)_4$; $M_r = 1.3 \times 10^6$ Da Narrow substrate specificity. Delta subunit is CaM. Catalytic activity in gamma subunit.	91-94
Myosin light chain kinase	Monomer of 130 kDa Narrow substrate specificity Requires CaM for activity	95

Ca^{++} phospholipid protein kinase "C"	Monomer of 80-85 kDa
	Broad substrate specificity
	Requires phosphatidylserine
	    diacylglycerol, and Ca^{++} for
	    activity
		13,96

Ca^{++}(CaM)-dependent protein kinase I	Monomer of 37-45 kDa Limited substrate specificity High in neural tissues Cytosolic	50

Ca^{++}(CaM)-dependent protein kinase II	Composed of α and β subunits,
	    ratio varies depending on
	    tissue; M_r=300-700 kDa
	Broad substrate specificity
	Broad tissue distribution
	Particulate and cytosolic
		22,24,26,27,29,
		30,32,34,47,48,
		50

PHOSPHATASE:		
Calcineurin	$\alpha\beta$; $M_r = 80,000$ Alpha subunit has catalytic activity and binds CaM Limited substrate specificity	97,98

Ca^{++}(CaM)-dependent glycogen synthase kinase, they were readily phosphorylated (23,30). As will become apparent, evidence now strongly indicates that isozyme forms of a general Ca^{++}(CaM)-dependent protein kinase exist in many mammalian tissues. It is suggested that this multifunctional protein kinase be designated as Ca^{++}(CaM)-dependent protein kinase II (see footnote 1).

PHYSICAL PROPERTIES

The Ca^{++}(CaM)-dependent protein kinase II has been highly purified from the high-speed supernatant fractions of liver (19-22), brain (24,26,27,29), skeletal muscle (32), and cardiac muscle (33,34). Although these preparations do not exhibit identical physical properties, they all have subunits in the M_r range of 50-60 kDA (Table 2).

The liver kinase II consists of equal stoichiometric amounts of 51 and 53 kDa subunits (Fig. 1). Upon incubation with magnesium and (γ-^{32}P]-ATP, both subunits become phosphorylated, in a calcium and calmodulin-dependent manner, and exhibit a characteristic decrease in mobility on gel electrophoresis in sodium dodecyl sulfate (21,35). Analysis of phosphoamino acids reveals that both serine and threonine are labeled in a ratio of approximtely 2:1, respectively. Both subunits bind CaM in a calcium-dependent manner using the gel overlay technique. The exact relationship between the 51 and 53 kDa subunits has not been established, and the possibilities of limited proteolysis or differential phosphate contents have not been completely excluded.

The Ca^{++}(CaM)-dependent kinase II purified from rat brain is comprised of 50 and 58-60 kDA subunits. When forebrain is the source of kinase II, the amount of the 50 kDA subunit predominates over the 60 kDa subunit by a factor of 3-6 on a molar basis. However, when the enzyme is purified from cerebellum, the 60 kDA polypeptide is the major subunit (31,36). Both of these polypeptides undergo intramolecular "autophosphorylation" and bind CaM (24,29). Surprisingly, when the two subunits are resolved using SDS/PAGE and subsequently incubated in buffer containing CaM, autophosphorylation of each resolved subunit can be demonstrated (37). This strongly indicates that both subunits contain catalytic activity. This is in contrast to the cAMP-dependent protein kinase where the catalytic subunit catalyzes intramolecular phosphorylation of the regulatory subunit (38).

Table 2. Properties of purified Ca^{++}(CaM)-dependent protein kinase II

Tissue	Native M_r	Subunit M_r	K_m (ATP)	$K_{0.5}$ (CaM)	Reference
Rabbit liver	300,000	51,000 53,000	27 μM	22 nM	21,22,35
Rat brain	650,000	50,000 58-60,000	60 μM	10 nM	24,26
Canine heart	600,000	55,000	--	--	34
Rabbit skeletal muscle	696,000	58,000 54,000	45 μM	--	32

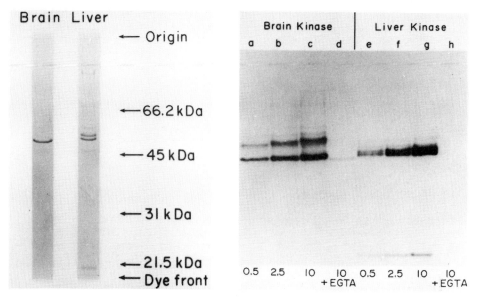

Fig. 1. Subunit composition and autophosphorylation of Ca^{++}(CaM)-dependent protein kinase II. LEFT. Rat brain and rabbit liver Ca^{++}(CaM)-dependent kinase II were analyzed by SDS/PAGE with Coomassie blue staining. RIGHT. Brain and liver Ca^{++}(CaM)-dependent kinase II were subjected to autophosphorylation conditions for the times (min.) indicated at the bottom of the figure. Lanes d and h shows samples where the phosphorylation reaction contained 0.5 mM EGTA. Samples were run on SDS/PAGE and subjected to autoradiography. The M_r of the major ^{32}P-labeled polypeptides were 51 and 62 kDa for the brain kinase and 51 and 53 kDa for the liver kinase. (Reproduced by permission from ref. 35).

Two dimensional peptide maps of the brain kinase II reveal marked similarities between the 50 and 60 kDa subunits although some differences also are apparent (29,39). Immunological studies also indicate that the two subunits are similar but not identical. Using a Western blot analysis, most antibodies cross react with both subunits, but specificity for a particular subunit can be obtained with some monoclonal and polyclonal antibodies (31,36). These specific antibodies have yielded useful information concerning the intramolecular subunit relationships. One possible arrangement would consist of mixtures of homo-oligomers containing just 50 kDa or 60 kDa polypeptides. Alternatively, one could have hetero-oligomers containing both subunits in a single molecule. This latter arrangement is favored by the immunological studies. When the native kinase II is immunoprecipitated, using the monoclonal antibody specific for the 50 kDa subunit, both subunits are precipitated in the same ratio as exists in the original kinase II preparation (36). A possible caveat in this experiment is that the specificity of the antibody for the 50 kDA subunit was demonstrated using denatured polypeptides in a Western blot analysis. It is possible that in the native structure the antibody crossreacts with both subunits.

Ca^{++}(CaM)-dependent protein kinase II with subunits in the M_r range of 50 to 60 kDa has been purified to varying degrees from skeletal muscle (32), cardiac muscle (33,34,40), pancreas (41), erythrocytes (42), Aplysia neurons (43), Torpedo electric organ (44), mammary tissue (45), and anterior pituitary (46). The similarities in physical and catalytic properties indicate the existence of isozyme forms of a multifunctional Ca^{++}(CaM)-dependent protein kinase II.

It should be pointed out that most of the above studies were performed on the Ca^{++}(CaM)-dependent protein kinases purified from the cytoplasmic fractions of tissue homogenates. However, in those tissues examined, a majority of the activity is associated with pellet fractions. In several cases, the membrane-associated Ca^{++}(CaM)-dependent kinase II has been solubilized using detergents or hypotonic lysis and purified (47,48). These preparations of kinase II exhibit properties similar or identical to the corresponding cytoplasmic kinase II. Several subcellular organelles which have significant amounts of the Ca^{++}(CaM)-dependent protein kinase II associated with them include nuclei (49), synaptic vesicles (50), and postsynaptic densities (51) from brain, and cardiac sarcoplasmic reticulum (34,52, 53). The widespread cellular and subcellular localization of this kinase II again is suggestive of diverse functions.

CATALYTIC PROPERTIES

The activity of the Ca^{++}(CaM)-dependent protein kinase II is completely dependent on the presence of CaM and calcium ion (Fig. 2). Half-maximal activation has been reported to require a CaM concentration of 10-20 nM (21,26,27,29,35,36). The pH maximum with glycogen synthase as substrate is at 7.8 (22), and the K$_m$ for ATP is approximately 40 μM (mean from several laboratories) (21,26,27,29,32,36). Kinase activity is completely inhibited by

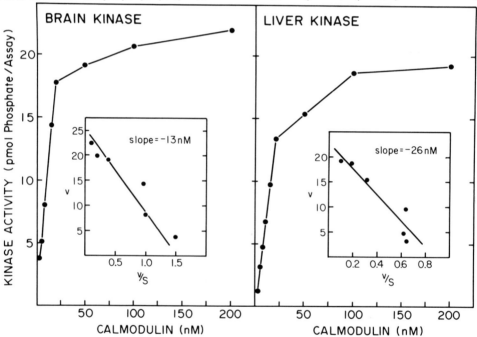

Fig. 2. Effect of CaM on CaM-dependent protein kinase II activity. Brain and liver CaM-dependent protein kinase II were assayed using glycogen synthase as substrate and the indicated concentrations of CaM.

the chelator EGTA and by NaF (35). Both of these inhibitions can be reversed by addition of sufficient calcium ion. The inhibition by NaF is important to recognize as many protein kinase assays, especially in tissue extracts, contain fluoride to inhibit phosphatases. Ca^{++}(CaM)-dependent protein kinase II activity is also inhibited by phenothiazine antipsychotics such as trifluoperazine (19).

146

As mentioned above, the Ca^{++}(CaM)-dependent protein kinase II exhibits autocatalytic phosphorylation on both serine and threonine residues (54,55). Two laboratories have reported that this autophosphorylation is associated with inactivation of the kinase (37,55). Several other experiments have suggested other possible functions of autophosphorylation, but these studies were performed in tissue extracts and are therefore more difficult to interpret. One report shows that incubation of a membrane-cytoskeletal complex with Mg-ATP in the presence of calcium and CaM results in dissociation from the complex of the kinase in a CaM-independent form (43). Another study reports that autophosphorylation results in increased kinase activity when assayed at low concentrations of CaM (56). More studies are needed to establish the putative role(s) of this autophosphorylation.

PROTEIN SUBSTRATES

The several laboratories which initially purified the Ca^{++}(CaM)-dependent protein kinase II utilized different protein substrates. Included among these proteins were glycogen synthase (19), synapsin (47), and tryptophan hydroxylase (26). Other proteins which appear to be potential substrates for the Ca^{++}(CaM)-dependent protein kinase II include microtubule-associated protein 2 (MAP-2) and tau protein (27,29,57), tubulin (29), tyrosine hydroxylase (58,59), myelin-basic protein (27,29), ribosomal protein S6 (41), phospholamban (38,53), pyruvate kinase (60), phenylalanine hydroxylase (61), and cyclic nucleotide phosphodiesterase (62). It should be stressed that most of these studies have been conducted in vitro with purified proteins. Whereas good evidence exists in the literature that many of these proteins are subject to calcium-dependent phosphorylation in vivo, extrapolation of in vitro data to in vivo situations can be fraught with dangers. For example, isolated smooth muscle myosin light chain is a rather good substrate for Ca^{++}(CaM)-dependent protein kinase II, but the myosin light chain is not phosphorylated in the intact myosin complex (22). Although the multifunctional Ca^{++}(CaM)-dependent protein kinase II is an attractive candidate as the catalyst for many of the above reactions, rigorous evidence for this in vivo role is generally lacking. The possible exception is the case of brain synapsin. The laboratory of Greengard has accumulated convincing evidence for the role of Ca^{++}(CaM)-dependent protein kinase II catalyzed phosphorylation of synapsin in the regulation of catecholamine secretion (63,64).

The phosphorylation of several purified enzymes by purified Ca^{++}(CaM)-dependent protein kinase has been investigated in our laboratory (Fig. 3). These enzymes were chosen because of data in the literature demonstrating a calcium-dependent phosphorylation in intact cell preparations. Whether these reactions are catalyzed in vivo by the Ca^{++}(CaM)-dependent kinase, as occurs in vitro, has yet to be established. The concentrations (based on 50 kDa subunit) of Ca^{++}(CaM)-dependent protein kinase used in these in vitro experiments varied between .02 and 0.2 μM. The estimated subunit concentration of the Ca^{++}(CaM)-dependent kinase in rat liver is 0.1 μM and in rabbit liver is 0.6 μM. Therefore, the in vitro reaction conditions are within the physiological range of kinase concentrations.

Glycogen Synthase: When isolated rat hepatocytes are incubated wih calcium mobilizing agents such as vasopressin, angiotensin II, or phenylephrine, there is inactivation of glycogen synthase (14,15,65). Although the studies utilizing intact cells were performed with liver, most studies on the purified enzyme have utilized skeletal muscle glycogen synthase. This is because the skeletal muscle enzyme has been very highly characterized and the liver synthase is relatively poorly understood. The activity of glycogen synthase, the rate-limiting enzyme in glycogen biosynthesis, is regulated in the cell by allosteric stimulation and inhibition and by covalent phosphorylation. The allosteric interactions allow

147

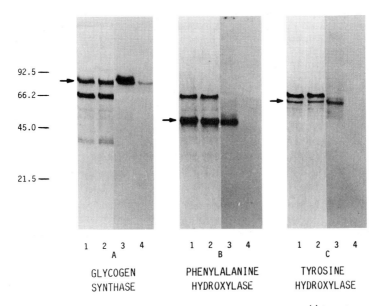

Fig. 3. SDS/PAGE analysis of proteins phosphorylated by Ca⁺⁺(CaM)-dependent protein kinase II. For each protein analyzed, lanes 1 and 2 were stained with Coomassie blue and lanes 3 and 4 are the autoradiographs. Reactions corresponding to lanes 1 and 3 were standard phosphorylations containing 0.5 mM $CaCl_2$; those represented by lanes 2 and 4 contained 0.1 mM trifluoperazine. Proteins analyzed, their concentrations in the phosphorylation reactions, and their subunit M_r were as follows: rabbit skeletal muscle glycogen synthase, 0.20 mg/ml, 90 kDa; rat liver phenylalanine hydroxylase, 0.18 mg/ml, 51 kDa; pheochromocytoma tyrosine hydroxylase, 0.04 mg/ml, 60 kDa. The protein band at 66 kDa present in all reactions is bovine serum albumin which was included to stabilize the kinase. The scale to the extreme left shows the migrations of proteins of the indicated M_r (kDa). (Reproduced by permission from ref. 30)

the enzyme to be responsive to changes in local cellular environment and metabolism. Covalent modifications usually are triggered in response to hormonal signals, thereby integrating glycogen synthesis with other physiological requirements of the animal. Skeletal muscle glycogen synthase is a tetrameric enzyme of 90 kDa subunits, each subunit containing at least seven serine residues which can be phosphorylated in vitro and in vivo by numerous protein kinases (for reviews see 8,9). In general, phosphorylation of most, but not all, sites results in inactivation due to either an increase in the K_m for the substrate UDP-glucose and/or in the K_a for the allosteric activator glucose-6-P. Ca⁺⁺(CaM)-dependent protein kinase from either liver or brain phosphorylates preferentially site 2 but also site 1b (for amino acid sequences, see Table 3). Phosphorylation of these sites, especially site 2, results in partial inactivation of glycogen synthase. The K_m of the Ca⁺⁺(CaM)-dependent protein kinase for glycogen synthase is approximately 7 μM (subunit concentration). The site specificity of the Ca⁺⁺(CaM)-dependent kinase for glycogen synthase (sites 2 and 1b) can be compared to several other kinases of interest. The cAMP-dependent protein kinase phosphorylates sites 1a, 2, 1b, and 3 (66,67). Phosphorylase kinase phosphorylates only site 2 (17), and protein kinase C catalyzes phosphorylation of sites 1a and 2 (68).

Phenylalanine Hydroxylase: Phenylalanine hydroxylase is important in liver since almost all metabolized phenylalanine is converted to tyrosine. It is well established that the activity of this enzyme is regulated by a cAMP-dependent phosphorylation, demonstrated in vitro (69), in vivo (70), and in isolated liver cells (71). Calcium-mobilizing hormones or agents produce a 2-fold increase in the phosphorylation state of phenylalanine hydroxylase (71). This effect is not mimicked by phorbol esters and occurs in liver phosphorylase kinase deficient rats, thereby suggesting that protein kinase C and phosphorylase kinase are probably not involved (72). We demonstrated that the Ca^{++}(CaM)-dependent protein kinase can catalyze incorporation of 0.5-0.7 mol phosphate per mole hydroxylase subunit (61). Not only could the kinase catalyze the forward reaction (i.e., phosphorylation), but it could also catalyze the reverse reaction (dephosphorylation) in the presence of high concentrations (5 mM) of ADP. The reversibility of kinase reactions under appropriate conditions is well known (73). The Ca^{++}(CaM)-dependent protein kinase and the cAMP-dependent protein kinase catalyze phosphorylation of the same serine residue in phenylalanine hydroxylase. This fact was established by two criteria. Both kinases catalyzed the reversal reaction of the other's forward reaction. Secondly, reverse-phase high performance liquid chromatography of peptic and tryptic ^{32}P-peptides proved the existence of a single, common phosphopeptide regardless of the catalytic protein kinase (61). Not surprisingly, equal stoichiometric phosphorylation by either kinase gave equivalent activation of the phenylalanine hydroxylase. What was unusual, however, was that phenylalanine, acting through an allosteric site, increased the rate of phosphorylation by the cAMP-dependent kinase but decreased the rate of phosphorylation of the same site by the Ca^{++}(CaM)-dependent kinase II (61,74). These results were interpreted to indicate that although the two kinases may share some common amino acid recognition determinants in their substrates, important differences also exist. This hypothesis has been confirmed, as presented below, using synthetic model peptides.

Pyruvate Kinase: Liver pyruvate kinase is one of several enzymes involved in the acute hormonal regulation of gluconeogenesis (75). Hormones such as glucagon, whose actions are mediated by elevations in intracellular cAMP, stimulate gluconeogenesis partly by inhibiting pyruvate kinase II. This inhibition of pyruvate kinase reduces substrate cycling by limiting the conversion of phosphoenolpyruvate to pyruvate and is brought about by phosphorylation catalyzed by the cAMP-dependent protein kinase (76). Because of evidence, largely from the laboratory of Garrison (14,71), demonstrating a calcium-mediated phosphorylation of pyruvate kinase, we tested this enzyme as a substrate for the Ca^{++}(CaM)-dependent protein kinase II. Indeed, the Ca^{++}(CaM)-dependent protein kinase does catalyze incorporation of up to 1.7 mol of phosphate per mol subunit of rat liver L-type pyruvate kinase (Fig. 4).

This phosphorylation is associated with an inactivation of pyruvate kinase due to a 3-fold increase in the $K_{0.5}$ for phosphoenolpyruvate (60).

Phosphorylation of pyruvate kinase by the Ca^{++}(CaM)-dependent kinase II occurs at two sites, one of which is the same serine phosphorylated by the cAMP-dependent protein kinase. This was determined as follows. If pyruvate kinase is phosphorylated with the Ca^{++}(CaM)-dependent protein kinase and then subsequently incubated with EGTA (to inhibit the Ca^{++}(CaM)-dependent kinase), 5 mM ADP and a high concentration of the cAMP-dependent protein kinase (kinase reversal conditions, see above), the cAMP-dependent kinase catalyzes removal of about 50-60% of the phosphate (Fig. 4). The remaining 40-50% of the phosphate is resistant to dephosphorylation by the cAMP-dependent protein kinase, indicative of a unique site of phosphorylation. Phosphoamino acid analysis shows that the Ca^{++}(CaM)-dependent protein kinase phosphorylates not only serine but also threonine residues in pyruvate kinase in a ratio of 2:1, respectively (60). Analysis of the CNBr phosphopeptides by reverse phase high performance liquid

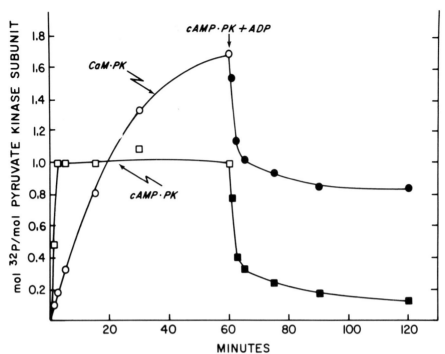

Fig. 4 Phosphorylation-dephosphorylation of pyruvate kinase. At zero time
either the CaM-dependent protein kinase II (CaM-PK) or the cAMP-dependent protein
kinase (cAMP-PK) was added to standard phosphorylation reactions containing
pyruvate kinase. At 60 min. there was addition to both reactions of 5 mM ADP, 10
mM glucose, 2 mM EGTA, 12 units/ml hexokinase, and 180 units/ml of the
cAMP-dependent protein kinase (all final concentrations). These latter
conditions promote the reversal reaction (dephosphorylation) catalyzed by the
added cAMP-dependent protein kinase. (Reproduced by permission from ref. 60)

chromatography showed a single peptide containing both the ^{32}P-serine and
^{32}P-threonine. This is the same peptide containing the site of phosphorylation
by the cAMP-dependent kinase. The sequence of this peptide is known (77,78), and
it contains only a single serine and two threonine residues. Currently, the
identity of which threonine is phosphorylated is being established. The K_m of
the Ca^{++}(CaM)-dependent protein kinase for pyruvate kinase is 12 μM (subunit)
compared to the K_m of 68 μM of the cAMP-dependent protein kinase. The subunit
concentration of pyruvate kinase in rat liver is about 0.5 to 1 μM (79). Whether
calcium-mobilizing agents promote phosphorylation of pyruvate kinase at threonine
in isolated rat hepatocytes is under investigation.

As with phenylalanine hydroxylase, calcium-mobilizing agents promote
phosphorylation of pyruvate kinase in rats deficient in liver phosphorylase
kinase, and phorbol esters have no effect in normal rats (72). Therefore, the
Ca^{++}(CaM)-dependent protein kinase II is an attractive candidate as the
catalyst of Ca^{++}-dependent phosphorylation of pyruvate kinase. The magnitude
of this phosphorylation and the resultant effect on gluconeogenesis is rather
small in rat liver (14,71,75), but it would be interesting to determine if the
effect were larger in a species, such as rabbit, which contains higher
concentrations of Ca^{++}(CaM)-dependent protein kinase II.

Tyrosine Hydroxylase: Tyrosine hydroxylase, the rate-limiting enzyme in catecholamine biosynthesis, is believed to be regulated in vivo by several mechanisms, including protein phosphorylation. Phosphorylation of the purified enzyme by the cAMP-dependent protein kinase is associated with activation due to a large decrease in the K_m for the cofactor 6-methyltetrahydropterin (80,81). Treatment of brain slices or adrenal chromaffin cells with agents which elevate cAMP concentrations result in activation due to a decreased K_m for cofactor (82,83). Depolarization of brain slices or pheochromocytoma cells (PC12 cells) with potassium causes a calcium-mediated phosphorylation and activation of tyrosine hydroxylase due to an increase in V_{max} (83,84). Protein kinase C phosphorylates the same site in tyrosine hydroxylase as does the cAMP-dependent protein kinase, causing a "K_m type" activation, and is therefore an unlikely candidate for the calcium-mediated V_{max} activation. Since we had demonstrated that the Ca^{++}(CaM)-dependent protein kinase phosphorylates phenylalanine hydroxylase, which is similar in many properties to tyrosine hydroxylase, we tested tyrosine hydroxylase as a substrate. The Ca^{++}(CaM)-dependent protein kinase does phosphorylate tyrosine hydroxylase which has been purified from PC12 cells (Fig. 3 and ref. 30) as well as from brain (59,85). Analysis of tryptic phosphopeptides established that phosphorylation occurs primarily on a serine residue which is unique from the serine phosphoryated by the cAMP-dependent protein kinase. However, phosphorylation by the Ca^{++}(CaM)-dependent protein kinase had little effect on the K_m for cofactor (59). In fact, phosphorylation by the Ca^{++}(CaM)-dependent kinase appears to have little direct effect on tyrosine hydroxylase activity (58,59,86), but it does confer upon the enzyme the ability to be activated by the addition of an activator protein (86). This heat-labile activator protein is a dimer of 35 kDa subunits (87). The activator protein was without effect on nonphosphorylated tyrosine hydroxylase or on the enzyme which had been phosphorylated by the cAMP-dependent protein kinase. The activator protein produced its effect through an increase in the V_{max} of tyrosine hydroxylase (59). Therefore, it would appear that the Ca^{++}(CaM)-dependent protein kinase is a good candidate to catalyze the depolarization induced, calcium-dependent increase in tyrosine hydroxylase V_{max}. Studies are being conducted in isolated striatal synaptosomes to test this hypothesis.

Synthetic Peptides: Studies on the cAMP-dependent protein kinase using synthetic peptides as substrates revealed the presence of certain amino acid recognition determinants in substrates (88,89). In particular, the presence of a basic residue, preferably arginine, three residues amino terminal from the phosphorylated serine was essential. The presence of a second basic amino acid two residues amino terminal also greatly enhanced the kinetic parameters (89). A synthetic peptide commonly used as a substrate for the cAMP-dependent protein kinase which conforms to these principles has the sequence Leu-Arg-Arg-Ala-Ser-Leu-Gly. This peptide was designed after the sequence of the phosphorylation site in pyruvate kinase (89).

The known sequences in proteins of sites phosphorylated in vitro by the Ca^{++}(CaM)-dependent protein kinase are given in Table 3. From these results, we proposed that the arginine three residues amino terminal from the phosphorylated serine was probably an important recognition determinant for the Ca^{++}(CaM)-dependent protein kinase (22). This was especially apparent upon examination of the myosin light chains from smooth muscle versus cardiac and skeletal muscles. The Ca^{++}(CaM)-dependent kinase readily phosphorylates the isolated smooth muscle, but not the cardiac or skeletal muscle, myosin light chains (19,22). All three of these proteins have very high homologies around their phosphorylation sites except for the presence of the arginine residue in the smooth muscle variant. Synthetic peptides were made, modeled after the smooth muscle myosin light chain sequence (Table 4). Peptide 1 had a K_m of

Table 3. Amino acid sequences of phosphorylation sites in proteins tested as substrates for the Ca^{++}(CaM)-dependent protein kinase II. MLC is an abbreviation for myosin light chain.

Protein	Substrate	Sequence	Reference
Glycogen synthase site 2	yes	-Arg-Thr-Leu-Ser-Val-Ser-Ser-Leu	17,90
Glycogen synthase site 1B	yes	-Arg-Ser-Asn-Ser-Val-Asp-Thr-Ser	99
Glycogen synthase site 1A	no	-Arg-Arg-Ala-Ser-Cys-Thr-Ser-Ser	99
Phosphorylase	no	-Lys-Glu-Ile-Ser-Val-Arg-Gly-Leu	100
Phenylalanine hydroxylase	yes	-Arg-Lys-Leu-Ser-Asp-Phe-Gly-Glu	101
Pyruvate kinase	yes	-Arg-Arg-Ala-Ser-Val-Ala-Gln-Leu	102
Smooth muscle MLC	yes	-Arg-Ala-Thr-Ser-Asn-Val-Phe-Ala	103
Skeletal muscle MLC	no	-Glu-Gly-Ser-Ser-Asn-Val-Phe-Ser	104
Cardiac muscle MLC	no	-Glu-Gly-Ala-Ser-Asn-Val-Phe-Ser	104

approximately 27 μM and a V_{max} of approximately 1 μmol/min/mg. The ratio of V_{max} K_m indicates how well the substrate is phosphorylated by the Ca^{++}(CaM)-dependent protein kinase II. Deletion of the arginine residue (peptide 2) decreased this ratio from 100 to 0.2, confirming the absolute requirement for this arginine. Addition of a second arginine (peptide 3) or lysine (peptide 4),

Table 4. Ca^{++}(CaM)-dependent protein kinase II amino acid recognition determinants.
Peptides were synthesized and tested as substrates for the liver Ca^{++}(CaM)-dependent protein kinase II. All peptides had the same sequence as peptide 1 except as shown for the individual peptides. The ratio of the V_{max}/K_m, an index of the relative suitability of the peptide as a substrate for the kinase, was given a relative value of 100 for peptide 1.

Peptide	Sequence	Relative V_{max}/K_m
1	Lys-Lys-Ala-Pro-Gln-Arg-Ala-Ala-Ser-Asn-Val-Phe-Ala-Met	100.0
2	-Ala-Ala-Ala-Ser-	0.2
3	-Arg-Arg-Ala-Ser-	6.6
4	-Arg-Lys-Ala-Ser-	5.8

which would greatly enhance catalysis by the cAMP-dependent kinase, was unfavorable for the Ca^{++}(CaM)-dependent protein kinase.

These model peptide studies have suggested some interesting concepts which appear to be correct. The fact that the first arginine residue is a strong positive determinant for both the cAMP-dependent and Ca^{++}(CaM)-dependent protein kinases suggests there will be significant overlaps in substrate sites for these two kinases (Fig. 5). In fact, both kinases phosphorylate sites 2 and 1b in glycogen synthase, and the same site in phenylalanine hydroxylase. The fact that the presence of the second arginine residue is a positive determinant for the cAMP-dependent kinase but a negative determinant for the Ca^{++}(CaM)-dependent kinase suggests that there will also be unique sites or substrates for these two kinases (Fig. 5). For example, site 1a (see Table 3)

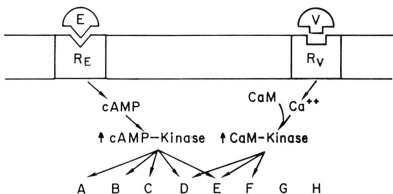

Fig. 5. Hypothetical model depicting substrate specificities of Ca^{++}(CaM)-dependent protein kinase II and cAMP-dependent protein kinase. Hormones which elevate intracellular Ca^{++} (hormone V) or cAMP (hormone E) act through separate plasma membrane receptors. Activation of the CaM-dependent protein kinase II or cAMP-dependent protein kinase results in phosphorylation of the indicated proteins (A through H), thereby altering the physiological functions of these proteins. The CaM-dependent protein kinase II and cAMP-dependent protein kinase phosphorylate some of the same residues in proteins (D and E) because of common recognition determinants for substrates and in other cases unique proteins because of differences in substrate recognition determinants. See text for a discussion of these recognition determinants.

is the best site in glycogen synthase for the cAMP-dependent kinase (90) but is not phosphorylated by the Ca^{++}(CaM)-dependent kinase (22,66). Pyruvate kinase (Table 3) is a very good substrate for the cAMP-dependent kinase and a mediocre substrate for the Ca^{++}(CaM)-dependent kinase. Sequences of the phosphorylation sites in tyrosine hydroxylase are not yet known. We would predict that the site phosphorylated by the cAMP-dependent kinase would have two or more basic residues amino terminal from the serine. It should be pointed out that the Ca^{++}(CaM)-dependent protein kinase does incorporate some phosphate (about $10^0/o$) into this site (Neil Richtand, Janet Atkinson, T.R. Soderling, unpublished data).

It should also be emphasized that although these studies with synthetic peptides support the importance of amino acid recognition determinants in the primary structure immediately surrounding the phosphorylated residue, other important amino acid recognition sites probably exist and depend on the secondary and tertiary structures of the intact proteins. Two examples will tend to verify this statement. Firstly, the synthetic peptides are generally poorer substrates than the intact protein. Their kinetic values are usually an order of magnitude or more lower than those of the corresponding proteins. The second example was already discussed in relation to phenylalanine hydroxylase. The presence of phenylalanine, acting through an allosteric site (i.e., a conformational change), increases the rate of phosphorylation by the cAMP-dependent protein kinase but also decreases the rate of phosphorylation of this same site by the Ca^{++}(CaM)-dependent protein kinase (61,74).

In summary, it appears that just as the cAMP-dependent protein kinase mediates the intracellular effects of cAMP, so also the Ca^{++}(CaM)-dependent protein kinase II may mediate several of the effects of elevated intracellular calcium ion. Current efforts are directed towards establishing the in vivo relevance of many of these reactions catalyzed in vitro by the Ca^{++}(CaM)-dependent protein kinase II.

REFERENCES

1. Rosen, O.M., and Krebs, E.G.(1981): In Protein phosphorylation. Cold Spring Harbor Conference on Cell Proliferation, Vol 8.
2. Corbin, J.D., and Hardman, J.G.(1983): Hormone Action: Protein kinases. Methods in Enzymology, Vol 99.
3. Krebs, E.G.(1972): Protein Kinases. Current Topics in Cellular Regulation 5, 99-133.
4. Krebs, E.G., and Fischer, E.H.(1956): Biochim. Biophys. Acta 20, 150-157.
5. Walsh, D.A., Perkins, J.P., and Krebs, E.G.(1968): J. Biol. Chem. 243, 3763-3765.
6. Robison, G.A., Butcher, R.W., and Sutherland, E.W.(1971): Cyclic AMP. New York, Academic Press.
7. Krebs, E.G., and Beavo, J.A.(1979): Annu. Rev. Biochem. 48, 923-959.
8. Cohen, P.(1982): Nature (Lond). 296, 613-620.
9. Soderling, T.R., and Khatra, B.S.(1982): In Calcium and Cell Function, ed W.Y. Cheung, Vol 3, pp 189-221. New York: Academic Press.
10. Hunter, T., and Cooper, J.A.(1985): Annu. Rev. Biochem. 54, 897-930.
11. Cohen, P.(1980): In Calcium and Cell Function, ed W.Y. Cheung, Vol 1, pp 183-199. New York: Academic Press.
12. Klee, C.B., Crouch, T.H., and Richman, P.G.(1980): Annu. Rev. Biochem. 49, 489-515.
13. Ashendel, C.L.(1985): Biochim. Biophys. Acta 822, 219-242.
14. Garrison, J.C., Borland, M.K., Florio, V.A., and Twible, D.A.(1979): J. Biol. Chem. 254, 7147-7156.
15. Strickland, W.G., Blackmore, P.F., and Exton, J.H.(1980): Diabetes 29, 617-622.
16. Roach, P.J., DePaoli-Roach, A.A., and Larner, J.(1978): J. Cyclic Nucleotide Res. 4, 245-257.
17. Soderling, T.R., Sheorain, V.S., and Ericsson, L.H.(1979): FEBS Lett. 106, 181-184.
18. Walsh, K.X., Millikin, D.M., Schlender, K.K., and Reimann, E.M.(1979): J. Biol. Chem. 254, 6611-6616.
19. Payne, M.E., and Soderling, T.R.(1980): J. Biol. Chem. 255, 8054-8056.
20. Soderling, T.R., and Payne, M.E.(1981): In Protein Phosphorylation. Cold Spring Harbor Conference on Cell Proliferation, Vol 8, pp 413-423.
21. Ahmad, Z., DePaoli-Roach, A.A., and Roach, P.J.(1982): J. Biol. Chem. 257, 8348-8355.
22. Payne, M.E., Schworer, C.M., and Soderling, T.R.(1983): J. Biol. Chem. 258, 2376-2382.
23. Kennedy, M.B., and Greengard, P.(1981): Proc. Natl. Acad. Sci. USA 78, 1293-1297.
24. Bennett, M.K., Erondu, N.E., and Kennedy, M.B.(1983): J. Biol. Chem. 258, 12735-12744.
25. Yamauchi, T., and Fujisawa, H.(1980): FEBS Lett. 116, 141-144.
26. Yamauchi, T., and Fujisawa, H.(1983): Eur. J. Biochem. 132, 15-21.
27. Fukunaga, K., Yamamoto, H., Matsui, K., Higashi, K., and Miyamoto, E.(1982): J. Neurochem. 39, 1607-1617.
28. Goldenring, J.R., Gonzalez, B., and DeLorenzo, R.J.(1982): Biochem. Biophys. Res. Commun 108, 421-428.
29. Goldenring, J.R., Gonzalez, B., McGuire, J.S., and DeLorenzo, R.J.(1983): J. Biol. Chem. 258, 12632-12640.
30. Schworer, C.M., and Soderling, T.R.(1983): Biochem. Biophys. Res. Commun. 116, 412-416.
31. McGuinness, T.L., Lai, Y., and Greengard, P.(1985): J. Biol. Chem. 260, 1696-1704.

32. Woodgett, J.R., Davison, M.T., and Cohen, P.(1983): Eur. J. Biochem. 136, 481-487.
33. Palfrey, H.C.(1984): Fed. Proc. 43, 1466.
34. Iwasa, T., Inoue, N., and Miyamoto, E.(1985): J. Biochem. 98, 577-580.
35. Schworer, C.M., McClure, R.W., and Soderling, T.R.(1985): Arch. Biochem. Biophys. 242, 137-145.
36. Miller, S.G., and Kennedy, M.B.(1985): J. Biol. Chem. 260, 9039-9046.
37. Kuret, J., and Schulman, H.(1985): J. Biol. Chem. 260, 6427-6433.
38. Rangel-Aldao, R., and Rosen, O.M.(1976): J. Biol. Chem. 251, 7526-7529.
39. Kelly, P.T., McGuinness, T.L., and Greengard, P.(1984): Proc. Natl. Acad. Sci. USA 81, 945-949.
40. Kloepper, R.F., and Landt, M.(1984): Cell Calcium 5, 351-364.
41. Gorelick, F.S., Cohen, J.A., Freedman, S.D., Delahunt, N.G., and Gershoni, J.M., Jamieson, J.D.(1983) J. Cell Biol. 97, 1294-1298.
42. Palfrey, H-C., Lai, Y., and Greengard, P.(1984): In Proc. 6th Int. Conf. Red Cell Structure and Metab., ed G. Brewer, pp 291-308. New York: Liss.
43. Saitoh, T., and Schwartz, J.H.(1985): J. Cell Biol. 100, 835-842.
44. Palfrey, H.C., Rothlein, J.E., and Greengard, P.(1983): J. Biol. Chem. 258, 9496-9503.
45. Brooks, C.L., and Landt, M.(1985): Arch. Biochem. Biophys. 240, 663-673.
46. Hatada, Y., Munemura, M., Fukunaga, K., Yamamoto, H., Maeyama, M., and Miyamoto, E.(1983): J. Neurochem. 40, 1082-1089.
47. Kennedy, M.B., McGuinness, T., and Greengard, P.(1983): J. Neurosci. 3, 818-831.
48. Fukunaga, K., Yamamoto, H., Tanaka, E., and Miyamoto, E.(1984): Biomed. Res. 5, 165-176.
49. Sahyoun, N., Levine, H., III, and Cuatrecasas, P.(1984): Proc. Natl. Acad. Sci. USA 81, 4311-4315.
50. Nairn, A.C., Hemmings, H.C., Jr., and Greengard, P.(1985): Annu. Rev. Biochem. 54, 931-976.
51. Kennedy, M.B., Bennett, M.K., and Drondu, N.E. (1983): Proc. Natl. Acad. Sci. USA: 80, 7357-7361.
52. Campbell, K.P. and Mac Lennan, D.H. (1982): J. Biol. Chem. 257, 1238-1246.
53. Jett, M.F., and Soderling, T.R. (1984): Fed Proc 43, 1949.
54. Landt, M., and McDonald, J.M.(1984): Int. J. Biochem. 16, 161-169.
55. Yamauchi, T., and Fujisawa, H.(1985): Biochem. Biophys. Res. Commun. 129, 213-219.
56. Shields, S.S., Vernon, P.J., and Kelly, P.T.(1984): J. Neurochem. 43, 1599-1609.
57. Yamamoto, H., Fukunaga, K., Tanaka, E., and Miyamoto, E.(1983): J. Neurochem. 41, 1119-1125.
58. Vulliet, P.R., Woodgett, J.R., and Cohen, P.(1984): J. Biol. Chem. 259, 13680-13683.
59. Richtand, N.M., Atkinson, J., Schworer, C., Kuczenski, R., and Soderling, T.R.(1986): sudmitted for publication.
60. Schworer, C.M., El-Maghrabi, M.R., Pilkis, S.J., and Soderling, T.R.(1985): J. Biol. Chem. 260, 13018-13022.
61. Døskeland, A.P., Schworer, C.M., Døskeland, S.O., Chrisman, T.D., Soderling, T.R., Corbin, J.D., and Flatmark, T.(1984): Eur. J. Biochem. 145, 31-37.
62. Fukunaga, K., Yamamoto, H., Tanaka, E., Iwasa, T., Miyamoto, E. (1984): Life Sci 35, 493-499.
63. Huttner, W.B., Schiebler, W., Greengard, P., and Camilli, P.(1983): J. Cell Biol. 96, 1374-1388.
64. Llinas, R., McGuinness, T.L., Leonard, C.S., Sugimori, M., and Greengard, P.(1985): Proc. Natl. Acad. Sci. 82, 3035-3039.
65. Hutson, N.J., Brumley, F.T., Assimacopoulos, F.D., Harper, S.C., and Exton, J.H.(1976): J. Biol. Chem. 251, 5200-5208.

66. Juhl, H., Sheorain, V.S., Schworer, C.M., Jett, M.F., and Soderling, T.R.(1983): Arch. Biochm. Biophys. 222, 518-526.
67. Sheorain, V.S., Corbin, J.D., and Soderling, T.R.(1985): J. Biol. Chem. 260, 1567-1572.
68. Ahmad, Z., Lee, F.-T., DePaoli-Roach, A., and Roach, P.(1984): J. Biol. Chem. 259, 8743-8747.
69. Abita, J.-P., Milstien, S., Chang, N., and Kaufman, S.(1976): J. Biol. Chem. 251, 5310-5314.
70. Donlon, J., and Kaufman, S.(1978): J. Biol. Chem. 253, 6657-6659.
71. Garrison, J.C., and Wagner, J.D.(1982): J. Biol. Chem. 257, 13135-13143.
72. Garrison, J.C., Johnsen, D.E., and Campanile, C.P.(1984): J. Biol. Chem. 259, 3283-3292.
73. Flockhart, D.A.(1983): Methods Enzymol. 99, 14-20.
74. Carr, F.P.A., and Pogson, C.I.(1982): Biochem. J. 198, 655-660.
75. Claus, T., and Pilkis, S.J.(1981): Biochem. Action Horm. 8, 209-271.
76. Claus, T., Park, C.R., and Pilkis, S.J. (1983): In Handbook Experimental Pharmacology, Vol 66, I, ed P.J. Lefebre, pp 315-360. Berlin, Heidelberg: Springer-Verlag.
77. Humble, E.(1980): Biochim. Biophys. Acta 626, 179-187.
78. Hoar, C.G., Nicoll, G.W., Schiltz, E., Schmitt, W., Bloxam, D.P., Byford, M.F., Dunbar, B., and Fothergill, L.A.(1984): FEBS Lett. 171, 292-296.
79. Pilkis, S.J., El-Maghrabi, M.R., Coven, B., Claus, T.H., Tager, H.S., Steiner, D.F., Keim, P.S., and Heinrikson, R.L.(1980): J. Biol. Chem. 255, 2770-2775.
80. Joh, T., Park, D., and Reis, D.(1978): Proc. Natl. Acad. Sci. USA 75, 4744-4748.
81. Richtand, N.M., Inagami, T., Misono, K., and Kuczenski, R.(1985): J. Biol. Chem. 260, 8465-8473.
82. Meligeni, J.A., Haycock, J.W., Bennett, W.F., and Waymire, J.C.(1982): J. Biol. Chem. 257, 12632-12640.
83. El Mestikawy, S., Glowinski, J., and Hamon, M.(1983): Nature (Lond) 302, 830-832.
84. Yanagihara, N., Tank, A.W., and Weiner, N.(1984): Molecular Pharmacol. 26, 141-147.
85. Richtand, N., Schworer, C., Kuczenski, R., and Soderling, T.(1985): Fed. Proc. 44, 705.
86. Yamauchi, T., and Fujisawa, H.(1981): Biochem. Biophys. Res. Commun. 100, 807-813.
87. Yamauchi, T., Nakata, H., and Fujisawa, H.(1981): J. Biol. Chem. 256, 5404-5409.
88. Kemp, B.E., Benjamini, E., and Krebs, E.G.(1976): Proc. Natl. Acad. Sci. USA 73, 1038-1042.
89. Kemp, B.E., Graves, D.J., Benjamini, E., and Krebs, E.G.(1977): J. Biol. Chem. 252, 4888-4894.
90. Embi, N., Parker, P.J., and Cohen, P.(1981): Eur. J. Biochem. 115, 405-413.
91. Malencik, D.A., and Fischer, E.H.(1982): In Calcium and Cell Function, ed W.Y. Cheung, Vol 3, pp 161-188. New York: Academic Press.
92. Cohen, P.(1973): Eur. J. Biochem. 34, 1-14.
93. Cohen, P., Burchell, A., Foulkes, J.G., Cohen, P.T.W., Vanaman, T.C., and Nairn, A.C.(1978): FEBS Lett. 92, 287-293.
94. Skuster, J.R., Cahn, K.F.J., and Graves, D.J.(1980): J. Biol. Chem. 255, 2203-2210.
95. Adelstein, R.S., and Eisenberg, E.(1980): Annu. Rev. Biochem. 49, 921-956.
96. Nishizuka, Y. (2984): Nature (Lond) 308, 693-698.
97. Klee, C.B., Crouch, T.H., and Krinks, M.H.(1979): Proc. Natl. Acad. Sci. USA 76, 6270-6273.

156

98. Stewart, A.A., Ingebritsen, T.S., Manalan, A., Klee, C.B., and Cohen, P.(1982): FEBS Lett. 137, 80-84.
99. Parker, P.J., Aitken, A., Bilham, T., Embi, N., and Cohen, P.(1981): FEBS Lett. 123, 332-336.
100. Titani, K., Cohen, P., Walsh, K.A., and Neurath, H.(1975): FEBS Lett. 55, 120-123.
101. Wretborn, M., Humble, E., Ragnarsson, U., and Engstrom, E.(1980): Biochem. Biophys. Res. Commun. 93, 403-408.
102. Edlund, B., Andersson, J., Titanji, V., Dahlquist, U., Ekman, P., Zetterquist, O., and Engstrom, L.(1975): Biochem. Biophys. Res. Commun. 65, 1516-1521.
103. Northrop, R., and Kendrick-Jones, J.(1976): FEBS Lett 70, 229-234.
104. Matsuda, G., Maita, T., Kato, Y., Chen, J.E., and Umegane, T.(1981): FEBS Lett. 135, 232-236.

Résumé

La protéine kinase II Ca^{2+} (CaM) dépendante est largement distribuée dans plusieurs tissus et espèces et existe sous forme de plusieurs iosozymes qui diffèrent par leur Mr apparent et leurs sousunites dans une gamme de 50 à 60 kDa. Des études biochimiques avec une kinase et des substrats purifiés ont démontré que la protéine kinase II, Ca^{2+} (CaM) dépendante a une spécificité de substrat plutôt large. Parmi les protéines phosphorylées in vitro on trouve la synapsine de cerveau, la tyrosine hydroxylase, la tryptophane hydrolase, la phenylalanine hydroxylase, la glycogène synthetase, la pyruvate kinase, la tubuline, la protéine associée aux microtubules 2 (MAP$_2$) la protéine ribosomique S6 et le phospholambame cardiaque. Bien que plusieurs de ces mêmes protéines soient phosphorylées in vivo en réponse à des conditions de mobilisation du calcium la preuve définitive que la protéine kinase II Ca^{2+} (CaM)-dépendante soit impliquée catalytiquement reste à être établie dans la plupart des cas. Des études utilisant des peptides synthetiques indiquent que la protéine kinase II Ca^{2+} (CaM) dépendante et la protéine kinase cAMP dépendante partagent certains déterminants de reconnaissance de leurs substrats mais ont aussi des différences marquées. Par conséquent, on peut prédire que ces deux kinases pourraient avoir quelques superpositions de spécificité de substrat mais également certains substrats uniques et différents.

Hormones and cell regulation. Ed J. Nunez *et al.* Colloque INSERM/John Libbey Eurotext Ltd. © 1986. Vol. 139, pp. 159–180.

The use of cAMP analogs to study cAMP-dependent protein kinase-mediated events in intact mammalian cells

S.J. Beebe, P.F. Blackmore, D.L. Segaloff, S.R. Koch, D. Burks, L.E. Limbird, D.K. Granner and J.D. Corbin

Howard Hughes Medical Institute, Department of Molecular Physiology and Biophysics, and Department of Pharmacology, Vanderbilt University Medical Center, Nashville, TN 37232, USA.

Cyclic AMP analogs were used in several intact mammalian cell preparations to study physiological responses mediated by the cAMP-dependent protein kinase (cA PK). These included adipocyte lipolysis, hepatocyte glycogenolysis, H4 hepatoma cell phosphoenolpyruvate carboxykinase gene transcription and granulosa cell LH receptor induction and progesterone synthesis. The basic principles which determined the efficacy of the analogs in stimulating these responses were studied. These included the partitioning characteristic of the analog, the concentration of analog required for protein kinase activation in vitro and the susceptibility of the analogs to hydrolysis by phosphodiesterases. The efficacy of the analogs differed among the various cell types. For example, hepatocyte glycogenolysis was 100-10,000 times more sensitive to analog stimulation than was adipocyte lipolysis.

In order to determine if cA PK was responsible for a cAMP effect, advantage was taken of a unique property of cA PK. cA PK can be synergistically activated in vitro and in vivo by using pairs of cAMP analogs, each one selective for one or the other of two intrasubunit cAMP binding sites (Site 1 and Site 2) on cA PK [J. Biol. Chem. 259, 3639 (1984)]. Various Site 1- and Site 2-selective analogs were added alone (in the linear dose-response range) and in combination, both in vitro to Type I and/or Type II cA PK isolated from various cell types, and to intact cells. Correlations were then made between the extent of synergism of cA PK activation and the synergism of the various physiological responses. All analog pairs which resulted in a synergistic activation of the respective cA PKs resulted in a synergistic increase in all the respective intact cell responses mentioned above. For all responses tested, synergism occurred only when Site 1- and Site 2-selective analogs were combined, a cA PK-specific characteristic. Because the synergism of cA PK activation was strongly correlated with the synergism of the intact cell responses, cA PK could be entirely responsible for the cAMP activation of all the physiological responses tested.

(Key words: cyclic AMP, cAMP analogs, protein kinase, phosphodiesterases, adipocytes, hepatocytes, H4 hepatoma cells, granulosa cells, lipolysis, glycogenolysis, transcription)

INTRODUCTION

Hormones such as epinephrine, glucagon and follicle-stimulating hormone (FSH) bind to specific cell surface receptors, stimulate adenylate cyclase and lead to elevations in intracellular levels of cAMP (1). The cAMP binds to the regulatory subunit(s) of the cAMP-dependent protein kinase(s) (cA PK), causing dissociation and activation of the catalytic subunit, which phosphorylates various protein substrates leading to appropriate physiological responses [see (2) for a review]. Phosphodiesterases hydrolyze the cAMP, leading to a reassociation of the cA PK subunits, thereby inactivating the enzyme as part of the mechanism to terminate the hormone response. When analogs of cAMP are used instead of hormones, they cross the cell membrane, bypass the hormone receptor-adenylate cyclase complex and directly activate the cA PK (3,4). The low K_m cAMP phosphodiesterase(s) (PDE) hydrolyzes some but not all cAMP analogs (4).

The objective of this article is to review the effects of cAMP analogs on the isolated cA PK isozymes, the low K_m PDE and isolated, intact cells. A comparison of the potencies of cAMP analogs on several cell types will be presented and factors responsible for different analog potencies will be discussed. A method will be presented which is designed to show whether or not the cA PK mediates a given physiological response. It is based on the different selectivities of cAMP analogs for the two different cAMP binding sites on the regulatory subunit (5,6), the positive cooperativity of cyclic nucleotide binding (7) and the corresponding positive cooperativity of activation of the isolated cA PK isozymes (3,7). It will also be shown that it is possible to selectively activate, in a synergistic fashion, one or the other of the cA PK isozyme in the intact cell and thereby potentially determine if the isozymes have specific roles in vivo. This method has previously been used to demonstrate that the Type II isozyme regulates the lipolytic response in isolated rat adipocytes (3). In the present report, it is extended to show that the cA PK regulates glycogenolysis in hepatocytes, transcription of the gene for phosphoenolpyruvate carboxykinase (PEPCK) and induction of the the mRNA[PEPCK] in H4 hepatoma cells, and induction of the luteinizing hormone (LH) receptor and synthesis of progesterone in granulosa cells.

MATERIALS AND METHODS

Materials: cAMP analogs were purchased from Sigma, ICN, or were gifts from Drs. Jon Miller and Robert Suva of the Biomedical Research Laboratory, Life Science Division, SRI International, Menlo Park, CA (3,4). Other materials were as previously published (3,4).

Preparation of intact cells and determination of cell responses. Adipocytes were isolated by collagenase treatment of rat epididymal fat pads as previously described (3). Glycerol release was determined as a measure of lipolysis after 30-120 min of incubation at 20° or 37° C with various concentrations of cAMP analogs or epinephrine as previously described (3,4).

Hepatocytes were prepared according to the method of Blackmore and Exton (8). Phosphorylase activity was determined after a 5 min incubation with various concentrations of cAMP analogs or glucagon by [14C] G-1-P incorporation into glycogen by the reverse reaction as reported (4,8).

H4IIE hepatoma cells were continuously cultured in T-175 culture flasks as previously reported (9). Messenger RNA[PEPCK] was determined following a 3-hour treatment with cAMP analog or hormone and phosphoenolpyruvate carboxykinase gene transcription was determined in a run-off assay following a 30 min incubation with cAMP analog as described in (9).

Porcine granulosa cells were prepared as primary cultures as reported by Segaloff et al. (10). LH receptor levels were quantitated by ^{125}I-labeled human chorio-gonadotropin as described (11). Progesterone synthesis was determined by an RIA as reported (10). Both cell responses were determined after a 48-hour treatment with various concentrations of cAMP analogs, FSH or cholera toxin (CT).

The EC_{50} for the various intact cell responses is defined as the effective concentration of cAMP analog required to stimulate the respective cell responses by 50% of the maximum determined by the appropriate hormone or other agonist as indicated (4).

For synergism experiments, intact cells were incubated with cAMP analogs alone or in combination as described (3). Each individual analog was used at a concentration that stimulated cell responses by 5-15% of the maximal response. These concentrations were in the linear dose response range. In separate incubations, two analogs were combined. One of these analogs (C8 analog) was selective for Site 1 on the regulatory subunit of cA PK and the other analog (C6 analog) was selective for Site 2. Since these two sites interact with positive cooperativity, these pairs of analogs elicit a synergistic cellular response if it is mediated by the cA PK. The synergism was estimated by the synergism quotient, which is defined as the cellular response observed when a pair of analogs were present divided by the sum of the cellular response elicited from each individual analog. A synergism quotient greater than 1.0 indicates the presence of synergism. In control experiments, two Site 1 selective analogs or two Site 2 selective analogs were tested in the same way. Two analogs selective for the same site do not cause synergism of cellular responses if mediated by the cA PK. A more thorough description of this technique will be described elsewhere (12).

Synergism of purified cA PK isozymes. The isozymes of the cA PK were purified from adipocytes (Type II), hepatocytes (Type I and Type II) and H4 hepatoma cells (Type II), using NaCl gradient elution from DEAE-cellulose as previously described (13). The isozymes were assayed in a standard procedure (3,14) in the presence or absence of cAMP analogs by measuring ^{32}P-incorporation into the synthetic heptapeptide (Leu-Arg-Arg-Ala-Ser-Leu-Gly). All activity was inhibited by the heat-stable protein kinase inhibitor. The cA PK activity ratio was defined as the activity measured in the absence of cAMP divided by maximal activity measured in the presence of 5 μM cAMP. In all cases, the basal kinase activity ratio of 0.03-0.15 was subtracted. Each individual cAMP analog was tested at a concentration that caused a cA PK activation of less than 10% of the maximal response. In separate incubations, two cAMP analogs were combined. Each analog was selective for either Site 1 or Site 2 on the regulatory subunit. In control experiments, two Site 1 selective analogs or two Site 2 selective analogs were combined. Synergism of activation occurred only when the pair included a Site 1 selective and a Site 2 selective analog.

Other determinations. The K_{act} for the cA PK is defined as the concentration of cAMP or cAMP analog which caused a half-maximal activation under standard assay conditions as previously described (4,14). The low K_m cAMP phosphodi-esterase was partially purified as previously described (4). The PDE I_{50} value was defined as the concentraton of cAMP analog which inhibited 0.1 μM [^{3}H]cAMP hydrolysis by 50% as previously described (4). The partition coefficient was determined using butanol and phosphate buffer, pH 7.5, as described in (3).

RESULTS

Table 1 shows the effects of six representative cAMP analogs on various intact cell responses. The potencies of the analogs for the various cell responses were compared according to the concentration of analog required to elicit

161

50% of the maximal response (EC$_{50}$, see Materials and Methods). The potency of any particular analog to half-maximally stimulate phosphorylase in normal hepatocytes, induce mRNA^{PEPCK} in hepatoma cells and stimulate lipolysis in adipocytes was considerably different. Hepatocyte phosphorylase activation was the most sensitive cell response tested. Half-maximal phosphorylase activation in hepatocytes required 10-1000 times lower analog concentration than required for the same extent of mRNA^{PEPCK} induction in hepatoma cells and 100-10,000 times lower analog concentration than was required for half-maximal lipolysis in adipocytes. For all responses tested, analogs modified with a thio moiety at the 8-carbon of the adenine ring (C8-thio analogs) were more potent than C8 amino analogs. The selective cell sensitivities for various cAMP analogs

Table 1. Effects of cAMP analogs on various intact cells.

The EC$_{50}$ is the concentration of cAMP analog required for a half-maximal stimulation of the indicated cell response (see Methods and Materials). Values were determined from dose response curves at 37°C.

cAMP analog	EC$_{50}$ (μM)		
	Hepatocyte Phosphorylase Activation	H4 Hepatoma mRNA^{PEPCK}	Adipocyte Glycerol Release
8-thioparachlorophenyl-cAMP	0.1	25	1000
8-thiomethyl-cAMP	0.5	50	1000
N^6-benzoyl-cAMP	0.5	60	700
N^6-butyryl-cAMP	4.8	1900	2250
8-aminomethyl-cAMP	6.0	>5000	8000
8-aminohexylamino-cAMP	60.0	1000	8600

Table 2. Responses of intact cells to 8-thioparachlorophenyl-cAMP

Species	Cell	Response	8-thioparachlorpheyl-cAMP EC$_{50}$ (μM)[a]
rat	hepatocyte	phosphorylase	0.1
rat	cardiac myocyte	phosphorylase	10
rat	H4 hepatoma	mRNA^{PEPCK}	25
rat	neutrophil	phosphorylase	500
bovine	neutrophil	phosphorylase	500
rat	adipocyte	glycerol	1000

[a]The EC$_{50}$ is defined in Table 1.

162

demonstrated in Table 1 are further illustrated in Table 2, which shows
responses of five different cell types to 8-thioparachlorophenyl-cAMP, generally
the most potent analog tested. When phosphorylase activation was measured in
rat tissues, the order of sensitivity was hepatocyte > cardiac myocyte >
neutrophils > hepatoma cell (hepatoma cell data not shown). Bovine and rat
neutrophil phosphorylase activation showed similar sensitivities to this analog.

It is well known that many analogs of cAMP are much more potent than cAMP itself
when tested as agonists on isolated cells or intact tissue preparations. It has
been assumed that this is because the analogs are more permeable and less
susceptible to degradation by phosphodiesterases than is cAMP itself. In order
to more quantitatively investigate this assumption and to determine why some
analogs are more potent than others, various parameters were investigated. In
Table 3, several cAMP analogs were tested as lipolytic agents and listed in
order of potency according to the concentration of analog required for
half-maximal glycerol release. Other properties of the analogs which might be
expected to determine the potency of the analog, or any drug acting in a similar
manner, were studied in some detail. As a measure of lipophilicity or membrane
permeability, partition coefficients for cAMP analogs were determined. The
concentration of analog required for half-maximal activation (K_{act}) of the
adipocyte Type II cA PK was determined in vitro as another parameter which
should influence potency. The concentration of analog (I_{50}) required to
inhibit by 50% the hydrolysis of 0.1 μM [^{3}H]cAMP by the low K_m,
adipocyte PDE, which has been shown to be well-correlated with actual analog
hydrolysis (4), was taken as a measure of the route of analog elimination.

Cyclic AMP was one of the poorest lipolytic agents. While it was among the best
cyclic nucleotides for protein kinase activation, it had one of the lowest partition coefficients, suggesting that it was not lipophilic enough to readily pene-
trate the cell membrane. Perhaps more importantly, it had a low I_{50} value,
indicating that it was readily hydrolyzed by the phosphodiesterase. On the other

Table 3. Various parameters of cAMP analogs

The various parameters are defined in Materials and Methods. NL = No lipolysis
(glycerol release) detected at highest concentration tested (12 mM). EC_{50}
values for glycerol release were determined from dose response curves for each
analog. Incubations were for 30 mins at 37°C.

Cyclic Nucleotide	EC_{50} (mM) Glycerol Release	Partition Coefficient	Protein Kinase activation K_{act} (μM)	PDE I_{50} (μM)
N^6-Carbamoylpropyl-cAMP	0.5	4.9	1.9	500
N^6-Benzyl-cAMP	0.7	3.0	2.1	90
8-Thiobenzyl	1.7	19.4	0.6	50
N^6-Butyryl-cAMP	2.2	1.3	3.0	600
8-Bromo-cAMP	3.9	1.6	0.2	15
8-Amino-cAMP	6.1	0.4	0.2	40
8-Aminobenzyl-cAMP	>15	10.9	4.0	200
cAMP	>15	0.6	0.2	1
2-Trifluoromethyl-cAMP	NL	4.6	1.6	0.1

163

hand, 8-bromo-cAMP was more potent than cAMP for glycerol release. It had a higher partition coefficient, suggesting more favorable lipophilicity and membrane permeability as well as a higher I_{50} value, indicating a greater resistance to hydrolysis by PDE. Both of these factors suggest that 8-bromo-cAMP is more likely to reach effective concentrations at its site of action. The most potent lipolytic agents, N^6-carbamoylpropyl- and N^6-benzoyl-cAMP were relatively poor activators of the cA PK but readily attained effective intracellular concentrations since they had very favorable lipophilicity and were highly resistant to hydrolysis (3,4). 8-Thiobenzyl-cAMP appeared to be a less effective agonist because it was so highly lipophilic that it probably remained in the membrane and/or partitioned into the lipid droplet. 2-Trifluoromethyl-cAMP was an ineffective agonist at the highest concentration tested. Although it had a similar lipophilicity and apparent K_{act} for cA PK compared to the most potent analogs, it apparently never attained effective concentrations because it was so readily hydrolyzed.

One potential complication of using cAMP analogs in intact cells is that they may elevate intracellular cAMP levels to varying degrees via competitive inhibition of PDEs. However, data in Table 3 argue that this is not the case since cA PK activation (K_{act}) occurs at much lower analog concentrations than does PDE inhibition (I_{50}). Direct evidence that cAMP analogs do not elevate

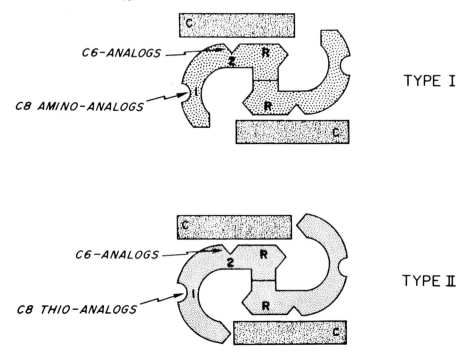

Fig. 1. Selectivities of cAMP analogs for Type I and Type II cA PK. Cyclic AMP analogs modified at the 6 carbon of the adenine ring (C6-analogs) are selective for Site 2 on the regulatory subunit of both Type I and Type II cA PK. Analogs modified with an amino group at the 8 carbon of the adenine ring (C8 amino-analogs) are generally selective for Site 1 on Type I and analogs modified with a thio group at the 8 carbon (C8 thio-analogs) are generally selective for Site 1 on Type II. The combination of a C6- plus a C8-amino analog is a Type I-directed analog pair and the combination of a C6-plus a C8 thio-analog is a Type II-directed analog pair.

but, in fact, lower hepatocyte endogenous cAMP levels via a cA PK-mediated mechanism has been recently reported (22). Results of more recent experiments from this laboratory indicate that this negative feedback regulation of cAMP levels is at least in part due to PDE stimulation. The effect is also observed in adipocytes and cardiac myocytes (unpublished). These and other results (3) indicate that the cAMP analog substitutes for cAMP as the intracellular cA PK activator.

There are two major isozymes of the cA PK, designated Type I and Type II (15) (Fig. 1). The latter type can be divided into the major subclasses referred to as Type IIA and Type IIB (13). Both isozymes are tetramers composed of two identical monomeric catalytic subunits and a single dimeric regulatory subunit. The catalytic subunits of both isozymes are identical while the regulatory subunits are similar but not identical. Both regulatory subunits contain two different cAMP binding sites per monomeric chain (16,17). Site 1 (or Site B) on both isozymes is characterized by a slow dissociation rate and a relative selectivity, in general, for cAMP analogs modified at the 8 carbon of the adenine ring (C8 analogs), while Site 2 (or Site A) on both isozymes has a rapid cAMP dissociation rate and a relative selectivity for analogs modified at the nitrogen of the 6 carbon of the adenine ring (C6-analogs) (Fig. 1). For both isozymes, binding at Site 1 stimulates binding at Site 2 and vice versa. This is correlated with a synergistic activation of both isozymes in vitro (3,7,18) and leads to a synergism of a cA PK-mediated physiological responses when appropriate pairs of analogs are added to intact cells (3). Although the regulatory subunits of both isozymes are homologous proteins and have the strongest homologies in the amino acid sequences determining their cAMP binding sites (19), Type I and II exhibit differences in cAMP analog specificities for binding. These differences, together with the positively cooperative interactions between these binding sites, can be expected to differentiate between the isozymes in vitro (3,7,18) and in the intact cell (2,3). These differences are illustrated in Fig. 1. Although both isozymes have a relative selectivity (compared to cAMP) at Site 1 for C8-analogs, the combination of a C6-analog with a C8 amino-analog stimulates binding and causes a synergistic activation of the Type I isozyme but not Type II (2,7,18). On the other hand, the combination of a C6-analog with a C8 thio-analog stimulates binding and causes a synergistic activation of the Type II isozyme but not Type I (2,3,7,18).

The cA PK can be shown to mediate a given response if incubations of the isolated cA PK and intact cells show a predicted pattern of synergism. This occurs when incubations with combinations of a Site 1- and a Site 2-selective analog cause a greater response (observed response) than the sum of responses (expected response) from incubations with each individual analog. A synergism quotient (observed response divided by expected response) greater than 1.0 indicates synergism. Incubation with two C6-analogs or two C8-analogs should not show synergism since these pairs of analogs are usually selective for the same site on the regulatory subunits. In the results and discussion that follow the relationships between the cAMP analogs and their site selectivity will be simplified in the following way (see Fig. 1):

C6-analogs + C8 amino-analogs are Type I-directed analog pairs
C6-analogs + C8 thio-analogs are Type II-directed analog pairs

The first experiments using cAMP analog pairs for synergism were carried out using rat adipocyte lipolysis (3). This system presented two major advantages. First, lipolysis had been shown, albeit indirectly, to be mediated by the cA PK. Second, the rat adipocyte contained predominantly, if not exclusively, the Type II isozyme (3,15). Therefore, only a Type II-directed analog pair would be expected to show synergism of lipolysis. Experiments were carried out by incubating cAMP analogs alone, and in combination, with intact adipocytes (Fig. 2, panel A) and with the isolated adipocyte Type II cA PK in vitro (Fig. 2, panel B). Experiments with the Type II-directed analog pair are illustrated by the set of bars on the left of each panel and Type I-directed analog pairs are

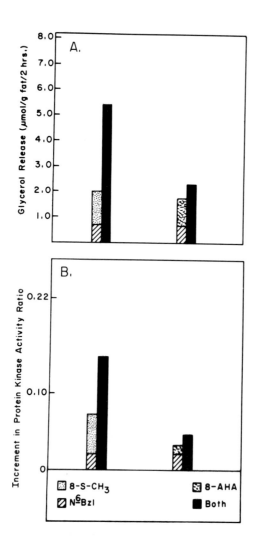

Fig. 2. Comparison of synergism of adipocyte lipolysis and synergism of adipocyte Type II cA PK activation. Synergism of glycerol release (A) was determined as a measure of lipolysis following 90 min incubations at 22°C and synergism of adipocyte Type II cA PK (B) was determined using the cA PK activity ratio as described in the text and in Materials and Methods for the respective responses. The effects of individual analog and analog pairs are indicated by the absolute height of the bars as indicated at the bottom of the Fig. N^6-benzoyl-(N^6-BZL) and 8-thiomethyl-cAMP (8-S-CH$_3$) is a Type II-directed analog pair and N^6-benzoyl- and 8-aminohexylamino-cAMP (8-AHA) is a Type I-directed analog pair.

shown by the right set of bars in each panel. In Fig. 2, panel A, when each analog is used alone, approximately 1 μmol glycerol/g fat/2 hrs was released. This is shown by the absolute height of the bars on the left of each set. Therefore, it would be expected, by summing the effects of each individual analog, that approximately 2 μmol glycerol/g fat/2 hrs would be released when the respective pairs of analogs was used. However, when a Type II-directed analog pair was used (Fig. 1, panel A, left set of bars), the observed glycerol released was 6 μmol glycerol/g fat/2 hrs (solid bars). This is almost three times the expected response. Thus, the synergism quotient (observed response/expected response) was nearly 3.0. When a Type I-directed analog pair was tested on intact adipocytes (Fig. 2, panel A, right set of bars), the results were different. The amount of glycerol released was only slightly more than expected. The synergism quotient was only 1.3. Similar results were obtained when the same experiments were carried out on the isolated adipocyte Type II cA PK (Fig. 2, panel B). Synergism of cA PK activation occurred only when a Type II-directed analog pair was used (left set of bars). The Type I-directed analog pair (right set of bars) produced no synergism.

To further demonstrate that only Type II-directed, and not Type I-directed pairs resulted in synergism of lipolysis other analog pairs were tested (Table 4). In these experiments, the Site 2 selective analog was N^6-butyryl-cAMP. For simplicity, only the synergism quotients (observed/expected) are shown. Generally synergism quotients significantly greater than 1.0 were observed only

Table 4. The effect of cAMP analog combinations on lipolysis in isolated adipocytes

The synergism quotient is defined in Materials and Methods. The indicated concentrations of cAMP analogs elicited 5-12°/o of the maximal response determined with epinephrine (53 nmol glycerol/g dry wt./min.). All incubations were at 22°C.

cAMP Analog	N^6-butyryl (0.3 mM)	8-thiobenzyl (1 mM)
	Synergism Quotient	
8-thiobenzyl-cAMP (1 mM)	2.9	1.1
8-bromo-cAMP (1 mM)	2.7	1.3
8-thiomethyl-cAMP (0.1 mM)	2.1	
8-thioparanitrobenzyl-cAMP (0.2 mM)	1.7	
cAMP (4 mM)	1.4	
8-amino-cAMP (1 mM)	1.4	
8-aminomethyl-cAMP (1 mM)	1.3	1.1
8-aminohexylamino-cAMP (2 mM)	1.3	
8-aminobenzyl-cAMP (2 mM)	1.2	
N^6-benzoyl-cAMP (0.5 mM)	1.2	
N^6 carbamoylpropyl-cAMP (1 mM)	0.8	

when N^6-butyryl-cAMP was combined with C8 thio-analogs (Type II-directed analog pairs). Synergism also occurred when N^6-butyryl-cAMP was combined with analogs such as 8-bromo-cAMP, 8-amino-cAMP and cAMP itself, but these latter cyclic nucleotides in combination with C6 analogs do not adequately differentiate between the isozymes (6). Little, if any, synergism of lipolysis occurred when N^6-butyryl-cAMP was combined with C8 amino-analogs (Type I-directed analog pairs). Small synergistic effects were also obtained when two analogs which are selective for the same site on the regulatory subunit were combined (two N^6-analogs or two C8-analogs).

Type I and Type II-directed analog pairs were also used in hepatocytes to measure the activation of phosphorylase as a measure of glycogenolysis. This represented a more complex situation because hepatocytes have approximately equal amounts of Type I and Type II cA PK. Figure 3 shows the results of these experiments in isolated, intact hepatocytes (panel A) and on the isolated hepatocyte cA PK isozymes (panel B). In Fig. 3, panel A, each cAMP analog was used alone to elicit less than 10% of the maximal phosphorylase activation. This is indicated by the absolute height of each bar. For the Type II-directed analog pairs (left set of bars), the sum of the effects of the individual analogs would be expected to cause approximately 10% of the maximal response. However, when hepatocytes were incubated with this pair of analogs, phosphorylase was activated three times more than expected. Thus, the synergism quotient was 3.0.

The Type II-directed analog pairs was also tested on the Type I and Type II cA PK (Fig. 3, panel B, left side). Individual analogs were tested such that they caused less than 10% of the maximal protein kinase activation. When the Type II-directed analog pair was tested on the hepatocyte Type I isozyme, no synergism of cA PK activation was observed (extreme left set of bars). However, the Type II isozyme was synergistically activated by this analog pair (set of bars second from left). Because the observed response was two times greater than expected, the synergism quotient was 2.0. Consequently, the synergism of phosphorylase activation observed in the intact hepatocyte with the Type II-directed analog pairs was likely due to the selective activation of the Type II isozyme.

The Type I-directed analog pair was also tested in the same way. Individual analogs were tested at concentrations that again elicited less than 10% of the maximal phosphorylase response (Fig. 3, panel A, right set of bars). The sum of the individual analog responses was expected to activate phosphorylase by about 10% of the maximum, but the pair of analogs actually caused a response 6-fold greater than expected. In other words, the synergism quotient was 6.0. Note that this synergism quotient was considerably greater than the synergism quotient obtained with a Type II-directed analog pair. When the Type I-directed analog pair was tested on the isolated hepatocyte cA PK isozymes (Fig. 3, panel B, right side), only the Type I isozyme showed a significant synergism of activation. Therefore, the synergism of phosphorylase activation induced by the Type I-directed analog pair was likely due to a selective activation of the Type I isozyme.

To further show that the Type I isozyme was not synergistically activated by the Type II-directed analog pairs and the Type II isozyme was not synergistically activated by the Type I-directed analog pair, further in vitro experiments were carried out with the partially purified isozymes. In one series of experiments, more detailed than those shown in Fig. 3B, a wide range of cAMP analog concentrations were used on the isolated isozymes to activate them (data not shown). In all experiments, the respective isozymes were only synergistically activated by the appropriate analog pair.

A further test was carried out in which various, defined proportions of partially

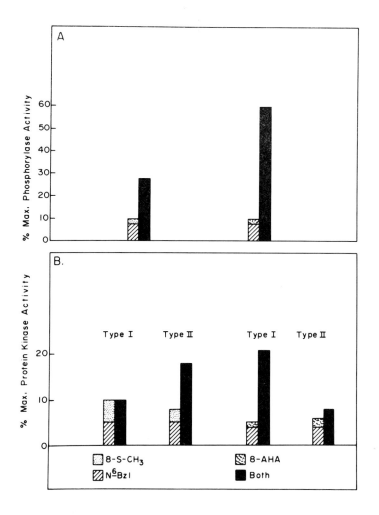

Fig. 3. Comparison of synergism of hepatocyte glycogenolysis and synergism of hepatocyte Type I and Type II cA PK activation in vitro. Synergism of hepatocyte phosphorylase activation (A) was determined as a measure of glycogenolysis following 5 min incubations and synergism of hepatocyte Type I and Type II cA PK activation (B) was determined using the cA PK activity ratio as described in the text and in Materials and Methods for the respective responses. The effects of individual analogs or analog pairs are indicated by the absolute height of the bars as indicated at the bottom of the Fig. The Type II-directed and Type I-directed analog pairs are defined in the legend to Fig. 2.

purified hepatocyte Type I and Type II isozymes were tested with both Type I and Type II-directed analog pairs (Fig. 4A and 4B). This design attempted to mimic in vivo situations in which the Type I/Type II ratio varies from one tissue to another. The cA PK activity ratio (see Materials and Methods) was used as a measure of isozyme activation and the synergism quotients. This response was plotted against the percentage of Type II isozyme present in the incubation. Fig. 4A shows the experiment with a Type II-directed analog pairs (the same Type II pair shown in Fig. 3). When only the Type I isozyme was present (zero percent Type II), no synergism was seen (synergism quotient = 1.0). As the amount of Type I was decreased and Type II increased proportionally, the synergism quotient increased in nearly linear proportion according to the amount of Type II isozyme present. When 100°/o Type II isozyme was present, the synergism quotient was slightly greater than 2.0. Fig. 3 shows the same experiment with a Type I-directed analog pair (the same Type I pair shown in Fig. 3). When all of the enzyme present was Type I isozyme (0°/o Type II), the Type I-directed analog pair showed synergism. However, as the amount of Type I isozyme was replaced with a corresponding amount of Type II, the synergism quotient decreased. When all the isozyme was Type II, no synergism was seen. These experiments show that the relative extent of synergism with a Type I- or Type II-directed analog pairs is proportional to the amount of each isozyme in a mixture.

It has recently been shown that agents which elevate intracellular cAMP in H4 hepatoma cells cause an induction of mRNA[PEPCK] and an increase of PEPCK gene transcription (9). While it is known that cAMP mediates acute changes in the activities of certain enzymes through direct phosphorylation of the enzymes by cA PK, regulation of enzyme activities by cAMP effects on mRNA levels or gene transcription, which changes the amount of enzyme, have been less well characterized. Cyclic AMP regulation of certain enzymes involved in sugar transport in E. coli is via cAMP and the catabolite gene activator protein (CAP) and not by the cA PK. The results presented in Figs. 2,3, and 4 and Table 4 indicated that an approach using cAMP analog pairs could establish whether or not the cA PK was responsible for PEPCK gene transcription and the induction of mRNA[PEPCK]. The experiments in Figs. 5 and 6 were designed like those in Figs. 2 and 3 except the Site 1 selective analog used in the Type II-directed pair was 8-thioparachlorophenyl-cAMP instead of 8-thiomethyl-cAMP. C6-analogs in combination with 8-thiomethyl-cAMP also caused synergism of mRNA[PEPCK] induction (data not shown). Figure 5A shows an experiment measuring mRNA[PEPCK] levels in H4 hepatoma cells using a Type II-directed analog pairs (left set of bars) and a Type I-directed analog pair (right set of bars) in the same format used in Figs. 2 and 3. Both the Type I-directed and the Type-II-directed analog pairs resulted in a synergistic induction of mRNA[PEPCK]. Figure 5B shows the corresponding in vitro experiments using the H4 cell Type II cA PK. Because the absolute cA PK levels in these cells were low (20) and because the Type II isozyme represented 75-85°/o of the activity recovered (recoveries were generally 60-80°/o) from DEAE-cellulose chromatography (data not shown), only the Type II isozyme was characterized in these studies. As indicated in Fig. 5B, the extent of cA PK activation expected from the sum of the individual analog effects was about 15°/o of the maximum. However, when both analogs were included, the observed enzyme activation was two times the expected amount (left set of bars). However, when the Type I-directed analog pair was tested on the Type II isozyme, no synergism was observed (right set of bars). Therefore, the H4 hepatoma cell Type II isozyme behaved like the adipocyte and hepatocyte Type II isozyme when tested in this way. Since the Type II-directed analog pair synergistically activated only the Type II isozyme, yet both Type I-directed and Type II-directed analog pairs synergistically induced mRNA[PEPCK] in the intact cell, it is likely that either isozyme can mediate the induction of mRNA[PEPCK].

170

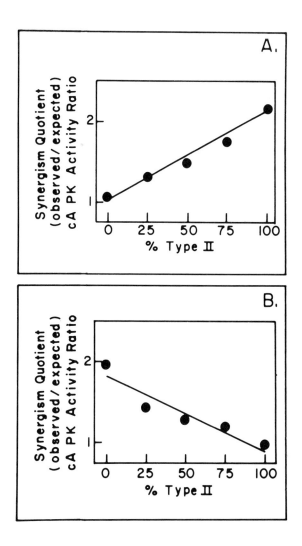

Fig. 4. Synergism of cA PK activation with various, defined proportions of hepatocyte Type I and Type II isozymes. The Type I and Type II isozymes were completely separated and partially purified from hepatocytes as described in Materials and Methods. The isozymes were diluted to equivalent activities and mixed in various proportions and indicated as the percent Type II isozyme. The various mixtures of isozymes were then tested for synergism of protein kinase activation as described in the text and in Materials and Methods. The Type II-directed analog pair (A) is N^6-benzoyl- and 8-thiomethyl-cAMP and the Type I-directed analog pair (B) is (N^6-benzoyl- and 8-aminohexylamino-cAMP. The extent of synergism is indicated by the synergism quotient.

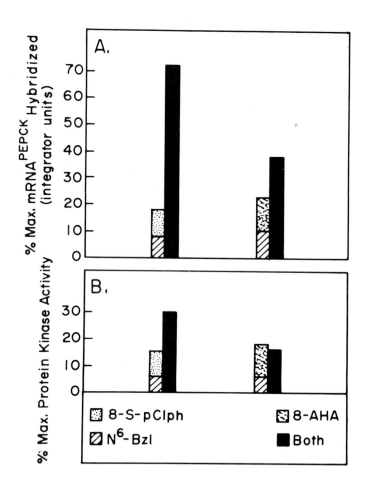

Fig. 5. Comparison of H4 hepatoma cell mRNA[PEPCK] induction and synergism of H4 hepatoma cell Type II cA PK activation. Synergism of mRNA[PEPCK] induction (A) and synergism of Type II cA PK activation using the activity ratio (B) were determined as described in the text and in Materials and Methods. The effects of individual analogs or analog pairs are indicated at the bottom of the Fig. N[6]-benzoyl-(N[6]BZL) and 8-thioparachlorophenyl-cAMP (8-S-pClph) is a Type II-directed analog pair and N[6]-benzoyl- and 8-aminohexylamino-cAMP (8-AHA) is a Type I-directed analog pair.

Although these results indicate that the cA PK is involved in mediating the induction of mRNA^{PEPCK} in H4 hepatoma cells, they did not establish the step at which the enzyme mediates these changes. For example, the increase in mRNA^{PEPCK} could be due to increases in mRNA synthesis through activation of transcription, or it could be due to stabilizing the mRNA and/or decreasing its breakdown. To determine if transcription of mRNA^{PEPCK} is mediated by cA PK, synergism experiments were carried out by measuring PEPCK gene transcription, using a run-off assay (see Materials and Methods). In Fig. 6, it can be seen that the observed responses for the Type II-directed analog pairs (left set of bars) was 13 times greater than expected and for the Type I-directed analog pair (middle set of bars) was five times greater than expected. It can also be seen that two analogs selective for the same site (right set of bars) caused little, if any, synergism.

Stimulation of ovarian granulosa cells with agents which elevate intracellular levels of cAMP is known to induce the LH receptor and increase progesterone synthesis in these cells (21). Therefore, cAMP analogs were used to directly determine if these events were mediated by the cA PK. Fig. 7A presents results of experiments in which Type I-directed and Type II-directed analog pairs were tested on LH receptor induction in primary cultures of porcine granulosa cells (see Materials and Methods). Follicle-stimulating hormone (FSH) and cholera toxin (CT), both of which are known to elevate cAMP, were used as positive controls. The Type II-directed analog pair produced an 11-fold greater increase in the LH receptor than expected from the sum of the individual analog effects (Fig. 7A, left set of bars). The Type I-directed analog pair also resulted in a significantly greater response than expected (Fig. 7A, middle set of bars), albeit less than that observed with the Type II-directed analog pair. Progesterone synthesis also showed synergistic responses with both Type I-directed and Type II-directed analog pairs. The results of a number of experiments measuring these responses indicated that the extent of synergism stimulated by the Type II-directed analog pair predominated over the synergism observed with the Type I-directed analog pair. Although synergism studies were not carried out for the isolated granulosa cell cA PK, 8-azido[^{32}P]-cAMP labeling experiments indicated that the Type II isozyme was the predominant isozyme present (data not shown). Generally speaking, the extent of synergism for the Type I and Type II-directed analog pairs seen with the intact cell responses approximated that which would be expected from the relative amounts of the two major isozymes present in the porcine granulosa cells.

DISCUSSION

The method of incubating intact cells with a combination of two cAMP analogs has proved to be a useful approach to determine if the cA PK mediates a given cell response. This technique is based on the finding that the two cAMP binding sites on the cA PK interact with positive cooperativity (7) and that certain cAMP analogs selectively bind to one or the other of these different binding sites (5,6). In other words, the binding of a Site 1 selective analog stimulates the binding of a Site 2 selective analog and vice versa. The positive cooperativity of binding is reflected in a synergistic activation of the cA PK. Because the regulatory subunit of the cA PK is the only known cAMP binding protein of physiological significance in mammalian cells, the demonstration of synergism of a cell response under these defined conditions proves that the cA PK mediates the response. To further establish the validity of the conclusion, the use of a combination of two Site 1 selective analogs or two Site 2 selective analogs does not cause either the stimulation of analog binding to the regulatory subunit (6), the synergistic activation of the cA PK (3,7) or the synergism of intact cell responses. Furthermore, the recent finding that

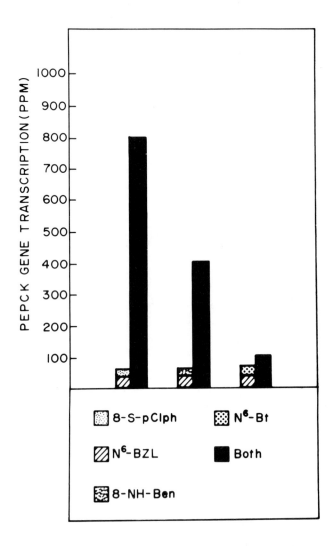

Fig. 6. Synergism of H4 hepatoma cell PEPCK gene transcription. PEPCK gene transcription was determined following incubation of H4 cells with individual analogs and analog pairs as described in Materials and Methods. The effects of individual analogs or analog pairs are shown by the absolute height of the bars as indicated at the bottom of the Fig. The pair of bars on the left side of the panel indicate a Type II-directed analog pair and the pair of bars in the middle of the panel indicated Type I-directed analog pair. The pair of bars on the right side of the panel indicate incubation of H4 cells with two analogs which are selective for the same site (Site 2) on the regulatory subunit.

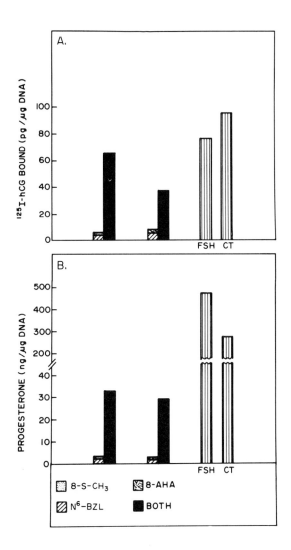

Fig. 7. Synergism of LH receptor induction and progesterone synthesis in primary cultures of porcine granulosa cells. Synergism of LH receptor induction (A) and progesterone synthesis (B) were determined as described in the text and Methods and Materials. The effects of individual analogs and analog pairs are indicated by the absolute height of the bars as indicated at the bottom of the Fig. The Type I-directed and Type II-directed analog pairs are as defined in the legend to Fig. 2.

cAMP analogs do not increase endogenous cAMP levels, but in fact decrease them (22), provides additional evidence that the cAMP analogs themselves are activating the cA PK. The method of using site selective cAMP analog combinations in intact cells has now been used to demonstrate that the cA PK mediates adipocyte lipolysis, hepatocyte glycogenolysis, H4 hepatoma cell PEPCK gene transcription and mRNA^{PEPCK} induction, and ovarian granulosa cell LH receptor induction and progesterone synthesis. In addition to these studies in isolated cells, Lefort, Van Sande et al. (manuscript in preparation) have recently used this approach in dog thyroid slices and have demonstrated that thyroid hormone secretion is mediated by cA PK. Consequently, these experiments indicate that this procedure is useful to study cA PK mediated responses and appears to be generally applicable.

There are several advantages to this technique. First, the hormone receptor - adenylate cyclase complex is bypassed and only the activation of the cA PK and events following this activation are involved. Second, because of the specificity of the analog pairs for the cA PK, the demonstration of synergism argues against cA PK-independent mechanism(s). Another advantage of using cAMP analog pairs in intact cells is that the cA PK isozyme which mediates the cell response can potentially be identified. This is possible because of a difference in analog specificity for the two cA PK isozymes. The C6 analogs, such as N^6-benzoyl-cAMP, are highly selective for Site 2 on both isozymes. However, C8 amino analogs are highly selective for Site 1 only for Type I and C8 thio analogs are highly selective for Site 1 only for Type II. Several lines of evidence suggest that this technique, in fact, can differentiate between isozymes. The adipocyte contains primarily, if not exclusively, the Type II isozyme and only Type II-directed analog pairs cause a significant synergism of glycerol release. A number of Type I-directed analog pairs were ineffective in causing synergism of lipolysis. The second line of evidence comes from in vitro cA PK activation. The Type II isozymes from rat adipocytes (3), rat hepatocytes, H4 hepatoma cells, rat heart (7), bovine heart (18), bovine neutrophils (unpublished) and solubilized rat liver membranes (unpublished) are synergistically activated only by Type II-directed analog pairs. The Type I isozymes from rat hepatocytes, rat heart (7), bovine neutrophils, (unpublished) and rabbit muscle (18) are synergistically activated only by Type I-directed analog pairs. Furthermore, when various mixtures of hepatocyte Type I and Type II isozymes were tested with Type I-directed and Type II-directed analog pairs (Fig. 4), linear changes in the extent of synergism occurred as expected when the proportions of isozymes were altered. A more conservative interpretation of these data may suggest that Type I-directed or Type II-directed analog pairs activate either isozyme in the intact cell. This is referred to as cross-over synergism of activation. The use of other cAMP analogs which have even greater site selectivities for one or the other isozymes may help to exclude this potential problem. Ogreid et al. (18) have recently described an usual analog, 8-piperidino-cAMP, which has a much greater potential to discriminate between cell responses mediated by Type I or Type II cA PK. 8-Piperidino-cAMP is selective for Site 1 on the Type II isozyme and for Site 2 on the Type I isozyme. This distinctive difference in site selectivities for the two isozymes provides a greater assurance that only a single isozyme is synergistically activated when 8-piperidino-cAMP is combined with an analog which is selective for the opposite site.

A third aspect of this technique is that it proves that the cA PK mediates a given response without the identification of putative phosphoproteins, which are presumably phosphorylated by the catalytic subunit of the cA PK. Consequently, this method is a short-cut to the criteria proposed by Krebs et al. (23) to prove that the cA PK mediates a physiological response. However, identification of putative phosphoproteins is of critical importance to futher understand cA PK action. While the phosphoproteins responsible for lipolysis and glycogenolysis have been identified, the events following cA PK activation responsible for PEPCK gene transcription, LH receptor induction and progesterone synthesis are less well understood.

Although the major consequence of cA PK activation is increased protein phosphorylation, it is possible that cAMP binding may confer some functional property to the regulatory subunit which then acts independently of or in conjunction with catalytic subunit action. This may be important in PEPCK gene transcription since the Type II regulatory subunit has recently been reported to have topoisomerase activity (24). This may also have relevance to the cAMP-stimulated elevation of the Type II regulatory subunit levels in granulosa cells reported in the present study and by others (25). In contrast to the findings of others (25), we observe an increase in catalytic subunit activity, although the fold increase is always greater for the regulatory subunit than for the catalytic subunit. Since for the responses measured in the present study both isozymes appear to mediate cAMP effects and since the catalytic subunit is common to both isozymes, it is reasonable to suggest that these responses are mediated at least in part by the catalytic subunit through phosphorylation.

Another advantage in using pairs of cAMP analogs is that the total concentration of analog required to elicit a response is decreased considerably. For example, the total analog concentration required for a given level of mRNA^{PEPCK} induction elicited by a Type II-directed analog pairs is as much as 30-fold lower than if the same response is elicited by either of the analogs alone. If cAMP analogs are ever used in drug therapy, this property may provide a advantage and reduce drug toxicity. Furthermore, even greater specificity can be achieved by activating one or the other isozyme, depending on the isozyme present in the target cell. Still greater specificity may be possible since some cell types, such as hepatocytes or cardiac myocytes, are more sensitive to cAMP analogs than other cell types, such as neutrophils or adipocytes.

ACKNOWLEDGEMENTS

The experiments using cardiac myocytes were done in collaboration with Bruce Redmon and experiments using neutrophils were done in collaboration with Dr. Ted Chrisman. We thank Penny Stelling and Becky Lawson for typing the manuscript and Bernard Bouscarel for the French translation of the resume.

REFERENCES

1. Robison, G.A., Butcher, R.W., and Sutherland, E.W. (1971): Cyclic AMP, New York and London: Academic Press.
2. Beebe, S.J. and Corbin, J.D. (1986, in press): Cyclic nucleotide-dependent protein kinases. In The Enzymes, Vol. 17, eds E.G. Krebs and P.D. Boyer, Orlando, Fla and London: Academic Press.
3. Beebe, S.J., Holloway, R., Rannels, S.R., and Corbin, J.D. (1984): Two classes of cAMP analogs which are selective for the two different cAMP-binding sites of Type II protein kinase demonstrate synergism when added together to intact adipocytes. J. Biol. Chem.. 259, 3539-3547.
4. Beebe, S.J., Redmon, J.B., Blackmore, P.F., and Corbin, J.D. (1985): Discriminative insulin antagonism of stimulatory effects of various cAMP analogs on adipocyte lipolysis and hepatocyte glycogenolysis. J. Biol. Chem. 260, 15781-15788.
5. Rannels, J.R., and Corbin, J.D. (1980): Two different intrachain cAMP binding sites of cAMP-dependent protein kinases. J. Biol. Chem. 255, 7085-7088.

6. Corbin, J.D., Rannels, S.R., Flockhart, D.A., Robinson-Steiner, A.M., Tigani, M.C., and Doskeland, S.O., Suva, R., and Miller, J.P. (1982): Effect of cyclic nucleotide analogs on intrachain Site 1 of protein kinase isozymes. Eur. J. Biochem. 125, 259-266.

7. Robinson-Steiner, A.M., and Corbin, J.D. (1983): Probable involvement of both intrasubunit cAMP binding sites in activation of protein kinase. J. Biol. Chem. 258, 1032-1040.

8. Blackmore, P.F., and Exton, J.H. (1985): Assessment of effects of vasopressin, angiotensin II, and glucagon on Ca^{2+} fluxes and phosphorylase activity in liver. Methods in Enzymol. 109, 550-558.

9. Sasaki, K., Cripe, T.P., Koch, S.R., Andreone, T.L., Peterson, D.D., Beale, E.G., and Granner, D.K. (1984): Multihormonal regulation of phosphoenolpyruvate carboxykinase gene transcription. J. Biol. Chem. 259, 15242-15251.

10. Segaloff, D.H., May, J., Schomberg, D.W., and Limbird, L.E. (1984): A model system for the biochemical study of luteinizing hormone/chorionic gonadotropin receptor synthesis. Biochim. Biophys. Acta 804, 31-36.

11. Segaloff, D.H., and Limbird, L.E. (1983): Luteinizing hormone receptor appearance in cultured porcine granulosa cells requires continual presence of follicle-stimulating hormone. Proc. Natl. Acad. Sci. USA 80, 5631-5635.

12. Beebe, S.J. and Corbin, J.D. (in press): Use of synergistic pairs of cAMP analogs in intact cells. Methods in Enzymol.

13. Robinson-Steiner, A.M., Beebe, S.J., Rannels, S.R., and Corbin, J.D. (1984): Microheterogeneity of Type II cAMP-dependent protein kinase in various mammalian species and tissues. J. Biol. Chem. 259, 10596-10605.

14. Beebe, S.J., and Corbin, J.D. (1984): Rat adipose tissue cAMP-dependent protein kinase: a unique form of Type II. Mol. and Cell Endocrinol. 36, 67-78.

15. Corbin, J.D., Keely, S.L., and Park, C.R. (1975): The distribution and dissociation of cAMP-dependent protein kinase in adipose, cardiac, and other tissues. J. Biol. Chem. 250, 218-225.

16. Corbin, J.D., Sugden, P.H., West, L., Flockhart, D.A., Lincoln, T.M., and McCarthy, D. (1978): Studies on the properties and mode of action of the purified regulatory subunit of bovine heart cAMP-dependent protein kinase. J. Biol. Chem. 253, 3997-4003.

17. Doskeland, S.O. (1978): Evidence that rabbit muscle protein kinase has two kinetically distinct binding site for cAMP. Biochem. Biophys. Res. Comm. 83, 542-549.

18. Ogreid, D., Ekanger, R., Suva, R.H., Miller, J.P., Sturm, R., Corbin, J.D., and Doskeland, S.O. (1985): Activation of protein kinase isozymes by cyclic nucleotide analogs used singly or in combination. Eur. J. Biochem. 150, 219-227.

19. Takio, K., Wade, R.D., Smith, S.B., Krebs, E.G., Walsh, K.A., and Titani, K. (1984): Guanosine cyclic 3',5'-phosphate dependent protein kinase, a chimeric protein homologous with two separate protein families. Biochem. 23, 4207-4218.

20. Granner, D.K. (1974): Absence of high-affinity adenosine 3',5'-monophosphate binding sites from the cytosol of three hepatic-derived cell lines. Arch. Biochem. Biophys. 165, 359-368.

21. Knecht, M., Amsterdam, A., and Catt, K. (1981): The regulatory role of cyclic AMP in hormone-induced granulosa cell differentiation. J. Biol. Chem. 256, 10628-10633.

22. Corbin, J.D., Beebe, S.J., and Blackmore, P.F. (1985): Cyclic AMP-dependent protein kinase activation lowers hepatocyte cAMP. J. Biol. Chem. 260, 8731-8735.

23. Krebs, E.G. and Beavo, J.A. (1979): Phosphorylation-dephosphorylation of enzymes. Ann. Rev. Biochem. 48, 923-959.

24. Constantinou, A.I., Squinto, S.P., and Jungmann, R.A. (1985): The phosphorylaion of the regulatory subunit RII of cyclic AMP-dependent protein kinase possesses intrinsic topoisomerase activity. Cell 42, 429-437.
25. Richards, J.S., and Rolfes, A.I. (1980): Hormonal regulation of cAMP binding to specific receptor proteins in rat ovarian follicles. J. Biol. Chem. 255, 5481-5489.

Résumé

Différents analogues de l'AMPc ont été utilisés sur des préparations de cellules intactes de mammifères afin d'étudier les
réponses physiologiques régulées par la protéine kinase dépendante de l'AMPc (cA PK). Ces réponses concernent notamment:
la lipolyse du tissu adipeux, la glycogénolyse hépatique,
la transcription génétique de la phosphoénolpyruvate carboxyl-
kinase de la cellule, d'hépatome H4, ainsi que la synthèse
de la progésterone et l'induction du récepteur LH sur la cellule
de la granulosa. Les études ont porté sur les principes de
base permettant de déterminer l'efficacité des analogues à
stimuler ces réponses. Elles incluent certaines caractéristiques
des analogues; leurs critères physicochimiques, leurs concentrations nécessaires pour activer in vitro la proteine kinase,
ainsi que leurs dégradations par la phosphodiesterase. En
outre l'efficacité des analogues diffère selon le type cellulaire. Par exeemple, la glycogénolyse hépatique est de 100
à 10 000 fois plus sensible à la stimulation de l'analogue
que la lipolyse du tissue adipeux.

Dans le but de déterminer si la cA PK est responsable de l'effet
de l'AMPc, il a été pris en considération une caractéristique
spécifique de la cA PK. Différents analogues de l'AMPc se
lient sélèctivement à l'un des deux sites de liaison de l'AMPc
(site 1 et site 2) d la sous-unité régulatrice de la cA PK.
Généralement les analogues modifiés sur le carbone 8 du cycle
de l'adénine (analogues C8) se lient sélèctivement avec le
site 1 et les analogues modifiés sur le carbone en position
6 (analogues C6), se lient sélèctivement avec le site 2. Il
résulte que la cA PK peut être activée synergiquement in vitro
et in vivo par différentes paires d'analogues de l'AMPc; chacun
étant sélectif de l'un ou l'autre des sites de liaison de
l'AMPc (J. Biol. Chem.,259, 3639 (1984)).
Plusieurs analogues sélèctifs des sites 1 ou 2 ont été ajoutés
seuls (dans une relation dose-effet linéaire) ou en combinaison,
à la fois in vitro avec la cA PK de type I et/ou de type II,
isolés à partir de différents modèles cellulaires et, avec
les préparations de cellules intactes. La correlation a été
faite entre le synergisme de l'activation de la cA PK et le
synergisme des différentes réponses physiologiques obtenues.
Toutes les paires d'analogues produisant une activation synergique des différentes cA PK, produisent également sur tous les
types de cellules intactes testées, une augmentation synergique
des réponses respectives, mentionnées précédemment. Pour toutes
les reponses testées, le synergisme n'apparaît que lorsque
les analogues sélèctifs du site 1 et du site 2 sont combinés;
ceci représente une caractéristique spécifique de la cA PK.
À cause de la forte correlation entre le synergisme de l'activation de la cA PK et celui des réponses des cellules intactes,
le cA PK pourrait être rendue entièrement responsable de l'activation par l'AMPc de toutes les réponses physiologiques testées.

Hormones and cell regulation. Ed J. Nunez *et al.* Colloque INSERM/John Libbey Eurotext Ltd. © 1986. Vol. 139, pp. 181–197.

Epidermal growth factor regulation of S6 phosphorylation during the mitogenic response

I. Novak-Hofer[1], J. Martin-Pérez[2], Gary Thomas[3] and George Thomas

Friedrich Miescher-Institüt, PO Box 2543, CH-4002 Basel, Switzerland, [1]Department Forschung, Kantonsspital, Lab.f.Biochemie-Endokrinologie, CH-4031 Basel, Switzerland, [2]Harvard University, Department of Cellular and Developmental Biology, The Biological Laboratories, 16 Divinity Avenue, Cambridge, MA 02138, USA and [3]Institute of Advanced Biomedical Research, Department of Biochemistry, Oregon Health Sciences University, 02181 SW Sam Jackson Park Road, Portland, OR 97201, USA.

Abstract

Addition of EGF to quiescent cells in culture leads to a rapid activation of protein synthesis accompanied by a number of alterations in the pattern of translation. These changes in protein synthesis are closely associated with the multiple phosphorylation of 40S ribosomal protein S6. Extracts from EGF-stimulated 3T3 cells are up to 10 times more potent in phosphorylating ribosomal protein S6 than extracts from quiescent cells. The maximum increase in kinase activity is observed between 15 and 30 min with only 25% activity remaining after 2 h. Phosphorylation of S6 <u>in vivo</u> follows a similar pattern of activation reaching a maximum between 30 and 60 min and then slowly returns to basal levels by approximately 3 h. The activation of protein synthesis is also rapid, however, in contrast to the transient activation of the S6 kinase and S6 phosphorylation it remains persistently high for at least 6 h following EGF treatment. Comparison of these events with EGF binding shows that about 50% of the cell surface receptor sites are lost within 10 min of exposure to EGF, binding sites then continue to decrease to about 25% of their maximum level by 2 h. Finally, sodium orthovanadate, which is known to mimic the mitogenic effect of EGF, also leads to activation of the S6 kinase, however, with distinct kinetics and by an apparent EGF receptor independent pathway.

INTRODUCTION

The growth of cells in numerous biological systems has been shown to be intimately associated with increased rates of protein synthesis (1-5). In quiescent animal cells in culture, the stimulation to proliferate by specific growth factors or serum leads to a 2 to 3 fold increase in the rate of protein synthesis within 2 h (6). This increase is regulated at the level of initiation of protein synthesis (7,8) and is closely associated with a number of specific changes in the pattern of translation (9,10). Though the activation and maintenance of high rates of protein synthesis are essential for cell growth (11,12) the biochemical events involved in controlling this process are poorly understood. However many of the translational components which take part in the activation process are structurally well characterized and in many cases their functions are known from detailed <u>in vitro</u> studies (for a review see, 13). Thus we have reasoned that the activation of protein synthesis may serve as a useful model for uncovering the intracellular regulatory pathways employed by the growth factor in eliciting the

181

mitogenic response.

Because of the great number of translational components taking part in protein synthesis it was obvious that they all could not be analyzed simultaneously. Therefore we initially limited ourselves to looking at the phosphorylation and dephosphorylation of ribosomal proteins. This was because ribosomal proteins had been shown to be phosphorylated under a variety of growth conditions (14) and because protein phosphorylation serves as a common mediator of intracellular regulatory signals (15). Having made that our goal we initially found that there was a 30 fold increase in the amount of ^{32}P incorporated into 40S ribosomal protein S6 30 min following the addition of IGF, or serum to quiescent cells in culture (16,17). Subsequently we carried out a number of studies with inhibitors (18), titration with serum and growth factors (19), and selection of phosphorylated derivatives into polysomes (19) to determine whether the multiple phosphorylation of S6 could be dissociated from the activation of protein synthesis. The results from the experiments above and those of others (20-22) support the hypothesis that S6 phosphorylation is a prerequisite for the activation of protein synthesis during this stage of the cell cycle.

Our present aim is to establish a causal relationship, either _in vitro_ or _in vivo_, between increased S6 phosphorylation and the activation of protein synthesis and to begin to uncover the regulatory mechanisms involved in controlling S6 phosphorylation in the cell. It will be our efforts to understand this latter topic which will be chiefly reviewed in the following manuscript.

RESULTS

EGF Induced S6 Phosphorylation

Previously we employed serum as a mitogenic agent, however, because the aim is to understand in molecular terms how the interaction of a mitogen with its receptor leads to increased S6 phosphorylation, we have switched from serum to a well defined growth factor, EGF (for a review see, 23). To follow the effect of EGF on S6 phosphorylation quiescent 3T3 cells, which had been labeled to equilibrium with ^{35}S-methionine, were incubated with increasing concentrations of EGF (19). After 2 h the cells were lyzed, the ribosomal proteins applied to a 2-dimensional polyacrylamide gel and the extent of S6 phosphorylation analyzed by fluorography. The more phosphorylated the protein becomes the slower its mobility in both dimensions of electrophoresis. The results show that as the concentration of EGF is raised from 10^{-12} to 10^{-10}M there is little effect on the electrophoretic mobility of S6, as compared to untreated controls (Figs. 1A-1D). However increasing the concentration to 10^{-9}M leads to sharp increase in the number of phosphorylated derivatives (Fig. 1E) and increasing the concentration to 10^{-7}M EGF has no further effect (Fig. 1F). It should also be noted that the extent to which S6 is phosphorylated under optimal conditions is only about 30% of the response induced by serum (19).

EGF Induced Protein Synthesis

To examine the comparable effect of EGF on protein synthesis, quiescent cells were stimulated with increasing concentrations of EGF and the amount of ^{35}S-methionine incorporated into nascent protein and the amount of ribosomes as polysomes was determined. The results are very comparable to one another and to those obtained for S6 phosphorylation. There is little effect of EGF until the concentration is raised to 10^{-9}M EGF and addition of more EGF has little further effect on the rate of protein synthesis (Fig. 2). Again it should be noted that the effect is only about 30% of the effect induced by serum (19).

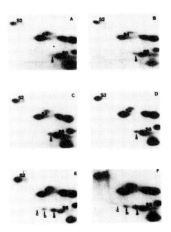

Fig. 1 Effect of increasing EGF concentration on extent of S6 phosphorylation. Cultures that had been labeled to equilibrium with [^{35}S]methionine were stimulated with increasing amounts of EGF and the extent of S6 phosphorylation was monitored by fluorographic analysis of two-dimensional polyacrylamide gels (A) 0.15M NaCl; (B) 10^{-12}M EGF, (C) 10^{-11} M EGF; (D) 10^{-10} M EGF; (E) 10^{-9} M EGF; (F) 10^{-7} M EGF. (Reprinted with permission, from ref. 19).

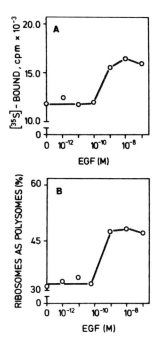

Fig. 2 Effect of increasing EGF concentration on initiation of protein synthesis. (A) Cultures were exposed to [^{35}S]methionine for 2 hr after addition of increasing concentrations of EGF and the amount of radioactivity incorporated into trichloroacetic acid-insoluble material was determined. (B) Cultures were

stimulated with increasing concentrations of EGF and the percentage of total cytoplasmic ribosomes as polysomes was determined. (Reprinted, with permission, from ref. 19).

EGF Induced Changes in the Pattern of Translation

Previously we showed that the addition of serum to quiescent cells leads to number of changes in the pattern of translation (9). These changes in the pattern of translation appear to be regulated at both the pretranslational and translational level (9). To ask whether EGF would also induce similar changes in the pattern of translation total cytoplasmic proteins from cells which had been treated with either 0.15 M NaCl, 10% serum, 10^{-10}M EGF or 10^{-9}M EGF were examined by NEPHAGE 2-dimensional gel electrophoresis (Fig. 3). The results show that all the major changes induced by serum (compare Figs. 3A and 3B) are also induced by EGF when the concentration is raised from 10^{-10}M to 10^{-9}M (compare Figs. 3C and 3D). Thus the effects induced on S6 phosphorylation, protein synthesis and the pattern of translation are very similar to those reported earlier for serum (9,19) and are consistant with the hypothesis that S6 phosphorylation is a prerequisite for the activation of protein synthesis during the transition of quiescent cells into the proliferative state of growth (18).

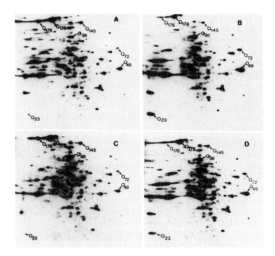

Fig. 3 Effect of serum and increasing EGF concentration on the pattern of translation. Cells were labeled with [^{35}S]methionine in the presence of the appropriate addition, followed by extraction of total cytoplasmic proteins and their analysis by fluorography of two-dimensional polyacrylamide NEPHAGE gels. (A) 0.15 M NaCl; (B) 10% serum; (C) 10^{-10} M EGF; (D) 10^{-9} M EGF.(Reprinted, with permission from ref. 10).

EGF Induced S6 Phosphopeptides

When the fully phosphorylated form of S6 is subjected to trypsin digestion and subsequent 2-dimensional thin layer analysis, 10 to 11 major phosphopeptides are observed (24). Furthermore these phosphopeptides appear in an ordered fashion (24). If instead cells are stimulated with 10^{-9}M EGF, the first 8 of 11 S6 phosphopeptides induced by serum are observed (Fig. 4). In addition those phosphopeptides which appear the earliest following serum stimulation, 7 through 11 are proportionally much more strongly labeled than those that appear later. This is the result that would be expected if EGF were following the same order of phosphorylation as serum. That EGF stimulates the same phosphopeptides as induced by serum is shown by mixing an equal amount of trypsin-digested S6 from either

EGF- or serum-stimulated cells together, and analyzing the mixture by 2-dimensional thin layer electrophoresis (Insert Fig. 4).

EGF activated S6 kinase in 3T3 cell extracts

Having established both the dose response curve for the activation of S6 phosphorylation in the intact cell by EGF and knowing the tryptic phosphopeptide pattern generated under optimal conditions, it was possible to search for a protein kinase with similar characteristics in cell extracts. When extracts from EGF treated cells were tested for their ability to phosphorylate S6 in vitro they were found to be up to 10 times more efficient than extracts derived from quiescent cells (Fig. 5). Indeed, when extracts were prepared from cells treated with increasing concentrations of EGF and subsequently tested for their ability to phosphorylate S6 in vitro they were found to closely mimic the dose response curve observed in vivo (compare Figs. 1 and 6). Furthermore if the S6 tryptic phosphopeptide pattern derived in vitro was compared with the in vivo map, the two maps were found to be essentially identical (compare Fig. 4 and 7). The results argue that the S6 kinase activity being followed in vitro is equivalent to the activity responsible for phosphorylating S6 in the intact cell.

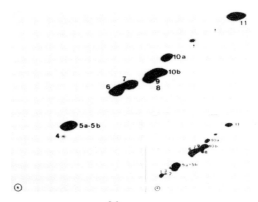

Fig. 4 Two-dimensional analysis of S6 ^{32}P-labeled tryptic phosphopeptides derived from cultures stimulated for 2 hr with 10^{-9} M EGF. The insert shows the same sample of S6 (2500 cpm) mixed with an aliquot of S6 (2500 cpm) derived from serum-stimulated cells. The direction of electrophoresis was from left to right in the first dimension at pH 1.5 and from bottom to top in the second dimension at pH 3.5 (O) Origin of sample; (↑) minor phosphopeptide. (Reprinted, with permission, from ref. 37).

Effect of phosphatase inhibitors on S6 kinase

In searching for an increased S6 kinase activity it was discovered that the presence of a phosphatase inhibitor, β-glycerol phosphate, and EGTA were required during the preparation of the extract in order to recover full activity (25). The effect of the phosphatase inhibitor was not on S6 but instead appeared to protect the S6 kinase. Following high speed centrifugation the inactivating phosphatase partitioned with the membrane fraction and the kinase remained in the soluble fraction (25). That β-glycerol phosphate was acting as a phosphatase inhibitor was shown by the fact that it could be substituted for in the extraction by a number of other phosphatase inhibitors. The most efficient of these inhibitors was phosphotyrosine followed by p-nitrophenylphosphate (PNPP), β-glycerol phosphate, phosphoserine and finally sodium orthovanadate (Fig. 8 and 26). Sodium fluoride was ineffective (Fig. 8). To further ensure that these agents are acting as phosphatase inhibitors total homogenates were incubated at 30°C in the absence of any phosphatase inhibitor (Fig. 9). At the indicated times PNPP was added and the amount of S6 kinase remaining in the homogenate was measured. The results show that in the absence of PNPP almost complete activity is lost within 10 min as compared to control homogenates incubated in the continuous presence of PNPP (Fig. 9). The results suggest that either the S6 kinase or a regulatory component of the kinase is controlled by phosphorylation.

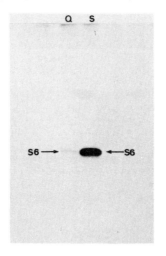

Fig. 5 Autoradiograph of S6 phosphorylating activity of cell-free extracts from Swiss 3T3 cells. Ribosomal 40S subunits from rat liver were incubated with [γ-^{32}P]ATP and soluble extracts from either quiescent cells (lane Q) or 10^{-9} M EGF-stimulated cells (lane S). After 30 min at 30°C, the reaction was stopped with concentrated SDS sample buffer, and the samples were subjected to SDS gel electrophoresis on a 15% acrylamide gel. The gel was then stained, dried, and autoradiographed.

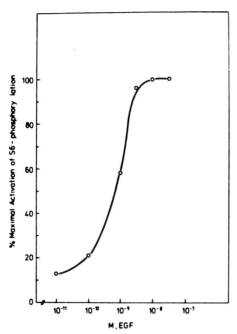

Fig. 6 Activation of S6 kinase by increasing EGF concentrations. Cells were stimulated with the indicated amounts of EGF, and extracts were prepared as described in Materials and Methods. Ribosomal 40S subunits from rat liver were incubated with each extract in the presence of [γ-^{32}P]ATP for 30 min at 30°C, and the incorporation of ^{32}P into S6 was determined by liquid scintillation spectroscopy. (Reprinted, with permission, from ref. 25).

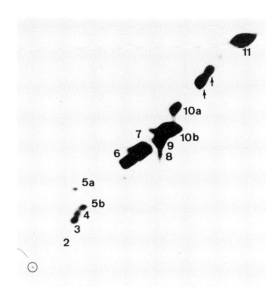

Fig. 7 Two-dimensional analysis of S6 ^{32}P-labeled tryptic phosphopeptides after _in vitro_ phosphorylation with extracts from cells that were stimulated with 10^{-8}M EGF for 30 min. 40S ribosomal subunits from rat liver were incubated as described in Materials and Methods with [^{32}P]ATP and extracts from cells that had been treated with EGF for 30 min. S6 was isolated by SDS-electrophoresis and tryptic phosphopeptide mapping was performed as described.

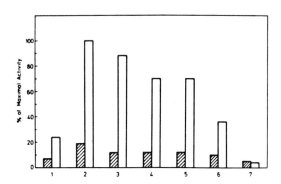

Fig. 8 Effect of phosphatase inhibitors on S6 kinase activity. Extracts from quiescent cells (hatched bars) or from cells stimulated for 30 min with 10^{-8} M EGF (open bars) were prepared in the presence of different phosphatase inhibitors and assayed for their ability to phosphorylate S6. Extracts (45 µg of protein) were added to 40S ribosomes (10 µg of protein) and ^{32}P incorporation into S6 was measured as described in Material and Methods. Results are means of three independent experiments and are expressed as percent of maximal activity. A 100% corresponds to 3950 cpm of ^{32}P incorporated into S6 per assay. 1, no phosphatase inhibitor present during cell extraction; 2, 40 mM DL-phosphotyrosine; 3, 40 mM p-nitrophenyl phosphate; 4, 80 mM β-glycerol phosphate; 5, 40 mM DL-phosphoserine; 6, 100 uM sodium orthovanadate; 7, 20 mM sodium fluoride. (Reprinted, with permission, from ref. 26).

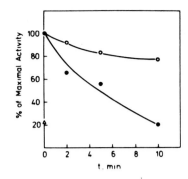

Fig. 9 Time course of inactivation of S6 kinase: effect of phosphatase inhibitor. Cells were treated for 30 min with 10^{-8} M EGF, homogenates were prepared as described in Material and Methods and incubated at 30°C for the indicated times in the presence (O) or absence (●) of 40 mM p-nitrophenyl phosphate. After incubation at 30 °C, samples were transferred to ice, and 40 mM p-nitrophenyl phosphate was added to homogenates not already containing it and then assayed for S6 kinase activity (see Material and Methods). Results are means of two independent experiments and are expressed as per cent of maximal activity. A 100% activity (samples in the presence of 40 mM p-nitrophenyl phosphate kept on ice) corresponds to 3126 cpm of ^{32}P incorporated into S6 per assay. △, basal activity of samples kept on ice and assayed in the absence of p-nitrophenyl phosphate. (Reprinted, with permission, from ref. 26).

Kinetics of EGF binding and kinase activation

The ability to isolate a stably activated form of the S6 kinase made it possible to follow its kinetics of activation following EGF treatment. The results show that the kinase is 80% activated within 5 min and reached a maximal activity between 15 and 30 min (Fig. 10). The activity then slowly declined reaching basal levels by 2h. In order to compare the kinetics of activation of the kinase to that of the down regulation of the EGF receptor, the kinetics of EGF-binding were investigated. Incubation of cells with EGF leads to a rapid loss of EGF binding sites (Fig. 11). The loss of EGF receptors preceded the activation of the kinase, such that by 10 min only 50% of EGF binding sites were remaining (Fig. 11), a time when the kinase is almost maximally activated (Fig. 10). In contrast if cells were kept at 4°C there was no detectable loss of EGF receptors (Fig. 11). Thus the level of EGF-occupied cell surface receptors seems to closely control the level to which the S6 kinase is activated.

Kinetics of EGF induced S6 phosphorylation and protein synthesis

In the intact cell the kinetics of S6 phosphorylation closely parallel the activation of the kinase, although temporally delayed. As observed by 2-dimensional gel electrophoresis within 15 min of EGF stimulation most of S6 has shifted from its native unphosphorylated position to derivative S6c (compare Figs. 12A and B), this increase continues, reaching a maximum between 30 and 60 min (Figs. 12C and D). Between 2 and 3 hrs S6 phosphorylation begins to decrease (Figs. 12E and F), this decrease continued such that no difference could be detected between quiescent cells and cells treated with EGF for 4 or 6 h (compare Fig. 12A with 12G and H). As shown previously for serum (17,19) the activation of S6 phosphorylation, closely parallels the activation of protein synthesis (Fig. 13). However unlike EGF receptors, the S6 kinase and S6 phosphorylation, protein synthesis does not appear to down regulate, remaining persistantly activated. It is interesting to also note that during the time S6 begins to become

188

dephosphorylated protein synthesis levels off. Thus as previously concluded (27) S6 phosphorylation appears to be required for the initial activation of protein synthesis but not for its maintenance.

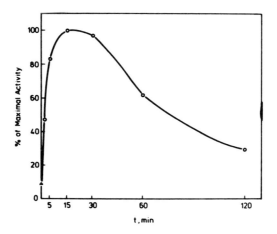

Fig. 10 Time course of activation of S6 kinase by EGF. Cells were treated with 10^{-8} M EGF for the indicated times, and extracts were prepared and assayed as described in Material and Methods. Results are means of three independent experiments and are expressed in per cent of maximal activity. A 100% corresponds to 3293 cpm of ^{32}P incorporated into S6 per assay.(Reprinted, with permission, from ref. 26).

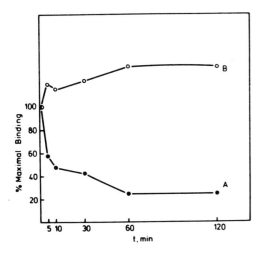

Fig. 11 Loss of ^{125}I-EGF binding after incubation with unlabeled EGF. 3T3 cells grown to confluency in 60-mm plates (2×10^6 cells/plate) were incubated in conditioned medium at 37°C (A) or on ice at 4°C (B) with 10^{-8}M EGF. At 0, 5, 10, 30, 60 and 120 min, cells were transferred to ice,washed and incubated with ^{125}I-EGF (4×10^{-9}M) as described in Material and Methods. Results are means of two independent experiments performed in triplicate and are expressed as per cent of maximal binding. 100% corresponds to 0.3 pmol of EGF bound per 2×10^6 cells. (Reprinted, with permission, from ref. 26).

189

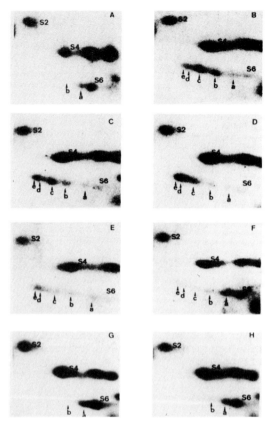

Fig. 12 Time course of activation of S6 phosphorylation by EGF. Cultures that had been labeled to equilibrium with 15 μCi/ml [^{35}S]methionine were stimulated for increasing times with 10^{-8} M EGF and the extent of S6 phosphorylation was monitored by fluorographic analysis of two-dimensional polyacrylamide gels: zero time (A), 15 min (B), 30 min (C), 1 h (D), 2 h (E), 3 h (F), 4 h (G), 6 h (H). (Reprinted, with permission, from ref. 26).

Activation of S6 kinase by Na orthovanadate

Sodium-orthovanadate has been reported to be mitogenic in Swiss 3T3 cells (28) and to mimick EGF effects on human fibroblasts without altering EGF binding and down regulation (29). While searching for agents that could activate the S6 kinase at a post-receptor level, it was found that vanadate was a potent activator of the enzyme (26). When 3T3 cells were incubated with Na orthovanadate S6 kinase was activated at concentrations as low as 10 μM with maximal activation at 4 mM (Fig. 14). In contrast to EGF, vanadate leads to a persistent activation of the S6-kinase, the onset of activation being slower and activity remaining at maximal level for at least 2 h (Fig. 15). Even in the presence of EGF, which leads to the down regulation of its receptors, vanadate can still activate the enzyme (Fig. 15, Table 1).

As shown above when the extent of S6 phosphorylation was measured in cells exposed to EGF for 4 h it was found that it had returned to its unphosphorylated form. Whereas addition of EGF at this point had no effect, vanadate could reactivate the S6 kinase and S6-phosphorylation (Table 1 and Fig. 16), demonstrating that it could bypass EGF-receptor regulated activation of the enzyme.

190

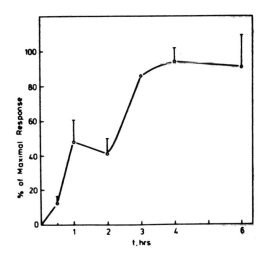

Fig. 13 Time course of activation of protein synthesis by EGF. 3T3 cells were grown to confluency in 60-mm plates (2 x 10^6 cells/plate), incubated for the indicated times with 10^{-8} M EGF, and pulse-labeled with [^{35}S]methionine as described in Material and Methods. Results are means ± S.E.M. of four indenpendent experiments performed in quadruplicate, except for the 3-h time point which is a mean of two experiments performed in quadruplicate. Results are expressed in per cent of maximal activation of protein synthesis, 100% corresponding to a 1.8 + 0.2 (n = 4)-fold increase above control levels. (Reprinted, with permission, from ref. 26).

Table 1. Effect of sodium orthovanadate on EGF-activated S6 kinase

Swiss 3T3 cells were stimulated for 30 min alone with either 10^{-8} M EGF or 1 mM sodium orthovanadate or together with both agents. Extracts were prepared and assayed for S6 kinase activity as described under "Materials and Methods". Results are means of two independent experiments expressed in per cent of maximal activity. A 100% corresponds to 2764 cpm of ^{32}P incorporated into S6 per assay.

Condition for stimulating cells	% of maximal activity of S6 kinase
30 min EGF	67
30 min vanadate	93
120 min EGF	28
120 min vanadate	100
120 min EGF followed by 30 min EGF	36
120 min EGF followed by 30 min vandate	88

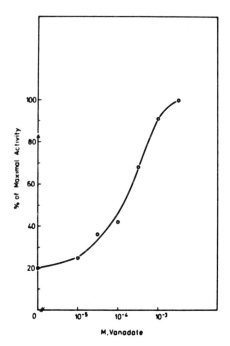

Fig. 14 Dose-response curve for activation of S6 kinase by sodium orthovanadate. Cells were treated for 30 min with the indicated amounts of sodium orthovanadate (○) or in the presence of 10^{-8} M EGF (●). Extracts were prepared and assayed for S6 kinase activity as described in Material and Methods. Results are means of two independent experiments expressed in per cent of maximal activity. 100% corresponds to 2422 cpm of ^{32}P incorporated into S6 assay. (Reprinted, with permission, from ref. 26).

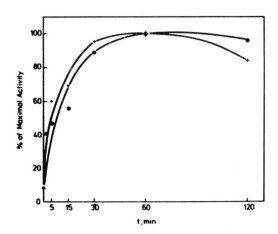

Fig. 15 Time course for activation of S6 kinase by sodium orthovanadate. Cells were treated with 1 mM sodium orthovanadate (●) or with 1 mM vanadate plus 10^{-8} M EGF (+) for the indicated times. Extracts were prepared and assayed for S6 kinase

activity as described in Material and Methods. Results are means of two independent experiments expressed in per cent of maximal activity. 100 % corresponds to 3982 cpm (0) and 4143 cpm (+) of ^{32}P incorporated in S6 per assay. (Reprinted, with permission, from ref. 26).

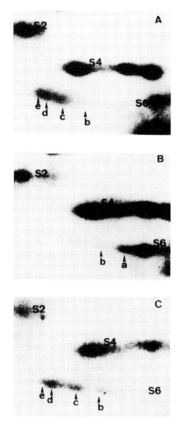

Fig. 16 Effect of EGF and sodium orthovanadate on S6 phosphorylation. Cultures that had been labeled to equilibrium with 15 μCi/ml [^{35}S]methionine were treated with 10^{-3} M sodium orthovanadate for 60 min (A), 10^{-8} M EGF for 4 h followed by a second addition of 10^{-8} M EGF for 30 min (B), or 10^{-8} M EGF for 4 h followed by 10^{-3} M sodium orthovanadate for 30 min (C). Cultures were then harvested and the extent of S6 phosphorylation was monitored on fluorograms of two-dimensional polyacrylamide gels. (Reprinted, with permission, from ref. 26).

DISCUSSION

Multiple phosphorylation of ribosomal protein S6 is associated with the activation of protein synthesis during tissue regeneration (30,31) development (27,32) cell growth (16,17) and transformation (23,24). From in vivo and in vitro studies it is argued that S6 phosphorylation exerts its effect on protein synthesis at the level of mRNA binding (19-22). The addition of EGF to quiescent 3T3 cells leads to a rapid burst in S6 phosphorylation (Fig. 1) followed by a 2 fold increase in the rate of protein synthesis (Fig. 2) which is accompanied by a number of specific changes in the pattern of translation (Fig. 3). In the intact cell EGF induces the phosphorylation of the first 8 of 11 peptides that are phosphorylated following serum stimulation of quiescent cells in culture (Fig. 4).

Knowing the dose response curve of activation of S6 phosphorylation by EGF and knowing the phosphopeptide pattern generated under optimal conditions we searched for a similar kinase activity in vitro. The results show that such a kinase activity is 10-fold stimulated in cells treated with EGF versus control cells (Fig. 5) and that the dose response curve of activation and the S6 phosphopeptides induced are equivalent to those observed in vivo (Figs. 6 and 7, respectively). A characteristic feature of the enzyme is that to recover full kinase activity in cell free extracts a phosphatase inhibitor must be present during extraction (Figs. 8 and 9), the most potent being phosphotyrosine. Following high speed centrifugation the kinase partitions in the soluble fraction and the inactivating phosphatase with membranes. At this point the phosphatase inhibitor can be removed by dialysis without loss of S6 kinase activity. Thus either the S6 kinase or a regulatory component of the kinase appears to be controlled by phosphorylation.

That a stably activated form of the S6 kinase could be extracted from cells made it possible to compare its kinetics of activation with receptor binding, S6 phosphorylation and protein synthesis following treatment of quiescent cells with EGF. The results show that all four processes are closely linked kinetically (Figs. 10-13). First EGF receptors down regulate reaching 50% of maximal binding in 10 min. The other three events reach their maximum at 1) 15-30 min, S6 kinase activation, 2) 1-2h, S6 phosphorylation, and 3) 2-3h protein synthesis. However unlike S6 kinase activation and S6 phosphorylation, protein synthesis does not down regulate within the first 6 h following EGF treatment. The results are compatible with the model that S6 phosphorylation is required for the initiation of protein synthesis but not for its maintenance (31). Thus once having triggered protein synthesis the signal is no longer required and is removed. In turn then the level and extent to which the S6 kinase and S6 phosphorylation are activated appears to be controlled by the number of EGF occupied cell surface receptors.

The latter argument above is supported by the fact that sodium orthovanadate, which others have shown to induce cell growth (28,29) and transformation (38) without effecting down regulation of EGF receptors (29), activates the S6 kinase in a persistant manner (Fig. 14) even in the presence of EGF (Fig. 15). Indeed if one first treats with EGF and allows EGF receptors to down regulate, it is no longer possible to activate either the S6 kinase or S6 phosphorylation in the intact cell with EGF (Figs. 15 and 16), however vanadate can still activate either event (Figs. 15 and 16). It has been proposed that sodium orthovanadate activates cell growth by either inhibiting tyrosine phosphatases (28,29), directly stimulating tyrosine-kinases (35) or activating the Na^+/H^+ exchanger (36). Regardless of the mechanism it is clear that sodium orthovanadate can activate the S6 kinase and S6 phosphorylation in EGF-receptor down regulated cells whereas EGF cannot, arguing that the down regulation of the S6 kinase and S6 phosphorylation are mediated by the down regulation of the receptor.

The argument that sodium orthovanadate can bypass the down regulation of the EGF receptor to activate the S6 kinase is based on the assumption that it is activating the same enzyme. Recent results from this laboratory show that both agents activate an enzyme which elutes from MonoQ (Pharmacia) in the identical position (I. Novak-Hofer, H. Luther, M. Siegmann, B. Friis and G. Thomas, in preparation). The enzyme has an apparent molecular weight of 70,000 and an apparent pI of 5.5-5.8 and is also activated by PDGF and insulin along with the oncogene sarc. It will now be necessary to purify this enzyme and determine its mode of activation.

MATERIALS AND METHODS

Cell growth

Cells were grown and maintained as previously described (19). Briefly cells were seeded at 3 x 10^5 cells per 15 cm dish (Falcon) on day 1, refed on day 4 and judged quiescent on day 8. At this point less than .05% population contained ^{3}H-labeled nuclei in a 24h radioactive labeling period.

Polyacrylamide gel electrophoresis

Labeling of cells with ^{35}S-methionine and 2-dimensional polyacrylamide gel electrophoresis for either the analysis of S6 or total cytoplasmic proteins was carried out as previously described (9,19, respectively). ^{32}P-labeling of S6 and its isolation by 1-dimensional SDS gel electrophoresis was as described by Martin-Pérez et al. (1984) and Novak-Hofer and Thomas (1984).

Phosphopeptide maps

Two dimensional analysis on thin layer plates of S6-tryptic phosphopeptides was carried out as described by Martin-Pérez and Thomas (1983) and Martin-Pérez et al.(1984). The conditions of incubation for the phosphopeptide map derived from S6 which had been incubated with the EGF extract were essentially identical to those previously described for serum (25).

Protein synthesis

Labelling cells with ^{35}S-methionine for determinating intact cell rates of protein synthesis were carried out as previously described (9).

Receptor binding and S6 kinase activity

The amount of EGF cell surface receptors and the level to which the S6 kinase is activated were determined as described by Novak-Hofer and Thomas (1984).

REFERENCES

1. Chauduri, S. and I. Liebermann (1968) J. Biol. Chem. 243: 29-33.

2. Turner, L.V. and P.J. Garlick (1974) Biochim. Biophys. Acta 563: 155-162.

3. Rudland, P.S. and Jimenez de Asua, L. (1979) Biochem. Biophys. Acta 560:91.

4. Wasserman, W., J.D. Richter and L.D. Smith (1982) Dev. Biol 89: 152.

5. Winkler, M.M., E.M. Nelson, C. Lashbrook and J.W.B. Hershey (1985) Devel. Biol. 107: 290-300.

6. Thomas, G. and J. Gordon (1979) Cell Biol.Int. Rep. 3: 307.

7. Stanners, C.P. and H. Becker (1971) J. Cell. Physiol 77: 31.

8. Rudland. P.S. (1974) Proc. Natl. Acad. Sci 71: 750.

9. Thomas, G., G. Thomas and H. Luther (1981) Proc. Natl. Acad. Sci. USA 78: 5712.

10. Thomas, G.,I. Novak-Hofer, J. Martin-Pérez, G. Thomas and M. Siegmann (1985) In Growth Factors and Transformation. Cold Spring Harbor Conference on Cancer Cells, 3, ed J. Feramisco, B. Ozanne and C. Stiles, pp.33-39. Cold Spring Harbor, New York: Cold Spring Harbor Press.

11. Brooks, R.F. (1976) Nature 260: 248.

12. Brooks, R.F. (1977) Cell 12: 311.

13. Perez-Bercoff, R. (1982) Protein biosynthesis in eukaryotes, New York: Plenum Press.

14. Rubin, C.S. and O.M. Rosen (1975) Ann. Rev. Biochem. 44: 831-887.

15. Krebs, E.G. and J.A. Beavo (1979) Ann. Rev. of Biochem. 48: 923-959.

16. Haselbacher, G.K., R.E. Humbel and G. Thomas (1979) FEBS Lett. 100: 185-189.

17. Thomas, G. M. Siegmann and J. Gordon (1979) Proc. Natl. Acad. Sci. USA 76: 3952.

18. Thomas, G., M. Siegmann, A.-M. Kubler, J. Gordon and L. Jimenez de Asua (1980) Cell 19: 1015.

19. Thomas, G., J. Martin-Pérez, M. Siegmann and A. Otto (1982) Cell 30:235.

20. Gressner, A.M. and E. Van de Leur (1980) Biochim. Biophys. Acta 608: 459.

21. Duncan, R. and E.H. McConkey (1982) Eur. J. Biochem. 123: 535-538.

22. Burkhard, S.J. and J.A. Traugh. (1983) J. Biol. Chem. 258: 14003.

23. Carpenter, G. and S. Cohen (1976) J. Cell. Biol. 71: 159.

24. Martin-Pérez, J., G. Thomas (1983) Proc. Natl. Acad. Sci. USA 80:926.

25. Novak-Hofer, I. and G. Thomas (1984) J. Biol. Chem. 259: 5995.

26. Novak-Hofer, I. and G. Thomas (1985) J. Biol. Chem. 260: 10134.

27. Nielsen, P.J., G. Thomas and J.L. Maller (1982a) Proc. Natl. Acad. Sci USA 79:2937.

28. Smith, J.B.(1983) Proc. Natl. Acad. Sci. USA 80:6162.

29. Carpenter, G. (1981) Biochem.Biophys. Res. Commun. 102:1115-1121.

30. Gressner, A.M. and I. Wool (1974) J. Biol. Chem. 249: 6917.

31. Nielsen, P.J., K.L. Manchester, H. Towbin, J. Gordon and G. Thomas. (1982b) J. Biol. Chem. 257:12316.

32. Ballinger, D. and Hunt, T.C. (1981) Dev. Biol. 87:277-285.

33. Decker, S. (1981) Proc. Natl. Acad. Sci. USA 78:4112-4115.

34. Blenis, J., Spivack, J.G., Erickson, R.L. (1984) Proc. Natl. Acad. Sci. USA 81:6408-6412.

35. Tamura, S., T.A. Brown, J.H. Whipple, Y. Fujita-Yamaguchi, R.E. Dubler, K. Cheng and J. Larner (1984) J. Biol. Chem. 259: 6650.

36. Cassel, D., Y.X. Zhuang and L. Glaser (1984) Biochem. Biophys. Res. Commun. 118: 675-681.

37. Martin-Pérez, J.,M. Siegmann, G. Thomas (1983) Cell 36:287-294.

38. Klarlund, J.K. (1985) Cell 41:707-717.

Résumé

L'addition d'EGF à des cellules quiescentes en culture conduit à l'activation rapide de la synthèse protéique qui est accompagnée par un certain nombre de modifications du processus de traduction. Ces changements de synthèse protéique sont étroitement associés à la phosphorylation multiple de la protéine S6 de la sous-unité ribosomale 40S. Des extraits de cellules 3T3 stimulées par EGF sont jusqu'à 10 fois plus actifs que les extraits correspondants de cellules quiescentes. L'augmentation maximale en activité kinasique est observée entre 15 et 30 min. avec seulement 25% d'activité résiduelle après 2 heures. La phosphorylation de S6 in vivo suit un type d'action similaire atteignant un maximum entre 30 et 60 minutes pour retourner ensuite lentement aux niveaux de base après 3 heures environ. L'activation de la synthèse protéique est aussi rapide; cependant contrairement à l'activation de la S6 kinase qui est transitoire la phosphorylation de S6 reste élevée d'une façon persistante pendant au moins 6 heures après traitement par EGF. La comparaison de ces évènements et de la liaison de EGF montre que près de 50% des sites récepteurs présents sur la surface cellulaire sont perdus dans les 10 minutes suivant l'addition de EGF; les sites de liaison continuent ensuite à diminuer jusqu'à 25% de leur niveau maximum dans les deux heures. Enfin l'orthovanadate de sodium, qui mime l'effet mitogénique de EGF produit aussi une activation de la S6 kinase mais avec des cinétiques différentes et par un processus indépendant du récepteur de l'EGF.

Hormones and cell regulation. Ed J. Nunez *et al*. Colloque INSERM/John Libbey Eurotext Ltd. © 1986. Vol. 139, pp. 199–217.

Protein sulfation of tyrosine

Wieland B. Huttner, Patrick A. Baeuerle, Ulrich M. Benedum, Evelyne Friederich, Annette Hille, Raymond W.H. Lee, Patrizia Rosa, Ulrich Seydel and Christian Suchanek

Max-Planck-Institute for Psychiatry, D-8033 Martinsried, and European Molecular Biology Laboratory, D-6900 Heidelberg, Federal Republic of Germany.

ABSTRACT

Protein sulfation on tyrosine has recently been recognized to be a widespread post-translational modification. The sulfation reaction is catalyzed by tyrosylprotein sulfotransferase and occurs on the luminal side of membranes of the trans Golgi network. Secretory proteins appear to be the predominant, if not exclusive, targets for tyrosine sulfation. Two tyrosine-sulfated secretory proteins, secretogranin I and secretogranin II, are of particular interest since they specifically occur in a wide variety of peptidergic endocrine and neuronal cells and presumably are involved in some aspect of peptide physiology, e.g., the packaging of peptides into secretory granules. Another well-characterized tyrosine-sulfated secretory protein, vitellogenin 2 of Drosophila, has been used, by means of gene transfer and site-directed mutagenesis, to investigate a possible role of tyrosine sulfation in the process of protein secretion.

INTRODUCTION

Modern DNA technology greatly facilitates the elucidation of the primary structure of proteins. However, one sometimes encounters the misconception that the knowledge of the amino acid sequence of a protein, as deduced by DNA sequencing, is sufficient to describe the detailed structure of a protein. One major reason why the knowledge of the primary structure is insufficient comes from the fact that most proteins are subject to post-translational modifications which affect their structure and thereby their function. The largest class of post-translational modifications comprises the covalent addition of chemical groups to the side chains of amino acid residues: most of the 20 "primary" amino acids used for protein synthesis can become modified on their side chains, often in many different ways. As a result, over 120 "secondary" amino acid derivatives have so far been found in proteins (1). Despite this great diversity of modifications, the principles by which most of these covalent additions affect protein structure can be classified as follows: (a) addition of charged groups (e.g., phosphorylation, sulfation, ADP-ribosylation); (b) neutralization of charged amino acid side chains by addition of uncharged groups (e.g. carboxymethylation, N-methylation); (c) addition of hydrophilic (but not necessarily charged) groups (e.g., N-glycosylation); (d) addition of hydrophobic groups (e.g., fatty acylation).

Covalent modifications not only can markedly alter the properties of amino acid side chains from the specifications encoded in the DNA but also represent an "indirect" transfer of genomic information into the phenotype of proteins. Whether or not a modification occurs, depends not only on the presence in the protein sequence of a recognition site for the modifying enzyme, but also on the presence and activity of the modifying enzyme, on the presence of the co-substrate that donates the modifying group, and on the accessibility of the protein to the modifying enzyme and the co-substrate, i.e. on a coordinated expression of several genes. Finally, one should realize that a seemingly small mutation in a protein's primary structure, say, from an asparagine to a glutamine residue, may affect the structure of the protein more than expected, since it may result in the loss of a site for modification, in this case N-glycosylation.

This article summarizes our current knowledge about the post-translational modification of proteins by sulfation of tyrosine residues. The interest of our laboratory in this modification developed from an original interest in protein phosphorylation. Following the discovery that tyrosine phosphorylation was likely to have a key role in cell transformation (2), we considered the possibility that other modifications of tyrosyl residues might also be involved in the control of cell function. The molecular resemblance of tyrosine sulfate and tyrosine phosphate was particularly intriguing. However, before tyrosine sulfation could be considered as a modification of general biological significance, it remained to be shown that this modification actually occurred widespread.

TYROSINE SULFATION - A WIDESPREAD POST-TRANSLATIONAL MODIFICATION

Tyrosine sulfate was first observed by Bettelheim more than 30 years ago in fibrinopeptide B (3). However, a search for tyrosine sulfate in proteins other than fibrinogen gave negative results (4), and subsequently only a few peptides, most of which belonged to the same family, were found to contain tyrosine sulfate (5-10). In view of the apparent scarcity of tyrosine sulfate in proteins, it was widely believed that sulfate incorporation into a protein was indicative of this being a glycoprotein, the sulfate being linked to carbohydrate residues.

The approach used in our laboratory to investigate the possible widespread occurrence of tyrosine sulfation differed from previous studies in that total cellular protein, separated by SDS-PAGE, rather than any specific protein, was analysed. Using radioactive sulfate labeling and highly sensitive methods for the detection of tyrosine sulfate in proteins, it turned out that tyrosine sulfation was a widespread protein modification (11,12).

We have found tyrosine-sulfated proteins in every animal species studied, from simple mollusks to man (11-19; Baeuerle and Huttner, unpublished data; Hille and Huttner, manuscript in preparation). Within a given animal species, tyrosine-sulfated proteins were found in most, if not all, tissues. For example, analysis of 20 different rat tissues showed that all contained proteins with sulfated tyrosine residues. When proteins of a given tissue were separated by SDS-PAGE, several tyrosine-sulfated proteins with different molecular weights were found. Different tissues contained different sets of tyrosine-sulfated proteins (11,14). Together, these results demonstrate that tyrosine sulfation is a widespread protein modification in animals.

In collaboration with Dr. Sumper's laboratory of the University of Regensburg, we have observed tyrosine sulfate in proteins of the green alga Volvox carteri, a simple multicellular organism (Huttner, Wenzl and Sumper, unpublished data). However, we have so far been unable to detect tyrosine-sulfated proteins in growing or aggregation-competent cells of Dictyostelium discoideum (20,21) and in single cell eucaryotes such as Paramecium and yeast (Huttner and Baeuerle, unpublished data). Whether or not these results indicate the appearance of tyrosine sulfation concomitantly with tissue-forming organisms during evolution remains to be determined. It will also be interesting to investigate whether tyrosine sulfation occurs in oocytes and, if not, when tyrosine sulfation appears during the development of an animal embryo.

Why was the widespread occurrence of protein tyrosine sulfation not noticed earlier? There are probably two major reasons. First, the ester bond in tyrosine sulfate is relatively acid-labile, and the sulfate group is readily hydrolyzed and thereby lost from tyrosine residues when proteins are analyzed by techniques involving strongly acidic conditions (12). The loss of the acidic sulfate group and the gain of an unmodified tyrosine can be missed easily, particularly when large proteins are studied. In fact, virtually all of the early observations on the occurrence of tyrosine sulfate were made with small peptides (3,5-10). Second, when sulfation is studied by labeling cells with radioactive sulfate, the incorporation of sulfate into carbohydrate residues of glycoproteins and, even more, into glycosaminoglycan chains of proteoglycans often masks the sulfation of tyrosine residues (12,14,16). Indeed, the demonstration of the widespread occurrence of protein tyrosine sulfation was largely based on (a) choosing hydrolysis conditions that liberated and preserved tyrosine sulfate of proteins and at the same time hydrolyzed and precipitated carbohydrate-linked sulfate, and (b) using a high-resolution thin-layer electrophoresis separation of the tyrosine sulfate-containing protein hydrolyzates (11,12).

At present, very little is known about the possible sulfation of other amino acid residues in proteins (e.g., serine, threonine, cysteine), but this may be an interesting topic for future investigations.

The widespread occurrence of tyrosine sulfation suggests important role(s) for this modification in cell function. As an approach towards understanding the significance of tyrosine sulfation, we have first investigated two questions. (a) What kind of proteins are the physiological targets for the tyrosine-sulfating enzyme(s) of cells? (b) In which subcellular compartment(s) and by which mechanism does tyrosine sulfation occur?

TARGET PROTEINS

We decided to investigate the first question, what kind of proteins are targets for tyrosine sulfation, from a cell biological point of view and to characterize the subcellular localization(s) of tyrosine-sulfated proteins. Two different approaches were considered:

(1) to take well-characterized proteins with established subcellular localizations, e.g., cytoplasmic proteins like actin and tubulin, nuclear proteins like histones, secretory proteins like immunoglobulins, etc., and to analyze which of them would be tyrosine-sulfated;

(2) to identify the major tyrosine-sulfated proteins of a cell and then to determine their subcellular localization.

Both approaches have so far provided the same answer: tyrosine sulfation appears to be largely, if not exclusively, a modification of secretory proteins.

Secretory proteins
Table 1 lists those identified proteins that have so far (October 1985) been shown to contain tyrosine sulfate. All of them have one property in common: they are secretory proteins. The sequence of events leading to the identification of tyrosine sulfate in these secretory proteins has been different. In most of the cases listed in Table 1, the secretory nature of the protein under study was known before the occurrence of sulfated tyrosine residues was shown; the latter were noticed either in the course of the characterization of the protein or as the result of a search for this modified amino acid in the protein. However, in other cases (e.g., yolk proteins (16), see below), proteins were initially studied as the major tyrosine-sulfated proteins of an organism, tissue, or cell, without knowing their secretory nature; the subsequent characterization of these tyrosine-sulfated proteins then revealed their identity with known secretory proteins.

202

Finally, in some cases (e.g., secretogranin I (13), see below), proteins were first identified by being the major tyrosine-sulfated proteins of cells and were subsequently demonstrated to be previously unknown secretory proteins.

TABLE 1 - IDENTIFIED TYROSINE-SULFATED PROTEINS

Proteins involved in blood clotting

      fibrinogen (3)
      hirudin (9)

Peptide hormones, neuropeptides, and related proteins

      gastrin (5)
      phyllokinin (6)
      caerulein (7)
      cholecystokinin (8)
      Leu-enkephalin (10)
      secretogranin I (13,18)
      secretogranin II (13,17,18)

Proteins of the immune system

      immunoglobulin G2a (15)
      immunoglobulin M (15,22)
      complement C4 (23)

Proteins of the extracellular matrix

      fibronectin (24,25)
      entactin (19)
      nidogen (19,26)
      procollagen type III (27)
      procollagen type V (28)

Proteins of the blood plasma

      alpha-2-macroglobulin (14)
      alpha-fetoprotein (29)

Storage proteins

      yolk protein 1 (16)
      yolk protein 2 (16)
      yolk protein 3 (16)

While Table 1 clearly makes the point that all tyrosine-sulfated proteins identified so far are secretory, it does not answer the question of whether tyrosine sulfation in general occurs preferentially in secretory proteins. We have investigated this question by either pulse-chase labeling or long-term labeling of cell cultures with radioactive sulfate, followed by analysis of the culture medium and various subcellular fractions (Hille and Huttner, manuscript in preparation). Pulse-chase labeling experiments with fibroblasts indicated that approximately 80% of all tyrosine-sulfated proteins labeled were secreted into the medium within 80 min. (The remaining 20% recovered in cell-associated form at the end of the chase will be discussed below.) This suggested a preferential occurrence of tyrosine sulfate in secretory proteins. However, the pulse-labeling of only a few minutes would have selected in favor of proteins with high rates of synthesis, and secretory proteins are often synthesized at high rates. Therefore, the possibility could not be excluded that the relative scarcity of tyrosine sulfate in non-secretory proteins resulted from suboptimal labeling conditions because of low rates of synthesis. This was investigated by long-term labeling of four different cell lines followed by subcellular fractionation. It was found that 90-99% of tyrosine sulfate detected in cell-associated proteins was present in soluble proteins and "peripheral" membrane proteins, both of which appeared to be mostly secretory proteins en route to the cell surface. (1-10% was found in apparently integral membrane proteins, see below). Thus, all the evidence obtained so far is consistent with the conclusion that tyrosine sulfation occurs predominantly in secretory proteins. This conclusion is further supported by experiments with whole rats which showed that plasma proteins, which are almost exclusively secretory in nature, contained much more frequently tyrosine sulfate than proteins of tissues (14).

Membrane proteins
The question of whether membrane proteins contain tyrosine sulfate residues was investigated in two ways. (a) After long-term labeling of cells, membranes were sequentially extracted with low ionic strength, high ionic strength, pH 11 and Triton X-114, followed by condensation of the Triton extract into an aqueous and a detergent phase. In various cells studied, the detergent phase, highly enriched in integral membrane proteins, was found to contain only 1-10% of the total cellular protein-bound tyrosine sulfate. SDS-PAGE of the detergent phase indicated that at least some of these tyrosine-sulfated proteins were secretory proteins that fractionated to a small extent into the detergent phase. Other tyrosine-sulfated proteins, however, did appear to be enriched in the detergent phase. Whether these proteins were indeed integral membrane proteins or whether they partitioned into the detergent phase for other reasons remains to be established (Hille and Huttner, manuscript in preparation). (b) A few bona fide integral membrane proteins, the hemagglutinin and neuraminidase of fowl plague virus and the G-protein of vesicular stomatitis virus, which are known to be sulfated, were investigated. No significant amounts of the incorporated sulfate could be recovered as tyrosine sulfate, indicating that essentially all of the sulfate was linked to carbohydrate residues (12; Huttner and Matlin, unpublished data). Thus it appears that tyrosine sulfation occurs only rarely

in integral membrane proteins, if at all.

Lysosomal proteins
Experiments carried out in collaboration with Dr. v. Figura's laboratory of the University of Muenster indicated that, after pulse-chase labeling of fibroblasts with radioactive sulfate and subcellular fractionation, only macromolecules with sulfated carbohydrates, but not protein-bound tyrosine sulfate, could be detected in the lysosomal fraction (Hille, v.Figura and Huttner, manuscript in preparation). Thus it appears that proteins destined for lysosomes do not undergo tyrosine sulfation.

Cytoplasmic, mitochondrial and nucleoplasmic proteins
As described in detail below, these three subcellular classes of proteins were found not to incorporate radioactive sulfate in sulfate labeling experiments of intact cells and tissues. Under the same labeling conditions, other proteins did incorporate sulfate, and the proteins not labeled with sulfate were found to incorporate radioactive amino acids. Therefore, the lack of radioactive sulfate incorporation did not appear to result from inappropriate experimental conditions and was taken to indicate lack of sulfation.

Cytoplasmic proteins. Several well-characterized cytoplasmic proteins such as actin, myosin, tubulin, clathrin and synapsin I were analyzed and were all found to be unsulfated. In addition, SDS-PAGE and fluorography of the soluble fraction of cell homogenates did not reveal any other sulfated proteins than secretory proteins that apparently were released from membrane vesicles by the homogenization (14; Hille and Huttner, manuscript in preparation).

Mitochodrial proteins. The crude mitochondrial fraction of a brain homogenate only contained sulfated proteins that were present in higher concentrations in the microsomal fraction and thus probably were non-mitochondrial proteins (Huttner, unpublished data). Consistent with the conclusion that cytoplasmic and mitochondrial proteins do not become sulfated are the observations that no protein sulfation could be detected after sulfate labeling of erythrocytes (which contain cytoplasm but lack other organelles) and intact synaptosomes (which contain cytoplasm and mitochondria but lack nucleus, ribosomes, rough endoplasmic reticulum and Golgi complex) (14; Huttner, unpublished data).

Nucleoplasmic proteins. No sulfation could be detected in histones, proteins that are known to receive many other post-translational modifications. Moreover, the crude nuclear fraction of a brain homogenate only contained sulfated proteins that were present in higher concentrations in the microsomal fraction and thus probably were non-nucleoplasmic proteins (Huttner, unpublished data).

The secretogranins represent examples of the experimental approach in which the study of the major tyrosine-sulfated proteins of cells has lead to the discovery of previously unknown secretory proteins. Secretogranin II was first described as the major sulfated secretory protein of the anterior pituitary by Rosa and Zanini (30,31) and was subsequently shown to be the major tyrosine-sulfated protein of this tissue, occurring in gonadotrophs, thyrotrophs, mammotrophs and corticotrophs (17). Secretogranin I was first identified in our search for the major tyrosine-sulfated proteins of PC12 cells, which revealed the presence of both secretogranin I and secretogranin II in these cells (13). In fact, the identity of the latter PC12 cell protein with the anterior pituitary protein was only recognized recently, in the course of a fruitful collaboration with the laboratory of Dr. A. Zanini from the University of Milan, and triggered a collaborative study with her and Dr. P. De Camilli on the cellular distribution of secretogranins I and II.

This study (18) revealed that secretogranins I and II occur in a wide variety of endocrine and neuronal cells, all of which have the property in common that they secrete peptide hormones/neuropeptides by the regulated pathway. Within these cells, the secretogranins take the same intracellular route as the peptides, becoming sorted into secretory storage vesicles and being released upon an appropriate stimulus. The widespread occurrence in secretory granules of peptidergic endocrine and neuronal cells has recently also been reported for chromogranin A (32-34), a well-characterized protein first observed in chromaffin granules (see 35). Interestingly, our biochemical and immunological characterization of secretogranin I, secretogranin II and chromogranin A showed that they are clearly distinct proteins but at the same time are sufficiently similar to be regarded as members of one protein class (see 18 and refs. therein). Secretogranin I, secretogranin II and chromogranin A are distinct proteins since they yield different tryptic fingerprints, since monospecific antisera can be raised against each one of them, and since they are translated from individual mRNAs. Nevertheless, the three proteins seem to share a certain extent of homology (36). Further evidence in favor of the three proteins being members of one class includes several properties that are common to secretogranin I, secretogranin II and chromogranin A (see 18). (a) All three proteins selectively occur in a wide variety of peptidergic endocrine and neuronal cells. (b) All three proteins are secretory, being sorted to the regulated pathway of secretion and being partially proteolyzed before exocytosis. (c) All three proteins have very low isoelectric points, containing high proportions of acidic amino acids and being post-translationally modified in the Golgi complex not only by sulfation on tyrosine and/or O-linked carbohydrate but also by phosphorylation on serine and to a lesser extent threonine. (d) All three proteins remain soluble upon boiling.

In view of these results, it seems likely that the three proteins serve a similar function and that this function is related to the

peptidergic nature of the endocrine and neuronal cells containing
them. While it is possible that these proteins are biologically
active after secretion, we have hypothesized (17,18) that this
protein class has a role in the process of peptide secretion. Due
to their large number of acidic amino acid residues, the
secretogranins will carry a large number of negative charges, even
at the acidic pH of the trans Golgi network (37). Therefore, these
proteins would be well suited to induce the aggregation of peptide
precursors in the trans Golgi network, which in turn might result
in the packaging and sorting of peptide precursors into vesicles
destined for the regulated pathway of secretion.

As an approach to investigate the possible function of the
secretogranins in peptide packaging, we have begun to isolate cDNA
clones coding for these proteins. In collaboration with Dr. J.
Mallet from the CNRS in Gif-sur-Yvette, we have screened cDNA
libraries from bovine adrenal medulla and rat PC12 cells with a
60-mer oligonucleotide synthesized according to published protein
sequences of chromogranin A (36). From the bovine adrenal cDNA
library, we have isolated a cDNA clone of approximately 2.1 kb
length that appears to code for chromogranin A and seems to be a
full length clone since it contains a poly(A) region at its 3´-end
and a sequence specifying the known aminoterminal protein sequence
of secreted chromogranin A near its 5´-end (Benedum, Mallet and
Huttner, manuscript in preparation). The latter sequence is
preceded by a hydrophobic sequence of 18 amino acids that starts
with methionine and is likely to represent the signal peptide of
chromogranin A. We are currently using this cDNA clone to isolate
cDNA clones coding for secretogranins I and II.

TYROSYLPROTEIN SULFOTRANSFERASE

We now turn to the second question, i.e. in which subcellular
compartment and by which mechanism tyrosine sulfation occurs.
Sulfation reactions of a variety of substances such as phenols,
steroids, and alcohols are catalyzed by specific enzymes that
transfer the sulfate from 3´-phosphoadenosine 5´-phosphosulfate
(PAPS) to the relevant acceptor substance (see 38,39). In our
efforts to characterize the mechanism and intracellular site of
tyrosine sulfation, we first searched for an enzymatic activity
capable of sulfating proteins on tyrosine residues. , Using
^{35}S-PAPS, we initially reported on such an enzymatic activity in
PC12 cells (13). In a cell-free system, this enzyme, referred to
as a tyrosylprotein sulfotransferase, was found to have the correct
substrate specificity since it catalyzed the tyrosine sulfation of
secretogranin I and secretogranin II, the major tyrosine-sulfated
proteins of PC12 cells.

For the characterization and purification of this enzyme,
sufficient quantities of purified substrate protein were required.
Although purified secretogranins were available, they were in the
sulfated form and therefore as such not suitable as substrates.
Chemical desulfation of the secretogranins by a short acid

hydrolysis removed not only the sulfate group but also led to partial cleavage of the polypeptide chain. Enzymatic desulfation by arylsulfatase yielded secretogranins in sulfatable form but was not a very efficient process (40). We therefore searched for other substrates that were suitable for the characterization of tyrosylprotein sulfotransferase.

In a few cases (e.g., fibrinopeptide B (41), gastrin (5), caerulein (7), cholecystokinin (8), hirudin (9)), the sequences surrounding sulfated tyrosine residues have been determined. Comparing these sequences we noted that the sulfated tyrosine residues were often located next or near acidic amino acid residues. Since this suggested that acidic amino acids might be involved in the recognition of substrate proteins by tyrosylprotein sulfotransferase, an acidic random polymer of tyrosine, consisting of 62% glutamic acid, 30% alanine and 8% tyrosine with an average mol. wt. of 30,000 and referred to as EAY, was tested for its ability to serve as substrate. EAY was found to be an excellent substrate for the enzyme and has been used for its characterization (40). It is also possible to measure tyrosylprotein sulfotransferase activity with other exogenous substrates such as tubulin (40) and various synthetic peptides modelled after caerulein (42), cholecystokinin (42) or the carboxyterminal nonapeptide of the cholecystokinin precursor (Domin, Adrian, Bloom, Lee and Huttner, manuscript in preparation). None of these substrates represents the physiologically relevant situation since tubulin is a cytoplasmic protein and the synthetic peptides are only fragments of the peptide hormone precursors which almost certainly are the physiological targets for sulfation.

Several lines of evidence are consistent with the conclusion or indicate that tyrosylprotein sulfotransferase is localized in the trans Golgi network. (a) Tyrosine sulfation of secretory proteins does not occur before the Golgi since the endoplasmic reticulum form of proteins that are secreted in tyrosine-sulfated form is unsulfated (43). (b) Tyrosine sulfation does not occur late after the Golgi since, upon subcellular fractionation of a bovine adrenal medulla homogenate, no significant tyrosylprotein sulfotransferase activity was detected in secretory granules (40). (c) Tyrosylprotein sulfotransferase activity is highest in the Golgi-enriched fraction, as determined by its co-enrichment with galactosyl transferase, a trans-Golgi marker enzyme (40). (d) Monensin inhibits protein tyrosine sulfation in intact cells (40). (e) In pulse-chase experiments with hybridomas secreting immunoglobulin M which contains tyrosine sulfate as well as N-linked oligosaccharides, sulfate is added to the protein at the same time during intracellular transport as galactose which is known to be added in trans-Golgi cisternae (22). (f) After short (5 min) pulse-labeling of immunoglobulin M-secreting hybridomas, sulfate is exclusively found in the form of immunoglobulin M that already contains complex-type N-linked oligosaccharides since the sulfate-labeled form is sensitive to neuraminidase (22). (g) Immunoglobulin M pulse-labeled with radioactive sulfate at $20^{\circ}C$ is not secreted when the chase of the label is performed at $20^{\circ}C$ but does become secreted when the chase is carried out at $37^{\circ}C$ (22). Treatment at $20^{\circ}C$ has been shown to block the exit from the trans

Golgi network of proteins that reach the cell surface by constitutive pathways (44).

Using EAY as substrate, tyrosylprotein sulfotransferase of Golgi membranes was characterized (40). As expected for an enzyme involved in the modification of secretory proteins, the catalytic center of tyrosylprotein sulfotransferase is oriented towards the lumen of the Golgi. The enzyme can be solubilized in active form from Golgi membranes by non-ionic detergent. Tyrosylprotein sulfotransferase requires divalent cations (Mg^{2+} or Mn^{2+}) for activity, is inhibited by salt concentrations above physiological values, and has a pH optimum between 5.3 and 7.0 with a peak between 6.0 and 6.5. The latter property fits well with the apparent trans-Golgi localization of the enzyme since the trans-cisternae of the Golgi are known to have a slightly acidic pH (37). Tyrosylprotein sulfotransferase has a very high affinity for EAY (K_m 0.3 uM) and an apparent K_m for the cosubstrate PAPS of 5 uM.

The observation that EAY serves as high affinity substrate for tyrosylprotein sulfotransferase, together with the available knowledge of the sequences surrounding tyrosine sulfate residues, suggests that an acidic sequence in the vicinity of a tyrosine residue can be sufficient for sulfation. It is conceivable that, at least in some cases, the consensus sequence for sulfation may be as simple as Glu-Tyr (EY) or Asp-Tyr (DY). Such a simple consensus sequence would not be without precedent since the consensus sequence for N-glycosylation consists of an appropriate arrangement of only two amino acid residues. However, it is also clear that sequences like EY or DY are not stringently required and that additional determinants are involved in the sulfation reaction since on the one hand tyrosine sulfate residues that are not located near acidic residues apparently can become sulfated in vivo (10) and on the other hand not all "acidic" tyrosines in secretory proteins undergo sulfation (Baeuerle, Lottspeich and Huttner; Friederich, Fritz and Huttner, manuscripts in preparation). In this context, it will be important to determine whether several tyrosylprotein sulfotransferases exist that recognize different protein substrates.

In our efforts to identify the tyrosylprotein sulfotransferase molecule, we have introduced direct photoaffinity labeling with radioactive PAPS as a method to identify PAPS-binding proteins (45). Using this method, we were able to demonstrate the presence of PAPS-binding proteins in various rat tissues (Seydel and Huttner, manuscript in preparation). However, in all cases these proteins were found to be soluble or peripherally associated with membranes, but not integral membrane proteins as expected for tyrosylprotein sulfotransferase. In the course of these studies, we noted an 85 kd membrane protein that was labeled with $3'-(^{32}P)PAPS$ also in the absence of UV irradiation. The 85 kd protein has been found in all tissues examined and has been shown to be an integral membrane protein, being enriched in Golgi membranes after subcellular fractionation. These results are consistent with the 85 kd protein being one of the Golgi sulfotransferases. Alternatively, it is also possible that the 85 kd protein is (part of) the PAPS translocator of Golgi membranes (Seydel and Huttner, manuscript in preparation).

POSSIBLE FUNCTION OF PROTEIN SULFATION ON TYROSINE

In discussing the biological significance of protein sulfation on tyrosine, one should first ask whether this modification has one, uniform, role or whether tyrosine sulfation can serve several functions. We find it likely that both is the case. First, the apparently predominant, if not exclusive occurrence of tyrosine sulfate in one topological class of proteins, the secretory proteins, suggests one primary role for this modification. Second, there are cases in which tyrosine sulfate residues serve very specific functions that cannot be generalized for other tyrosine-sulfated proteins. For example, the hormone cholecystokinin requires the sulfate on tyrosine for full biological activity (46), whereas the related hormone gastrin for some of its biological actions does not (47). We therefore believe that tyrosine sulfation may have one primary role and, in addition, can affect selected proteins in specific ways, for example with regard to their individual functions.

As to a possible uniform role of tyrosine sulfation of secretory proteins, does this modification affect proteins before or after they have been secreted from the cell? In answering this question, let us compare the proteins listed in Table 1. These proteins belong to different functional groups. Moreover, they also vary in other respects such as (a) stability to proteolysis, (b) quaternary structure, (c) solubility after secretion, or (d) uptake by receptor-mediated endocytosis, as described in the following examples. (a) Some (e.g., cholecystokinin) are subject to proteolytic processing before secretion, some (e.g., fibrinogen) are subject to proteolytic processing during circulation, some (e.g., vitellogenins) are destined to be degraded as nutrients after receptor-mediated endocytosis, and some (e.g., immunoglobulin M) are not subject to any specific proteolysis (except for the physiological turnover). (b) some (e.g., gastrin) are secreted as monomers while others (e.g., immunoglobulin M) are secreted as multimers. (c) Some (e.g., complement C4) circulate in soluble form while others (e.g., entactin) are incorporated into basement membranes. (d) Some (e.g., vitellogenins) are subject to receptor-mediated endocytosis while others (e.g., fibrinogen) are not. Thus it seems that the properties of tyrosine-sulfated proteins after secretion are diverse and that the one property common to all known tyrosine-sulfated proteins, namely to become secreted, is the only common denominator that is presently apparent. This has led us to propose, as a working hypothesis, that a uniform role of tyrosine sulfation could be related to the process of secretion of these proteins (14,15,17). (However, another uniform role of tyrosine sulfation is conceivable: although the various tyrosine-sulfated proteins as a whole behave differently with regard to proteolysis (see (a), above), tyrosine sulfation may affect specific domains of secretory proteins with

210

regard to their sensitivity to proteolysis.)

Regarding a possible role of tyrosine sulfation in protein secretion, it is interesting to note that the enzyme catalyzing the sulfation reaction, tyrosylprotein sulfotransferase, appears to be located in the trans Golgi network. This is the compartment in which secretory proteins are sorted into vesicles with specific destinations such as specific parts of the cell surface in the case of polarized cells or a storage vesicle in the case of typical secretory cells (see 48). It may be attractive to think of a role of tyrosine sulfation in the sorting of secretory proteins. One such role could be to serve as one (out of several) of the hypothesized secretion markers and to be recognized by a specific receptor in the trans Golgi network. Regarding such a role, mannose-6-phosphate provides a precedent for a post-translational modification being involved in the sorting of proteins to a specific subcellular compartment (in this case, the lysosome) (see 49). Another way in which tyrosine sulfation might affect the sorting of certain secretory proteins could be to serve as an uncoupling mechanism of secretory proteins from specific receptors once the secretory proteins have reached the correct vesicle-forming part of the membrane of the trans Golgi network. The latter model could be well reconciled with the fact that only a fraction of all secretory proteins are tyrosine-sulfated whereas others are glycosylated with sialic acid, sulfated on carbohydrate, or phosphorylated on serine and threonine. It is interesting to note that all these modifications have been shown or are likely to occur in the trans Golgi network (see 50,51; Rosa and Huttner, unpublished data) and make secretory proteins more acidic. This raises the possibility that the post-translational addition of negatively charged groups rather than any specific modification may be a critical step in the intracellular traffic of secretory proteins. In fact, we have obtained experimental evidence consistent with this possibility: a hybridoma line which secreted immunoglobulin G2a with sulfated N-linked carbohydrates in the Fab portion of the molecule was found to secrete this protein in the presence of tunicamycin not simply unglycosylated and unsulfated, but sulfated on tyrosine residues in the Fc portion (15).

A GENETIC APPROACH TOWARDS UNDERSTANDING THE FUNCTION OF TYROSINE SULFATION

In the course of studying tyrosine sulfation in various invertebrates, we also investigated the possible occurrence of this protein modification in Drosophila melanogaster (16). This showed that the three vitellogenins (yolk proteins), the major secretory proteins of female flies, were also the major sulfated proteins of Drosophila, the sulfate being essentially all linked to tyrosine. Using a novel method for the determination of the stoichiometry of tyrosine sulfation, it was demonstrated that vitellogenin 2 contained one mol of tyrosine sulfate per mol of polypeptide (16). We have purified vitellogenin 2 from sulfate-labeled flies and, in collaboration with Dr. F. Lottspeich from the Max-Planck-Institute for Biochemistry in Martinsried, have determined the sequence position of the sulfated tyrosine residue. This indicated that in virtually all vitellogenin 2 molecules, sulfate is linked to tyrosine 172 (Baeuerle, Lottspeich and Huttner, manuscript in preparation). Since the gene coding for vitellogenin 2 has been cloned (52,53), we decided to point-mutate tyrosine 172 to phenylalanine which closely resembles tyrosine but cannot be sulfated, to express the wildtype and the mutated vitellogenin 2 gene in a heterologous system (mammalian cell cultures), and to compare the secretion behavior of the wildtype and mutated protein.

We obtained the cloned gene for Drosophila vitellogenin 2 (52) from Dr. B. Hovemann from the University of Heidelberg. The 5´-flanking region of the gene, assumed to contain regulatory sequences that would hinder the expression in a heterologous system, was deleted and the remaining portion of the gene, including the entire coding sequence as well as one intervening sequence, was inserted into the expression vector pSV2, yielding pSV2-YP2(tyr). In collaboration with Dr. H. Garoff from the EMBL in Heidelberg, the expression of Drosophila vitellogenin 2 was tested by injecting pSV2-YP2(tyr) into BHK cells. This indicated that vitellogenin 2 was produced by these cells. Next, after transfection of pSV2-YP2(tyr) into L^{tk-} cells (mouse fibroblasts), stably transfected clones of these cells were selected and analyzed for synthesis, sulfation and secretion of vitellogenin 2. This indicated that the fly protein was not only synthesized and secreted by the mammalian cell, but was sulfated on tyrosine 172 as in the fly (Friederich, Hovemann, Garoff and Huttner, manuscript in preparation). Not only did these results establish the necessary basis for the site-directed mutagenesis, they also suggested that at least in this case, tyrosine sulfation is a highly conserved - and therefore presumably important - protein modification.

In collaboration with Dr. H.-J. Fritz from the Max-Planck-Institute for Biochemistry in Martinsried, tyrosine 172 has been mutated to phenylalanine, and the point-mutated vitellogenin gene has been inserted into pSV2, yielding pSV2-YP2(phe). Transient transfection of pSV2-YP2(phe) into L-cells revealed that a completely unsulfated vitellogenin 2 was produced (Friederich, Fritz and Huttner, manuscript in preparation). This result was not necessarily to be expected since vitellogenin 2 contains several other tyrosine residues in the vicinity of acidic amino acid residues (53) which might have become sulfated after mutating tyrosine 172 to phenylalanine. This result further supports the conclusion that tyrosine sulfation is a very specific post-translational modification. Most importantly however, it is now possible to precisely investigate the effect of the unsulfatability of vitellogenin 2 on its secretion behavior. Furthermore, by expressing the wildtype and mutated vitellogenin gene via the P-element in a specific strain of Drosophila which normally produces a variant form of vitellogenin 2, it should ultimately be possible to compare the wildtype and mutated protein during the entire life of the molecule in vivo.

ACKNOWLEDGEMENTS

We wish to thank Professor Hans Thoenen, Max-Planck-Institute for Psychiatry, for his interest in and support of our work. P.R. was supported by EMBO fellowships and W.B.H. was the recipient of grants from the Deutsche Forschungsgemeinschaft (Hu 275/3-1, Hu 275/3-2).

REFERENCES

1. Wold, F. (1981) Ann. Rev. Biochem. 50: 783-814.

2. Hunter, T. and Sefton, B.M. (1980) Proc. Natl. Acad. Sci. USA 77: 1311-1315.

3. Bettelheim, F.R. (1954) J. Am. Chem. Soc. 76: 2838-2839.

4. Jevons, F.R. (1963) Biochem. J. 89: 621-624.

5. Gregory, H., Hardy, P.M., Jones, D.S., Kenner, G.W. and Sheppard, R.C. (1964) Nature 204: 931-933.

6. Anastasi, A., Bertaccini, G. and Erspamer, V. (1966) Br. J. Pharmacol. Chemother. 27: 479-485.

7. Anastasi, A., Erspamer, V. and Endean, R. (1968) Arch. Biochem. Biophys. 125: 57-68.

8. Mutt, V. and Jorpes, J.E. (1968) Eur. J. Biochem. 6: 156-162.

9. Petersen, T.E., Roberts, H.R., Sottrup-Jensen, L., Magnusson, S. and Badgy, D. (1976) in Protides of Biological Fluids (Peeters, H., ed.) 23: 145-149, Pergamon Press, New York.

10. Unsworth, C.D., Hughes, J. and Morley, J.S. (1982) Nature 295: 519-522.

11. Huttner, W.B. (1982) Nature 299: 273-276.

12. Huttner, W.B. (1984) Methods Enzymol. 107: 200-223.

13. Lee, R.W.H. and Huttner, W.B. (1983) J. Biol. Chem. 258: 11326-11334.

14. Hille, A., Rosa, P. and Huttner, W.B. (1984) FEBS Lett. 117: 129-134.

15. Baeuerle, P.A. and Huttner, W.B. (1984) EMBO J. 3: 2209-2215.

16. Baeuerle, P.A. and Huttner, W.B. (1985) J. Biol. Chem. 260: 6434-6439.

17. Rosa, P., Fumagalli, G., Zanini, A. and Huttner, W.B. (1985) J. Cell Biol. 100: 928-937.

18. Rosa, P., Hille, A., Lee, R.W.H., Zanini, A., De Camilli, P. and Huttner, W.B. (1985) J. Cell Biol. 101: 1999-2011.

19. Paulsson, M., Dziadek, M., Suchanek, C., Huttner, W.B. and Timpl, R. (1985) Biochem. J. 231: 571-579.

20. Stadler, J., Gerisch, G., Bauer, G., Suchanek, C. and Huttner, W.B. (1983) EMBO J. 2: 1137-1143.

21. Hohmann, H.-P., Gerisch, G., Lee, R.W.H. and Huttner, W.B. (1985) J. Biol. Chem. 260: 13869-13878.

22. Baeuerle, P.A. and Huttner, W.B., in preparation.

23. Karp, D.R. (1983) J. Biol. Chem. 258: 12745-12748.

24. Paul, J.I. and Hynes, R.O. (1984) J. Biol. Chem. 259: 13477-13487.

25. Liu, M.-C. and Lipmann, F. (1985) Proc. Natl. Acad. Sci. USA 82: 34-37.

26. Paulsson, M., Hille, A., Huttner, W.B. and Timpl, R., in preparation.

27. Jukkola, A., Risteli, J., Niemela, O. and Risteli, L., Eur. J. Biochem., in press.

28. Fessler, L.I., Brosh, S., Chapin, S. and Fessler, J.H., submitted for publication.

29. Liu, M.-C., Yu, S., Sy, J., Redman, C.M. and Lipmann, F. (1985) Proc. Natl. Acad. Sci. USA 82: 7160-7164.

30. Rosa, P. and Zanini, A. (1981) Mol. Cell. Endocrinol. 24: 181-193.

31. Zanini, A. and Rosa, P. (1981) Mol. Cell. Endocrinol. 24: 165-179.

32. O'Connor, D.T., Burton, D. and Deftos, L.J. (1983) Life Sci. 33: 1657-1663.

33. Cohn, D.V., Elting, J.J., Frick. M. and Elde, R. (1984) Endocrinology 114: 1963-1974.

34. Somogyi, P., Hodgson, A.J., DePotter, R.W., Fischer-Colbrie, R., Schober, H., Winkler, H. and Chubb, I.W. (1984) Brain Res. Rev. 8: 193-230.

35. Winkler, H. (1976) Neuroscience 1: 65-80.

36. Settleman, J., Fonseca, R., Nolan, J. and Angeletti, R.H. (1985) J. Biol. Chem. 260: 1645-1651.

37. Anderson, R.G.W. and Pathak, R.K. (1985) Cell 40: 635-643.

38. Mulder, G.J. (1981) Sulfation of Drugs and Related Compounds, CRC Press, Boca Raton, Florida.

39. Mulder, G.J., Caldwell, J., Van Kempen, G.M.J. and Vonk, R.J. (1982) Sulfate Metabolism and Sulfate Conjugation, Taylor and Francis, London.

40. Lee, R.W.H. and Huttner, W.B. (1985) Proc. Natl. Acad. Sci. USA 82: 6143-6147.

41. Dayhoff, M.O. (1972) in Atlas of Protein Sequence and Structure (National Biomedical Research Foundation) Maryland, vol. 5: pp. D87-D97.

42. Vargas, F., Frerot, O., Dan Tung Tuong, M. and Schwartz, J.C. (1985) Biochemistry 24: 5938-5943.

43. Kudryk, B., Okada, M., Redman, C.M. and Blombaeck, B. (1982) Eur. J. Biochem. 125: 673-682.

44. Matlin, K.S. and Simons, K. (1983) Cell 34: 233-243.

45. Lee, R.W.H., Suchanek, C. and Huttner, W.B. (1984) J. Biol. Chem. 259: 11153-11156.

46. Bodanszky, M., Martinez, J., Priestley, G.P., Gardner, J.D. and Mutt, V. (1978) J. Med. Chem. 21: 1030-1035.

47. Rehfeld, J.F. (1982) Adv. Biochem. Psychopharm. vol. 33, Reulatory Peptides: From Molecular Biology to Function, E. Costa and M. Trabucchi, eds., pp. 405-411, Raven Press, New York.

48. Kelly, R.B. (1985) Science 230: 25-32.

49. Sly, W.S. and Fischer, H.D. (1982) J. Cell. Biochem. 18: 67-85.

50. Farquhar, M.G. (1985) Ann. Rev. Cell Biol. 1: 447-488.

51. Griffiths, G. and Simons, K. (1986) Science, in press.

52. Hovemann, B. and Galler, R. (1982) Nucleic Acid Res. 10: 2261-2274.

53. Hung, M.-C. and Wensink, P. (1983) J. Mol. Biol. 164: 481-492.

Résumé

Récemment, il a été reconnu que la sulfatation des protéines sur la tyrosine est une modification post-traductionnelle largement répandue. La réaction de sulfatation est catalysée par la tyrosylprotéine sulfotransférase et a lieu du côté luminal des membranes du réseau du trans Golgi. Il apparaît que ce sont les protéines sécrétrices qui sont la cible majeure, sinon exclusive, de la sulfatation de la tyrosine. Deux protéines secrétrices, sulfatées sur la tyrosine, la sécrétogranine I et la sécrétogranine II, sont d'un intérêt particulier. En effet, ces deux protéines se trouvent dans une large variété de cellules peptidergiques, endocrines et neuronales, et sont probablement impliquées dans certains aspects de la physiologie des peptides, par exemple dans l'empaquetage des peptides dans les granules de sécrétion. Une autre protéine bien caractérisée, sécrétrice et sulfatée sur la tyrosine, la vitellogénine 2 de Drosophile, a été utilisée au niveau du processus de la sécrétion des protéines, grâce au transfert de gêne et à la mutagénèse spécifique.

Ion channels

Canaux ioniques

Hormones and cell regulation. Ed J. Nunez *et al.* Colloque INSERM/John Libbey Eurotext Ltd. © 1986. Vol. 139, pp. 221–234.

Ion channels in secreting cells of exocrine glands

A. Marty

Laboratoire de Neurobiologie, 46, rue d'Ulm, 75005 Paris, France.

ABSTRACT

Three types of Ca-dependent channels are present in acinar cells of exocrine glands. They are respectively selective for K, for Cl and for monovalent cations. These channels control the resting permeability of the cells, and they are involved in the electrical response to secretion-inducing agents such as acetylcholine or noradrenaline. Transcellular fluid secretion may be understood on the basis of the activation of the Ca-dependent channels.

KEY WORDS
Ca-dependent channels. Fluid secretion. Exocrine glands.

INTRODUCTION

It has been known since the nineteenth century that exocrine gland secretion is controlled by the autonomic nervous system; either sympathetic or orthosympathetic stimulation lead to secretion. More recently, it appeared that direct application of neurotransmitters to the secreting cells of exocrine glands, the acinar cells, results in changes of ion conductance similar to those occurring in nervous and muscular tissue (see reviews 1,2,3). The pharmacology of electrical responses obtained in acinar cells when applying acetylcholine or noradrenaline is the same as that of muscarinic or α-adrenergic responses in these other tissues. Thus, the early steps of the cellular response may be identical in exocrine glands and in neurones, in the heart or in smooth muscle. Later steps of the cell responses are obviously different. Acinar cells of exocrine glands function as a neurotransmitter-responsive epithelium, and secrete both proteins and salt when stimulated.

Fluid transport across this structure can follow three routes. Some secretion results from exocytosis, but this probably amounts to a small fraction of the total volume which is extruded. Secondly, a major part of secretion is thought to occur via a transcellular ionic transport involving ion channels and ion carriers. Third, "tight junctions" regulating a paracellular route certainly play a role in secretion, since their permeability has been shown to increase markedly following neurotransmitter stimulation (4,5,6).

In the following, the membrane permeability of isolated acinar cells is discussed. More specifically, recent advances in the knowledge of ion channel properties in these cells will be reviewed. It is hoped that such studies will shed light on the transcellular part of fluid secretion.

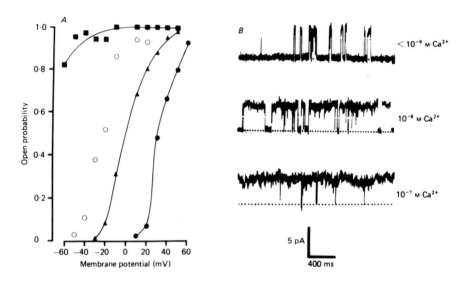

Fig. 1 . The effect of Ca upon the open probability of single K channels.
A: The variation of the probability of the open state of a single K channel as a function of membrane potential. Inside-out patch with high K solutions on both sides of the membrane, and 10^{-9}M Ca (●), 10^{-8}M Ca (▲) and 10^{-7}M Ca (■) on the internal side of the membrane. Also shown is the open state probability of this patch 'in situ', that is before obtaining the isolated patch (○).
B: Recordings at various Ca concentrations from the same patch. Membrane potential +20 mV. Dotted lines represent the channel-closed current level. (Reproduced with permission from ref. 9)

222

1) Resting conditions

       Conventional microelectrode studies showed that acinar cells have a resting potential around -40 mV for exocrine pancreas and lacrimal glands, and -70 mV for salivary glands; this potential depends on the bath K concentration (2). Current-voltage relations obtained with this technique were linear (2), but this result may have been biased to a large extent by the presence of electrical connections linking adjacent cells (2) and by ion accumulation in the extracellular spaces between cells.

       With the advent of patch-clamp techniques (7), it became possible to reexamine the permeability of the cells without interference from gap junction conductances or from ion accumulation. Cell-attached recordings (7) allow to probe for ion channels contributing to the cell conductance with a minimal disturbance to the cell environment. Such experiments revealed in many exocrine glands the presence of a particular K channel with large unitary conductance; "inside-out" recordings (see 7) soon revealed that this channel was sensitive to the internal Ca concentration, Ca_i (8,9,10,11; Fig.1). The same channel, called BK channel, had been previously found in a variety of glandular, muscular of nervous tissues (see for reviews 12,13). BK channel activity was found in rodent salivary glands (8) and lacrimal glands (9,10), and in pig exocrine pancreas (11).

       The results from cell-attached and inside-out experiments may be corroborated with those of whole-cell recordings, which allow to obtain the "macroscopic" current flowing through the entire cell surface (7). As shown in Fig. 2, the average current obtained by giving repetitive voltage jumps to an isolated patch containing a small number of BK channels (in the present case, 2) has the same kinetics as the macroscopic current obtained with a single jump on an entire cell. Furthermore, unitary currents in isolated patches are blocked by the same substances (notably tetraethylammonium, Fig.2) as the cell currents. These observations lead to conclude that the major part of the cell conductance is due to BK channels (9). Noise analysis confirmed this conclusion, and in addition gave an estimate of the total number of BK channels (50 to 150 per cell; ref. 9). A similar situation also pertains in pig pancreatic cells (11,14).

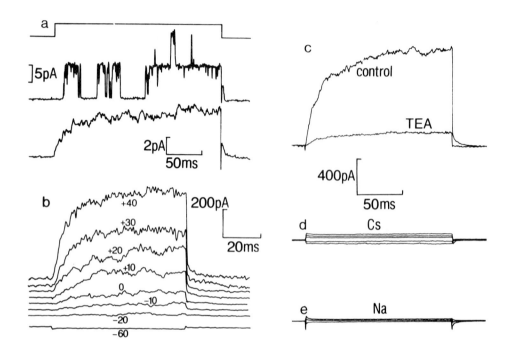

Fig. 2 : BK channels in isolated lacrimal gland cells.

a: In response to depolarizing voltage jumps from -30 to +10 mV (upper record), an isolated patch containing two BK channels displayed voltage-dependent single channel events (middle record). An average of such traces (lower record) yields the time course of the probability of opening of the channels.

b: Whole-cell recording in the same ionic conditions as in a (high internal K, high external Na).

c: Effect of tetraethylammonium (ET_4N^+, 2 mM) on whole-cell current. Test potential 0mV. Holding potential -40 mV.

d, e: Lack of voltage dependent currents in two cells dialysed with a high Cs (d) and high Na (e) solution respectively. Test voltages range between -110 and +50 mV. (Reproduced with permission from ref. 9)

2) Effects of hormones and neurotransmitters

BK channels were not found in all exocrine glands. In fact, the first investigations using patch-clamp techniques, which were conducted in rodent exocrine pancreas, revealed a totally different kind of Ca-dependent channels (15,16). These channels are selective for monovalent cations, but do not discriminate between Na and K. Their unit conductance is between 25 and 30 pS. Very similar channels were previously found in cardiac muscle (17) and in neuroblastoma cells (18).

Maruyama and Petersen (16) showed that application of cholecystokinin activates these channels by a mechanism involving an intracellular second messenger, probably Ca. Interestingly, no significant channel activity was observed in the absence of stimulation. Thus, we still do not know what kind of permeability regulates the resting conductance of these cells.

In rodent lacrimal glands, where mainly BK channels are recorded at rest, application of muscarinic agonists was found to increase the opening probability of these K channels (9,19,20). Noise analysis demonstrated that it is the opening probability of individual channels, and not the total number of channels available, which is augmented by agonist application (9). The response could be blocked by internal EGTA (9) and mimicked by application of the Ca ionophore, A23187 (19), showing that it was due to a rise of Ca_i. The response may be obtained in whole-cell recording or cell-attached modes, where Ca_i is not kept constant (unless a high EGTA internal solution is chosen in the former case). On the other hand, outside-out patches are not responsive (9), thus showing that the channels are not directly stimulated by ACh.

Whereas responses to moderate agonist doses (up to 0.5 μM carbamylcholine, or 0.2 μM ACh) can be entirely accounted for by the effect on BK channels, more complex conductance changes are elicited at higher concentrations (19,20). These are due to the additional activation of two other classes of Ca-dependent channels: Cl channels with a low unit conductance of 1-2 pS, and the monovalent cation selective channels described earlier in rodent exocrine pancreas, with a unit conductance of 25-30 pS. Cl channels are turned on above 10^{-7}M Ca_i, whereas the 25-30 pS channels need larger Ca_i values, in the micromolar range (19). When the latter channels are activated, BK channels undergo a voltage and Ca-dependent inactivation (19).

Examples of complex electrical responses obtained with a relatively high dose of carbamylcholine (CCh, 2 μM) are shown in Fig. 3. In the recordings shown (from three different cells), the holding potential was -60mV, and repetitive

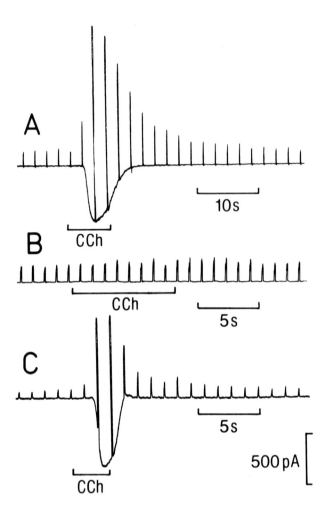

Fig. 3 : Whole-cell recordings during bath applications of 2μM CCh.
A: Response in a cell dialysed with a solution containing a low concentration of EGTA (0.5 mM). The 0-current level is undistinguishable from the holding current (holding potential -60 mV). In the presence of CCh, outward K current relaxations obtained in response to depolarizing voltage pulses (upward deflections) are greatly increased. In addition, an inward Cl current is observed at the holding potential.
B: Lack of response in another cell dialysed with a higher concentration of EGTA (5 mM).
C: Normal response in a third cell in the absence of external Ca ions. Bath and test solutions contained 0.5 mM EGTA and no added Ca. (Reproduced with permission from ref. 19)

depolarizing voltage pulses to OmV were given. Before CCh application, the holding current at the resting potential was very small. The depolarizing voltage pulses elicited outward relaxations of K currents similar to those illustrated in Fig. 2. (The relaxations are too fast to be seen on the time scale of Fig. 3). In the presence of CCh, an inward Cl current developped at -60mV, and the outward K current increased (Fig. 3A,C). No response was observed in cells dialysed with a high EGTA solution (Fig. 3B), but normal responses were obtained if Ca was removed from the external medium (Fig. 3C). This shows that the responses are mediated by a rise of internal Ca, and that the source of Ca is intracellular.

3) Localization of the various channel types

Only limited information is available concerning the repartition of the various Ca-dependent channels between basolateral and luminal membranes. Many of the studies used isolated cells, where the distinction between the two membrane domains is problematic or impossible. However, some of the cell-attached recordings were performed on cell clusters, and were thus most probably obtained from basolateral membrane. These studies showed that BK channels and monovalent cation selective channels are both located on the basolateral membrane (8,15).

4) Comparison between different exocrine glands

Here also, information is still scarce, and it is premature to generalize the findings obtained on a few preparations to all exocrine glands. Very different secretion patterns are found depending on the gland, the animal species, and the mode of stimulation (reviewed in 21). The responses to neurotransmitter, measured with labeled ion flux studies or conventional microelectrode recordings, show similar variations (reviewed in 2). It seems nevertheless plausible that this variability results from various combinations of the three basic elements found so far. As mentioned earlier, patch-clamp results in pig pancreas and in rodent salivary and lacrimal glands are very similar. On the other hand, exocrine pancreas from rodents displays a different pattern, since in these glands, the monovalent cation selective channel is directly activated without previous stimulation of the BK channel. (No activation of Cl channels was reported, but this may be due to the small size of elementary Cl currents in cell-attached recordings.)

5) Ca permeability

The three types of Ca-dependent channels may account for the increase of membrane permeability to K, Cl, and Na as revealed by conventional microelectrode recordings (2). Labeled flux studies also revealed, however, an increase in Ca permeability (22,23,24). The basis of this effect is still unknown. Petersen and Maruyama (25) proposed that the cation-selective channel of 25-30 pS could conduct Ca to some extent. No experimental evidence was found to support (or refute) this hypothesis.

6) Ca release mechanism

There is general agreement that Ca release is due to a rise of the intracellular concentration of inositoltrisphosphate (IP_3). IP_3 is produced by agents inducing fluid secretion (26), and it releases Ca from permeabilized cells (27) and from isolated endoplasmic reticulum vesicles (28). The mechanism linking receptor occupation to IP_3 production is still largely unknown. It was recently shown that, in insect salivary glands, GTP analogues are

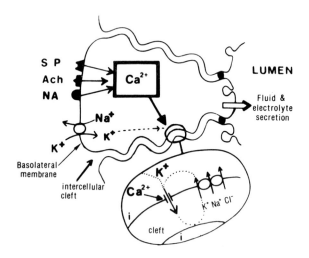

Fig. 4 :Secretion model of Petersen and Maruyama (1984).
An acinar cell is depicted with its basolaters1 membrane (in contact with the blood) on the left and its luminal membrane (in contact with the secreted fluid) on the right. Activation of receptors on the basolateral membrane leads to a rise of internal Ca, to the opening of BK channels, and to K accumulation in intercellular clefts. The local rise of K concentration serves as a driving force for the K-Na-Cl cotransport which carries Na and Cl ions into the cell. These ions are then extruded on the luminal side. (Reproduced with permission from ref. 13)

able to enhance the production of IP$_3$ (29). This observation adds up to the well known fact that GTP affects the binding of muscarinic and α-adrenergic agonists (reviewed in 26) to indicate that a GTP binding protein is most probably the missing link between receptor occupation and IP$_3$ production.

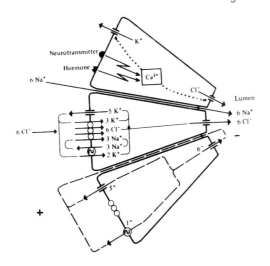

Fig. 5 : Secretion model of Suzuki and Petersen (1985).
This is a variant of the model of Fig. 4, where only Cl movements are transcellular. Na ions are secreted via leaky intercellular barriers. (Reproduced with permission from ref. 30)

7) Models of secretion

The Ca-dependent channels have been incorporated in several models of secretion (13,19,30).

Petersen and Maruyama (13) noted that secretion is blocked by replacement of Na with Li and by replacement of Cl with NO$_3$. To explain these observations, they suggested that a Na-K-Cl$_2$ cotransport system operates at the basolateral membrane. In this system, which has been studied in many transporting epithelia, Li cannot replace Na, nor can NO$_3$ replace Cl. In Petersen and Maruyama's model, inflow of salt by the cotransport system is driven by an accumulation of K ions elicited in intercellular clefts by the opening of BK channels (Fig. 4). Salt is then secreted in the lumen by an increase in Na and Cl conductance.

In a recent variant of this model, Suzuki and Petersen (30) proposed that the Na ions entering the cell at the basolateral membrane are extruded by the

Na-K pump. The net movement accross the basolateral membrane is therefore an entry of Cl ions. This is followed by extrusion of Cl on the luminal side by opening of Ca-dependent Cl channels. Na ions are supposed to reach the lumen through the tight junctions, so that the sum of transcellular and paracellular ion movements is a transfer of NaCl (Fig. 5).

As an alternative to the Na-K-Cl$_2$ cotransport, Marty, Tan and Trautmann (19) suggested that a combination of K, Cl, and monovalent cation selective channels are activated by internal Ca on the basolateral membrane. Because of the counteracting activity of the Na-K pump, the net movement is an entry of KCl. On the luminal membrane, opening of K and Cl selective channels leads to the extrusion of KCl. The model thus accounts for a transcellular movement of KCl, and supposes that any transfer of NaCl is paracellular and occurs through tight junctions (Fig. 6).

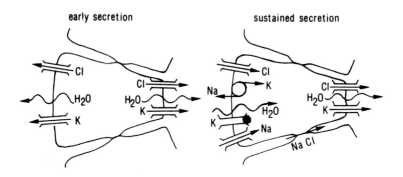

Fig. 6 : Secretion model of Marty, Tan and Trautmann (1984).
This model distinguishes an early phase of secretion, where the cell secretes KCl on both sides, and a sustained phase of secretion, where a transcellular KCl movement is established. Ca-dependent K and Cl channels are supposed to be present both on the basolateral and on the luminal membrane. Ca-dependent cationic channels are only present on the basolateral membrane. During the early phase of secretion, only K and Cl selective channels are activated. As internal Ca further rises, the cationic channels are also activated, and BK channels are blocked on the basolateral side. This leads to an inversion of Cl and K movement through the basolateral membrane. NaCl secretion, if present, is supposed to take place through cell to cell junctions. (Reproduced with permission from ref. 19)

CONCLUSION

A rather simple picture of the conductance systems of cells from exocrine glands is emerging from patch-clamp studies. Not too surprisingly, no voltage-gated Na, Ca or Cl channels were found, so that the cells are unable to generate the self-reinforcing depolarization leading to an action potential in other tissues. On the other hand, the cells have three kinds of Ca-dependent channels, which are selective for K, for Cl, and for monovalent cations. These Ca-sensitive channels are, together with the Na-K pump and possibly with the Na-K-Cl$_2$ cotransport, responsible for the movement of salt through the secreting cells.

1.Putney, J.W., Jr. (1979) Stimulus-permeability coupling : role of calcium in the receptor regulation of membrane permeability. Pharmacol. Rev. 30 , 209-245.
2.Petersen, O.H. (1980) The electrophysiology of gland cells. Academic Press.
3.Ginsborg, B.L. & House, C.R. (1980) Stimulus-response coupling in gland cells. Ann. Rev. Biophys. Bioeng. 9 , 55-80.
4.Burgen, A.S.V. (1956) The secretion of non-electrolytes in the parotid saliva. J. Cell. comp. Physiol. 48 , 113-138.
5.Martin, K. & Burgen, A.S.V. (1962) Changes in the permeability of the salivary gland caused by sympathetic stimulation and by catecholamines. J. Gen. Physiol. 46 , 225-243.
6.Mazariegos, M.R., Tice, L.W. & Hand, A.R. (1984) Alteration of tight junctional permeability in the rat parotid gland after isoproteranol stimulation. J. Cell Biol. 98 , 1865-1877.
7.Hamill, O.P., Marty, A., Neher, E., Sakmann, B. & Sigworth, F.J. (1981)

Improved patch-clamp techniques for high-resolution current recording from cells and cell-free membrane patches. Pflügers Arch. $\underline{391}$, 85-100.

8.Maruyama, Y., Gallacher, D.V. & Petersen, O.H. (1983) Voltage and Ca^{2+}-activated K^+ channel in baso-lateral acinar cell membranes of mammalian salivary glands. Nature $\underline{302}$, 827-829.

9.Trautmann, A. & Marty, A. (1984) Activation of Ca-dependent K channels by carbamylcholine in rat lacrimal glands. Proc. Nat. Acad. Sci. USA $\underline{81}$, 611-615.

10.Findlay, I. (1984) A patch-clamp study of potassium channels and whole-cell currents in acinar cells of the mouse lacrimal gland. J. Physiol. $\underline{350}$, 179-195.

11.Maruyama, Y., Petersen, O.H., Flanagan, P. & Pearson, G.T. (1983) Quantification of Ca^{2+}-ativated K^+ channels under hormonal control in pig pancreas acinar cells. Nature $\underline{305}$, 228-232.

12.Marty, A. (1983) Ca-dependent K channels with large unitary conductance. Trends in Neurosci. $\underline{6}$, 262-265.

13.Petersen, O.H. & Maruymama, Y. (1984) Calcium-activated potassium channels and their role in secretion. Nature $\underline{307}$, 693-696.

14.Maruyama, Y. & Petersen, O.H. (1984) Control of K^+ conductance by cholecystokinin and Ca^{2+} in single pancreatic acinar cells sltudied by the patch-clamp technique. J. Memb. Biol. $\underline{79}$, 293-300.

15.Maruyama, Y. & Petersen, O.H. (1982a) Single-channel currents in isolated patches of plasma membrane from basal surface of pancreatic acini. Nature $\underline{299}$, 159-161.

16.Maruyama, Y. & Petersen, O.H. (1982b) Cholecystokinin activation of single-channel currents is mediated by internal messenger in pancreatic acinar cells. Nature $\underline{300}$, 61-63.

17.Colquhoun, D., Neher, E., Reuter, H. & Stevens, C.F. (1981) Inward current channels activated by intracellular Ca in cultured cardia cells. Nature $\underline{294}$, 752-754.

18.Yellen, G. (1982) Single Ca++-activated non-selective cation channels in neuroblastoma. Nature $\underline{296}$, 357-359.

19.Marty, A., Tan, Y.P. & Trautmann, A. (1984) Three types of calcium-dependent channel in rat lacrimal glands. J. Physiol. $\underline{357}$, 293-325.

21.Schneyer, L.H., Young, J.A. & Schneyer, C.A. (1972) Salivary secretion of electrolytes. Physiological Reviews $\underline{52}$, 720-777.

22.Kanagasuntheram, P. & Randle, P.J. (1976) Calcium metabolism and amylase release in rat parotid acinar cells. Biochem. J. $\underline{160}$, 547-564.

23.Keryer, G. & Rossignol, B. (1976) Effect of carbachol on ^{45}Ca uptake and protein secretion in rat lacrimal gland. Amcr. J. Physiol. 230 , 99-104.

24.Putney, J.W., Jr. (1977) Muscarinic, alpha-adrenergic and peptide receptors regulate the same calcium influx sites in the parotid gland. J. Physiol. 268 , 139-149.

25.Petersen, O. H. & Maruyama, Y. (1983) What is the mechanism of the calcium influx to pancreatic acinar cells evoked by secretagogues? Pflügers Arch. 396 , 82-84.

26.Berridge, M. J. & Irvine, R. F. (1984) Inositoltrisphosphate, a novel second messenger in cellular signal transduction. Nature 312 , 315-321.

27.Streb, H., Irvine, R. F., Berridge, M. J. & Schulz, I. (1983) Release of Ca from a nonmitochondrial intracellular store in pancreatic acinar cells by inositol-1,4,5-trisphosphate. Nature 306 , 67-69.

28.Prentki, M., Biden, T. J., Janjic, D., Irvine, R. F., Berridge, M. J. & Wollheim, C. B. (1984) Rapid mobilization of Ca from rat insulinoma microsomes by inositol-1,4,5-trisphoshate. Nature 309 , 562-564.

29.Litosch, I., Wallis, C. & Fain, J. N. (1985) 5-Hydroxytryptamine stimulates inositol phosphate production in a cell-free system from blowfly salivary glands. J. Biol. Chem. 260 , 5464-5471.

30.Suzuki, K. & Petersen, O. H. (1985) The effect of Na and Cl removal and of loop diuretics on acetylcholin-evoked membrane potential changes in mouse lacrimal acinar cells. Quat. J. Exp. Physiol. 70 , 437-445.

Résumé

Les propriétés de conductance membranaire de cellules isolées à partir de glandes exocrines ont été étudiées au cours des 3 dernières années à l'aide des techniques de patch-clamp. En absence de stimulation chimique, les cellules possèdent une conductance potassique activable par une dépolarisation membranaire. Cette conductance est due à un canal sensible au Ca intracellulaire et possèdant une grande conductance élémentaire (canal BK).

Si on stimule les cellules par application externe d'acétylcholine, on observe que la probabilité d'ouverture des canaux BK augmente (glandes lacrymales de Rat; pancréas de Porc). Cet effet peut être bloqué en augmentant le pouvoir tampon du milieu intracellulaire vis à vis du Ca. Au contraire, des réponses d'amplitude normale sont obtenues après élimination du Ca du milieu intracellulaire. Cette réponse est sans doute due à une augmentation de la concentration cytosolique de Ca par suite de la libération de Ca de stocks calciques intracellulaires.

Dans certaines glandes exocrines (pancréas exocrine de Rongeurs), les canaux BK n'ont pas été observés au repos. Quand on stimule ces glandes par une application de cholécystokinine, on observe une activation de canaux de sélectivité cationique. Ces canaux sont activables par le Ca, et ont une conductance élémentaire de 25 à 30 pS.

Un troisième type de canaux dépendants du Ca a été décrit dans les glandes lacrymales de Rat. Ces canaux sont sélectifs aux ions Cl et ont une conductance élémentaire faible (1 - 2 pS).

Les concentrations de Ca nécessaires à l'activation des trois types de canaux sont différentes, de telle sorte que les canaux sélectifs pour le K, le Cl et les cations monovalents sont successivement activés quand la concentration en Ca cytosolique augmente.

La libération de Ca provient de l'action d'inositoltrisphosphate, qui libère le Ca du réticulum endoplasmique. Une protéine liant le GTP pourrait servir de lien entre l'activation des récepteurs et la production d'inositoltrisphosphate.

Plusieurs modèles de sécrétion tentent de rendre compte de la sécrétion d'électrolytes à partir des divers éléments moléculaires révélés par la technique du patch-clamp. Ces modèles supposent que le mouvement d'ions résulte d'une combinaison de l'activation de canaux sensibles au Ca, de transporteurs (pompe Na/K et/ou cotransporteur Na/K/Cl$_2$) et de transport paracellulaire au travers des jonctions "tight".

234

Hormones and cell regulation. Ed J. Nunez *et al.* Colloque INSERM/John Libbey Eurotext Ltd. © 1986. Vol. 139, pp. 235–248.

The cardiac calcium channel: properties and regulation

V. Flockerzi, P. Ruth, J. Oeken, B. Jung and F. Hofmann

Physiologische Chemie, Medizinische Fakultät der Universität des Saarlandes, D-6650 Homburg/Saar, Federal Republic of Germany.

RÉSUMÉ

Voltage dependent slow calcium channels are modulated by cAMP-dependent phosphorylation. The physiology and biochemistry of these channels will be discussed.

INTRODUCTION

An increase in the free calcium concentration within excitable cells initiates important cellular processes such as muscle contraction, the synthesis and secretion of transmitters and hormones, the regulation of enzyme activities and the control of membrane permeability. In unstimulated cardiac cells calcium is usually lower than 0.05 µM whereas calcium is about 2 mM in the extracellular fluids. Upon electrical stimulation the intracellular calcium concentration increases by 1 to 2 orders of magnitude as a result of calcium influx through voltage dependent calcium channels followed by an release of calcium from intracellular stores (1–3). Recently, three distinct potential dependent calcium channels have been described (4–10). The most common, which has been observed in many cells, produces a slow but long lasting inward current (9). This channel was the first ion channel for which modulation was suggested through phosphorylation/dephosphorylation (11,12). Calcium conductance through this channel is inhibited by the calcium channel blockers (13), a diverse chemical group of compounds which includes the phenylalkylamines, the dihydropyridines and the benzothiazepines . The second type of channel opens at much more negative membrane potentials and produces a transient inward current . Both types of channels are present in cardiac cells (8,9). The third type of calcium channel, found in sensory neurons, can only be activated by large depolarization steps from a very negative membrane potential (10). In addition to these voltage dependent channels, receptor operated calcium channels have been postulated, which are regulated directly or indirectly by occupation of a receptor protein by its specific ligand. This article will describe some regulatory and biochemical aspects of the cardiac voltage dependent slow calcium channel.

Stimulation of the β-adrenergic receptor increases the amount of calcium entering the cardiac muscle cell during depolarization. It was proposed that this increase in calcium conductance is caused by cAMP-dependent phosphorylation of the calcium channel or a protein closely related to the channel (11,15,16,). This prediction was later confirmed by experiments in which the catalytic subunit of cAMP-dependent protein kinase was injected into single cardiac muscle cells (17,18). Injection of the catalytic subunit of cAMP-dependent protein kinase increased about 3fold the calcium conductance of voltage clamped cells. These experiments ruled out the possibility that cAMP may modify the calcium conductance by an indirect mechanism such as changing the potassium permeability. However in these initial experiments rather high concentrations of the protein kinase were used. This raised the possibility that the observed increase in the calcium conductance was caused by nonphysiological concentrations of the enzyme which are not relevant to the physiological situation. Recent experiments (19) in which single cardiac cells were perfused with various concentrations of the catalytic subunit of cAMP-dependent protein kinase show that already 1μM catalytic subunit increases halfmaximally the calcium conductance. This enzyme concentration is almost identical to the concentration of catalytic subunit measured in cardiac muscle (20). The calcium conductance was raised maximally 3.3fold by 10 μM catalytic subunit. Similar changes are obtained either by perfusion of a single cell with cAMP or by the addition of a maximal concentration of isoproterenol to the perfusion chamber, suggesting that the catalytic subunit elicited the physiological response at a physiological concentration. The phosphorylation of the cardiac calcium channel increases the opening propability of the channel during depolarization by decreasing the number of zero sweeps during single channel recording (21-24). In an intact cell the change in the opening propability increases the number of channels which are opened during depolarization and enhances therefore the calcium conductance. Perfusion of a single cell with cAMP-free regulatory subunit of cAMP-dependent protein kinase decreases the calcium conductance, if the cell is stimulated by isoproterenol (17,25). This indicates that removal of the free catalytic subunit increases the rate of dephosphorylation of the channel and decreases thereby its propability to open. However, the calcium conductance can not be abolished completely by perfusion of a cell with high concentration of regulatory subunit. The same effect on calcium conductance was observed when a single cell was perfused with the heat stable inhibitor of cAMP-dependent protein kinase followed by the addition of a halfmaximal concentration of isoproterenol to the perfusion chamber (25). Taken together these results indicate that β-adrenergic stimulation of calcium conductance is mediated by cAMP-dependent protein kinase. However, about one third of the total number of calcium channels can still be opened in the absence of cAMP-dependent phosphorylation, indicating that two functionally different populations of channels are present in cardiac cells, namely phosphorylated and nonphosphorylated.

The experiments carried out in vivo suggested that phosphorylation of a membrane protein changes the properties of the calcium channel. In vitro a large number of different membrane proteins is phosphorylated after addition of cAMP-dependent protein kinase (26). To establish the physiological significance of these phosphorylations it is necessary to correlate functional changes observed with the phosphorylation of a specific sarcolemmal peptide. β-adrenergic stimulation of intact hearts results in the phosphorylation of a 15 kDa (27) and 22 to 27 kDa cardiac membrane protein (28,29). A possible correlation was found between the phosphorylation of the 15 kDa peptide and the increase in the maximal rate of developed tension in intact hearts after administration of isoproterenol(27). However, the abundance of the 15 kDa peptide in the sarcolemma and its high phosphate incorporation (100-300 pmol/mg) argues against it being a component of the sarcolemmal calcium channel which has a density of 0.3-0.6 pmol/mg sarcolemmal protein (see below). cAMP dependent phosphorylation of a 22 to 27 kDa peptide called phospholamban has been correlated with acceleration of the rate of calcium-uptake by the sarcoplasmic reticulum and with an increase in the rate of myocardial relaxation (307).However, there are other studies postulating that the phosphorylation of the sarcolemmal peptide might mediate the positive inotropic effect of β-adrenergic agonists (31,32). In support of this, Rinaldi et al.(33)suggested that phosphorylation of phospholamban (termed "calciductin")in sarcolemmal vesicles in vitro might open slow calcium channels. This study used potassium/sodium polarized vesicles of cardiac sarcolemma which accumulate "voltage"-dependent calcium if the extravesicular potassium concentration was raised (34). However, as shown by Flockerzi et al.(26), phosphorylation of this peptide does not increase the amount of calcium taken up by sarcolemmal vesicles after "depolarization". The difference between the results of both groups has been cleared up. In the experiments of Rinaldi et al. the catalytic subunit of cAMP-dependent protein kinase probably hydrolyzed ATP to adenosine and phosphate. The phosphate precipitated together with calcium and lanthanum on the filter. The study by Flockerzi et al. avoided this artefact and showed that the amount of calcium taken up by the vesicles is within the range which can passively enter these vesicles. However, this calcium uptake was not inhibited by the organic calcium channel blockers suggesting that this method is not sufficient to identify in vitro cardiac calcium channels. In addition there is now little doubt that sarcolemmal phospholamban and sarcoplasmic phospholamban are identical proteins(35) suggesting that they might serve similar functions (calcium ATPase) in both locations. Recently, an increased phosphorylation of a peptide of M_r 42.000 in cardiac membranes was described (36) after stimulation by isoproterenol or the calcium channel blocker nitrendipine. However, the exact role that phosphorylation of this peptide plays in modulating cardiac calcium channels has yet to be elucidated.

IDENTIFICATION OF CALCIUM CHANNELS IN SUBCELLULAR FRACTIONS OF BOVINE HEART.

The organic compounds which block calcium channels represent a diverse group of chemical structures. In general these substances

Table 1

Distribution of cardiac muscle binding sites for calcium channel blockers

fraction	dihydropyridine	phenylalkylamine	strophantin	oxalat supported calcium uptake
	fold(pmol/mg protein)			$(nmol \times min^{-1} \times mg^{-1})$
homogenate	1 (0.035)	1 (0.028)	1 (6.0)	1 (7.5)
microsomes	17.2	3.4	2.7	12.7
sarcolemma	12.3	3.3	9.8	5.7
SR, "junctional"	17.3	2.2	2.6	16.0
SR, free	-	10.1	2.0	53.5

Membranes were prepared from fresh bovine cardiac muscle as described by Jones and Cala (59). The concentration of radioligand was 1 nM (nimodipine, (-)desmethoxyvera-pamil) and 1 uM (strophantin). Dihydropyridine binding was measured in the presence of 0.1 uM nitrobenzylthioinosine. SR, sarcoplasmatic reticulum. Values are the average of 4 to 10 different membrane preparations. For further detail see Oeken et al. (46).

238

are amphiphiles that contain a large hydrophobic region- usually one or more ring structures- and a hydrophilic portion such as an amine or other nitrogen-containing group. It is unclear whether these drugs block calcium channel conductance by interacting directly with the channel protein or by disordering or otherwise modifying specific regions of the lipid bilayer. Tritiated derivatives of the organic calcium channel blockers have been used to identify binding sites in broken cell preparations in different tissues including skeletal muscle, heart, smooth muscle and brain. High affinity binding sites of proteinous nature have been studied in several laboratories (for review see 37-42), and it has been suggested that separate binding sites exist in broken cell preparations which are specific for the dihydropyridines, the phenylalkylamines and the benzothiazepine diltiazem (42-44). Evidence for the physiological significance of these high affinity binding sites has not been established beyond doubt. Skeletal muscle , e.g., is the richest and most convenient source of calcium channel blocker binding sites. However it was concluded from a comparison of the number of in vivo dihydropyridine binding sites with electrophysiological measurements of calcium channels that less than a few percent of the binding sites represent functional calcium channels.(45).These uncertainties made it mandatory to evaluate carefully the distribution and properties of calcium channel blocker binding sites in broken cell preparations.

In bovine cardiac muscle the distribution of the binding sites is studied in subcellular fractions (46) mainly containing sarcolemma, junctional sarcoplasmic reticulum and free sarcoplasmic reticulum (table 1). The origin of these distinct membrane fractions is supported by the distribution of the glycoside receptor and the calcium-uptake activity. The binding sites for strophantin, i.e. the Na/K-ATPase, are enriched 10fold in sarcolemma whereas the oxalate supported calcium-uptake activity sediments with free sarcoplasmic reticulum. As expected intermediate activities are associated with "junctional" sarcoplasmic reticulum. The binding sites for the dihydropyridines copurify with the sarcolemma, whereas only a minority of the binding sites for phenylalkylamines are localized in this membrane fraction. In contrast, the majority of the phenylalkylamine binding sites copurify with a fraction containing mainly free sarcoplasmic reticulum. Scatchard analysis of the binding data in sarcolemma for dihydropyridines and phenylalkylamines results in downward concave plots suggesting the presence of a high and a low affinity binding site with K_D values in the nmolar and µmolar range (44). Dihydropyridines bind optimally to the high affinity site only in the presence of mmolar calcium, whereas the binding of phenylalkylamines to its high and low affinity sarcolemmal sites is inhibited by a free calcium concentration of above 1 mM (44). The high affinity sites for ligands of both groups are specific for the respective group of calcium channel blockers and select with some exceptions the (-)isomers over the (+)isomers. However, binding of dihydropyridines is allosterically modified by occupancy of the phenylalkylamine site and also by the occupancy of a site specific for d-cis-diltiazem. Binding of a dihydropyridine to its high affinity site inhibits allosterically the binding of phenylalkylamines. In addition binding of phenylalkylamines is inhibited by diltiazem by a noncompetitive mechanism, suggesting that binding to the high affinity site of each group of calcium channel blocker affects allosterically the binding of the other two groups (44). These results indicate that cardiac sarcolemma contains at least a separate site for each group

Table 2

Summary of the properties of the cardiac muscle binding sites for calcium channel blockers

site	localization	K_D	B_{max}	interaction with	stereo-specifity	suggested identity
		uM	pmol/mg			
high affinity						
nimodipine	SL/junctional SR	0.35	0.3	PAA, allost. Diltia., allost.	yes	calcium channel
(-)desmethoxy-verapamil	SL/junctional SR	1.4	0.16	DHP, allost. Diltia., allost.	yes	calcium channel
low affinity						
nimodipine	SL/junctional SR	33	8.2	NBTI	yes	nucleoside transporter
(-)desmethoxy-verapamil	free SR	191	34.5	PAA	no	

SL, sarcolemma; SR, sarcoplasmatic reticulum; PAA, phenylalkylamine; DHP, dihydropyridine; Diltia, diltiazem; allost., allosteric; NBTI, nitrobenzylthioinosine.

of calcium channel blockers. The density of these high affinity binding sites in the homogenate is 20 to 40 fmol/mg protein. This value corresponds well with the number of functional calcium channels (3.000 - 16.000 per cell) measured by electrophysiological methods in a single cardiac cell (1,12). This number is equivalent to a density of 13 fmole/mg protein for functional calcium channels if the total cell volume, the specific density of the cell, and the protein content are taken as 10 pl, 1.1 and 20% of the wet weight. The similarity of values suggest that the high affinity binding sites for dihydropyridines and phenylalkylamines in cardiac sarcolemma represent part of the calcium channel. Their high affinity is not in contrast to this interpretation, since it was demonstrated recently that nitrendipine blocks the inactivated calcium channel with an apparent K_d of 0.34 nM whereas the apparent K_d for the resting channel is about 700 nM (47). It is very likely that the channel present in muscle homogenates is in its inactivated, high affinity state (table 2).

Although these considerations may be valid their plausibility depends on the identification of the physiological function of the sarcolemmal low affinity sites. Binding of dihydropyridines to the sarcolemma is inhibited by 50% at nmolar concentrations of nitrobenzylthioinosine and hexobendine (44). These compounds bind with high affinity to the low affinity site of dihydropyridines and do not affect binding of dihydropyridines to their high affinity site or the binding of phenylalkylamines. Both compounds are well characterized inhibitors of the nucleosid transporter. The dihydropyridine nimodipine inhibits binding of radioactive nitrobenzylthioinosine in cardiac membranes with an IC_{50} value of 44 nM which value is identical with the K_d value of 33nM determined for the low affinity site of nimodipine (48). In addition the density of the low affinity site for nimodipine is close to that for the nucleoside transporter suggesting that the low affinity site for dihydropyridines is related to the nucleoside transporter and has nothing to do with the calcium channel.

The low affinity phenylalkylamine binding site localized in free sarcoplasmic reticulum and sarcolemma is not related to the voltage dependent calcium channel. (-)desmethoxyverapamil binds to a single site in free sarcoplasmic reticulum with an apparent K_D value of 191 nM and a density of 34.5 pmol/mg protein (46). This low affinity site bound other phenylalkylamines, but stereospecific binding was not observed. Dihydropyridines did not affect the binding of phenylalkylamines to this site, suggesting that the low affinity site differs considerably from the high affinity sarcolemmal site. Some but not all compounds referred to as "calmodulin antagonists" which bind by hydrophobic interactions to membranes and calmodulin, inhibit binding of (-)desmethoxyverapamil to its low affinity site, suggesting similar properties of this low affinity site with binding sites present in calmodulin(46). The failure of the phenylalkylamines as well as of the other calcium channel blockers to affect calcium efflux from cardiac sarcoplasmic reticulum (49) indicates that the low affinity phenylalkylamine binding site localized in sarcoplasmic reticulum is probably not localized at a calcium release channel. However, the appearance of similar binding sites in brain, skeletal muscle (50) and plants(51) suggests some importance of this site for living cells.

PARTIAL PURIFICATION OF THE CARDIAC NITRENDIPINE BINDING SITE AND
COMPARISON WITH THE SKELETAL MUSCLE NITRENDIPINE BINDING SITE

The results described above indicate that the high affinity
binding site for the dihydropyridines and the phenylalkylamines are
part of the slow calcium channel.These sites are solubilized by
different detergents which include chaps, sucrosemonolaurylester
and digitonin (52).Chaps solubilizes about 40% of the membrane
protein and of the binding sites whereas the other two detergents
solubilize more protein than binding sites. Solubilization decreased the affinities for the high affinity binding sites 2fold
(nimodipine) and 10fold (desmethoxyverapamil) without changing
their densities. In contrast to the particulate state the low
affinity desmethoxyverapamil binding site is not measurable in the
detergent extract. The solubilized binding sites for dihydropyridines, phenylalkylamines and d-cis-diltiazem interfer with the
binding of each other in an allosteric manner. This suggests that
these sites are localized either on the same protein or on subunits
of the same structure which bind tightly together (52). The solu-

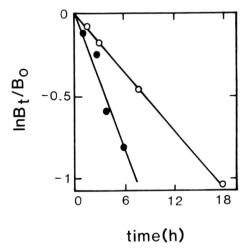

time(h)

Figure 1: Dissociation of nitrendipine from cardiac and skeletal
muscle binding sites.Sarcolemmal (open symbols) and T-tubule
binding sites (closed symbols) were incubated with 2.5nM (SL) and
6nM (T-tubule) (^{3}H)-nitrendipine in the presence of 100 µM
diltiazem.The binding sites were solubilized by 1%
digitonin.Aliquots of the solubilized (^{3}H)-nitrendipine-receptor
complex were assayed for bound nitrendipine by polyethylene glycoll
precipitation at the indicated times after 10fold dilution into
binding medium containing 2µM nitrendipine and 100µM diltiazem at
4°C.Specifically bound (^{3}H)-nitrendipine was determined from the
difference between samples incubated in the presence and absence of
2µM unlabelled ligand.

bilized site of cardiac sarcolemma is not identical to that
solubilized from skeletal muscle T-tubule membranes.The
dissociation rates of (^{3}H)-nitrendipine for both receptors differed
by a factor of 2 being 0.0011 min^{-1} and 0.0024 min^{-1}

(fig.1).Prelabelled sites from both sources sedimented with a similar S$_{20,w}$ value of 20 S in sucrose density gradients (fig.2).However,in agreement with others (53), the receptor from skeletal muscle sedimented with a broad shoulder suggesting heterogenety of this material.

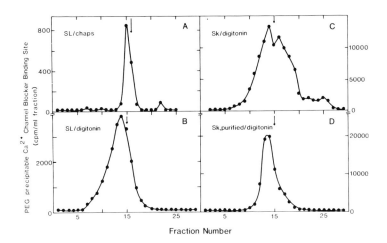

Figure 2: Sucrose gradient sedimentation of the solubilized cardiac and skeletal muscle dihydropyridine binding site. A: 0.9 ml undiluted chaps solubilized sarcolemmal membranes (1.55mg/ml) were layered onto a 39 ml linear gradient of 5-20% sucrose in 0.4% chaps, 48% glycerol and sedimented for 15 h at 210.000xg in a Ti 50 vertical rotor. 1.56 ml fractions were collected from the bottom of the tube and aliquots of each fraction were tested for polyethylene glycoll precipitable nimodipine binding. B,C,D: 2 ml containing the solubilized (^{3}H)-nitrendipine-receptor complex from cardiac sarcolemma (B), skeletal T-tubule (C) and the receptor-ligand complex, purified according to Curtis and Catteral (56) (D), were layered on a 39 ml linear sucrose gradient from 5-20% sucrose containing 0.1% digitonin and sedimented for 1.5 h at 210.000xg in a Ti 50 vertical rotor. 1.34 ml fractions were collected and aliquots were assayed for polyethylene glycoll precipitable (^{3}H)-nitrendipine. The migration of thyroglobulin (19.2 S) is indicated by the arrow.Bottom of the tube was at fraction one.

The solubilized cardiac high affinity site for calcium channel blockers was partially purified on a sepharose column substituted with a phenylalkylamine (54).Binding sites are eluted with high salt and have been characterized further by (^{3}H)-nitrendipine binding using the Hummel and Dreyer technique (55).Equilibrium binding experiments with four different preparations yielded a K$_d$ value of 1.2±.2 nM and a density of 146±4 pmol/mg.The specific density represents a 2,000fold purification of the high affinity site which may be about 10% pure.The major peptides detected by SDS gel electrophoresis had a molecular mass of 90, 58, 45 and 30 kDa. Binding of nitrendipine to this fraction was inhibited by other dihydropyridines in nmolar concentrations but only in μmolar concentrations by phenylalkylamines.The reason for the apparent loss of the allosteric interaction between the different sites is not clear but may be caused either by separation of these sites during purification or by a different susceptibility of these sites against proteases. The second explanation is supported by the observation that complete hydrolysis of the membrane bound phenylalkylamine binding site requires 1 µg trypsin/mg protein, whereas the nimodipine binding sites are destroyed completely only in the presence of 0.1mg trypsin/mg protein (44). The binding site eluted with high salt from the phenylalkylamine column was further purified on a DEAE-cellulose column. The fractions eluting between 50 and 100 mM sodium chloride bound nitrendipine. This step yielded an additional 3fold purification on the expence of recovered binding sites. The fraction

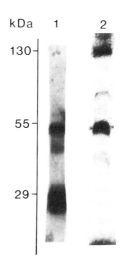

Figure 3:Phosphorylation of the partially purified dihydropyridine binding site from cardiac (lane 1) and skeletal muscle (lane 2). Binding sites from cardiac and skeletal muscle were purified as described in the text and according to Curtis and Catterall (56), respectively. Aliquots of the peak fractions of the final purification step were incubated at 37°C in a 40 mM Tris/MES buffer containing 5 mM MgCl$_2$, 0.5 mM (^{32}P)-ATP and 0.05 µM catalytic subunit of cAMP-dependent protein kinase. Total incubation volume was 100 µl. The reaction was started by the addition of (^{32}P)-ATP and terminated after 10 min by the addition of 33 µl NaDodSO$_4$ stopping solution containing 160 mM dithiothreitol. The tubes were boiled for 5 min. Peptides were separated in a polyacrylamide gel (7.5%) in the presence of 0.1% NaDodSO$_4$ using the TRIS/glycine system of (57). Gels were silver stained, dried and exposed to an X-ray film at -70°C for several days.

with the highest binding density was incubated together with (^{32}P)-ATP and the catalytic subunit of cAMP-dependent protein kinase. The audioradiogram after electrophoresis in the presence of SDS showed a major phosphorylated peptide of M$_r$ 55 000-60 000. This peptide may be involved in the regulation of the calcium channel. Its molecular weight is very similar to a peptide phosphorylated in skeletal muscle transveral tubule membranes , which is supposed to be part of the calcium channel in this tissue (fig.3).In contrast to the receptor purified from skeletal muscle T-tubule membranes phosphorylation of the M$_r$ 130.000 peptide was not observed with the cardiac receptor.The phosphorylation of these two peptides was demonstrated previously by Curtis and Catteral (56,57),which suggested that only the phosphorylation of the 55 kDa peptide might be of physiological significance.

CONCLUSIONS

These and other studies provide evidence that catecholamines increase the calcium conductance of cardiac muscle cells by cAMP-dependent phosphorylation of the voltage dependent calcium channel or a closely related protein. The identity of this protein has not been established so far, but it may be identical with the 55-60 kDa peptide phosphorylated in the partially purified preparation of cardiac and skeletal muscle binding sites for dihydropyridines. The cardiac voltage dependent slow calcium channel can be identified in vitro by tritiated organic calcium channel blockers. However, these compounds also bind to sites unrelated with the calcium channel.The channel contains at least three differnt binding sites: one for the dihydropyridines,one for the phenylalkylamines and one for diltiazem.Each site interacts allosterically with the other binding sites.These properties of the binding sites are similar to those present in the T-tubule membranes of skeletal muscle (42).The identity of these binding sites has yet to be shown since the physiological role of the calcium channels in these two tissues is different.The cardiac channel serves to initiate muscle contraction, whereas the channel from skeletal muscle may serve to replenish intracellular calcium stores.Preliminar evidence suggests that the peptide composition of the cardiac channel differs from that of the channel from skeletal muscle (60),although the regulatory peptide may be the same.However,confirmation of this suggestion has to wait until the channel from heart has been purified and reconstituted.

Acknowledgement
We would like to thank Dr. Hoffmeister from Bayer AG,Drs. Hollmann and Kretzschmar from Knoll AG and Drs.Satzinger and Marmé from Goedecke AG for providing some of the organic calcium channel blockers used in this study.This work was supported by grants from DFG and Fonds der Chemischen Industrie.

REFERENCES:
1)Reuter,H. (1983) Nature 301, 569-574
2)Reuter,H.(1984)Ann.Rev.Physiol.46,473-484
3)Fabiato,A.(1983)Am.J.Physiol.245,C1-4
4)Hagiwara,S.Byerly,L.(1983)Trends Neurosci.6,189-193
5)Carbone,E.,Lux,H.D. (1984) Nature 310,501-502
6)Yoshii,M.Tsunoo,A.,Narahashi,T.(1984)Biophys.J.47,433a
7)Armstrong,C.M.,Matteson D.R. (1985) Science 227,65-67
8)Nilius,B.,Hess,P.,Lansman,J.B.,Tsien,R.W.(1985)Nature316,443-446
9)Bean,B.P. (1985) J.Gen.Physiol. 86,1-30
10)Nowycky,M.C.,Fox,A.P.,Tsien,R.W.(1985)Nature316,440-443
11)Reuter,H.(1974)J.Physiol. 242,429-451

12) Trautwein,W.,Pelzer,D. (1985) in:Marmé,D. (ed) Calcium and Cell
    Physiology,Springer,54-93.
13) Fleckenstein,A. (1977) Ann.Rev.Pharmacol.Toxicol. 17, 149-166
14) Reuter,H. (1979)Ann.Rev.Physiol. 41,415-424
15) Tsien,R.W. (1973) Nature 245,120-122
16) Trautwein,W.,Taniguchi,J.Noma,A. (1982)Pfluegers
    Arch.Eur.J.Physiol.392,307-314
17) Osterrieder,W.,Brum,G.,Hescheler,J.,Trautwein,W.,Flockerzi,V.,
    Hofmann,F. (1982) Nature 298,576-578
18) Brum,G.,Flockerzi,V.,Hofmann,F.,Osterrieder,W.,Trautwein,W. (1983
    Pfluegers Arch.Eur.J.Physiol.398,147-154
19) Kameyama,M.,Hofmann,F.,Trautwein,W. (1985)Pfluegers
    Arch.Eur.J.PHysiol.in press
20) Hofmann,F.,Bechtel,P.J.,Krebs,E.G. (1977)J.Biol.Chem.252,1441-
    1447
21) Cachelin,A.B.,de Peyer,J.E.,Kokubun,S.,Reuter,H. (1983)Nature
    304,462-464
22) Hess,P.,Lansman,J.B.,Tsien,R.W. (1984) Nature 311,538-544
23) Brum,G.,Osterrieder,W.,Trautwein,W. (1984)Pfluegers
    Arch.Eur.J.Physiol.401,111-118
25) Kameyama,M.,Hescheler,J.,Trautwein,W.,Hofmann,F.(1985) submitted
26) Flockerzi,V.,Mewes,R.,Ruth,P.,Hofmann,F. (1983) Eur.J.Biochem.
    135, 131-142
27) Presti,C.F.,Jones,L.R.,Lindemann,J.P. (1985) J.Biol. Chem. 260,
    3860-3867
28) LePeuch,C.J.,Guilleux,J.-C.Demaille,J.G. (1980)FEBS Lett.114,165-
    168
29) Lindemann,J.P.,Jones,L.R.,Watanabe,A.M. (1980)Clin.Res.28,471
30) Tada,M.,Katz,A.M. (1982)Ann.Rev.Physiol.44,401-423
31) Huggins,J.P.,England,P.J. (1983) FEBS Lett. 163, 297-302
32) Iwasa,Y.,Hosey,M.M. (1983)J.Biol.Chem.258,4571-4575
33) Rinaldi,M.L.,Le Peuch,C.J.,Demaille,J.G. (1982) FEBS Letters
    129,277-281
34) Bartschat,D.K.,Cyr,D.L.,Lindenmayer,G.E. (1980) J.Biol.Chem.
    255,10044-10047
35) Jones,L.R.,Simmerman,H.K.B.,Wilson,W.W.,Gurd,F.R.N.,Wegener,A.D.
    (1985) J.Biol.Chem. 260, 7721-7730
36) Horne,P.,Triggle,D.J.,Venter,J.C. (1984)Biochem.Biophys.Res.
    Commun.121,890-898
37) Bellemann,P.,Ferry,D.,Luebbecke,F.,Glossmann,H. (1981)Arzneim.
    Forsch.31,2064-2067
38) Glossmann,H.Ferry,D.,Luebbecke,F.,Mewes,R.,Hofmann,F. (1982)
    Trends Pharmacol.Sci. 3,431-437
39) Janis,R.A.,Triggle,D.J. (1983)J.Med.Chem.26,775-785
40) Schwartz,A.,Triggle,D.J. (1984)Ann.Rev.Med.35,525-339
41) Snyder,S.H. (1984)Science (Wash.DC)224,22-31
42) Glossmann,H.,Ferry,D.Goll,A. (1985)Proc.9th.Int.Congr.Pharma
    col.2,329-335
43) Murphy,K.M.M.,Gould,R.J.,Largent,B.L.,Snyder,S. (1983)Proc.Natl.
    Sci.80,860-864
44) Ruth,P.,Flockerzi,V.,v. Nettelbladt,E.,Oeken,J.,Hofmann,F.
    (1985) Eur.J.Biochem.150,313-322
45) Schwartz,L.M.,McCleskey,E.W.,Almers,W. (1985)Nature 314,747-751
46) Oeken,J.,v.,Nettelbladt,E.,Zimmer,M.,Flockerzi,V.,Ruth,P.,Hof-
    mann, F. (1985) submitted
47) Bean,B.P. (1984) Proc.Natl.Acad.Sci.USA 81,6388-6392
48) Marangos,P.J.,Finkel,M.S.,Verma,A.,Maturi,M.F.,Patel,J.,Patter-
    son, R.E.(1984) Life Sciences 35,1109-1116

49) Chamberlain,B.K.,Volpe,P.,Fleischer,S. (1984) J.Biol.Chem.259, 7547- 7553
50) Reynolds,I.J.,Gould,R.J.,Snyder,S.H. (1983) Eur.J.Pharmacol.95, 319-321
51) Andrejauskas,E.,Hertel,RJ.,Marme,D.(1985)Biol.Chem.260,5411-5414
52) Ruth,P.,Flockerzi,V.,Oeken,J.,Hofmann,F.(1985)Naunyn-Schmiedebergs Arch.Pharmacol.331,suppl.143
53) Borsotto,M.,Norman,R.I.,Fosset,M.,Lazdunski,M.(1984) Eur.J.Biochem. 142, 449 -455
54) Flockerzi,V.,Ruth,P.,Hofmann,F(1983)Hoppe-Seyler's Z. Physiol. Chem.364,1124
55) Hummel,J.P.,Dreyer,W.J.(1962)Biochim.Biophys.Acta 63,532-534
56) Curtis,B.M.,Catterall,W.A. (1984) Biochemistry 23,2113-2118
57) Curtis,B.M.,Catterall,W.A. (1985) Proc.Natl.Sci.USA 82,2528-2532
59) Jones,L.R.,Cala,S.E.(1981) J.Biol. Chem. 256,11009-11018
58) Laemmli,U.K. (1970)Nature227,680-685
60) Rengasamy,A.,Ptasinski,J.,Hosey,M.M.(1985)Biochem.Biophys. Commun.126,1-7

Résumé

Les catécholamines augmentent la conductance du
calcium des cellules musculaires cardiaques par une phos -
phorylation cAMP dépendante du canal calcium ou d'une
protéine qui lui est étroitement associée. La conductance
calcique à travers ce canal est inhibée par un groupe de
composés chimiques, y compris les bloqueurs de ce canal ou
des antagonistes du calcium.

Des dérivés tritiés de ces bloqueurs du canal
calcium ont servi à identifier "in vitro" plusieurs sites
de liaison dans des préparations de cellules cassées du
muscle cardiaque. Seuls les sites de liaison de haute
affinité pour les dihydropyridines et les phenyl -
alkylamines, qui sont modulés par Ca^{2+} font partie du
canal calcium voltage dépendant.

Ces sites de liaison sont spécifiques pour les
différents groupes de bloqueurs du canal calcium. Chaque
site interagit allostériquement avec les autres sites de
liaison. Ces propriétés des sites de liaison sont sem -
blables à celles des sites des membranes des tubules T du
muscle squelettique. Les sites de liaison des deux tissus
peuvent être solubilisés par plusieurs détergents. La
protéine-kinase cAMP dépendante phosphoryle un peptide de
50-60 kDa dans les préparations partiellement purifiées de
muscle cardiaque et squelettique. Cependant la réponse à
la question selon laquelle les canaux du muscle cardiaque
et squelettique sont identiques doit attendre que l'on
dispose d'une préparation purifiée des sites de liaison
cardiaques.

Hormones and cell regulation. Ed J. Nunez *et al.* Colloque INSERM/John Libbey Eurotext Ltd. © 1986. Vol. 139, pp. 249–259.

Calcium ions and the hormonal control of intramitochondrial metabolism

Richard M. Denton and James G. McCormack*

*Department of Biochemistry, University of Bristol Medical School, University Walk, Bristol BS8 1TD and *Department of Biochemistry, University of Leeds, Leeds LS2 9JT, UK.*

Three key dehydrogenases from mammalian mitochondria have been found to be activated by Ca^{2+} ions with half-maximal effects at about $1\mu M$. They are pyruvate dehydrogenase, NAD^{+}-isocitrate dehydrogenase and oxoglutarate dehydrogenase. Activation of these enzymes can also be demonstrated within intact mitochondria when extramitochondrial Ca^{2+} is increased in the range of concentrations generally considered to occur in the cytcplasm of mammalian cells. It is argued that the main role of the calcium transport system in the inner-membrane of mammalian mitochondria is to relay changes in the cytoplasmic concentration of Ca^{2+} into the mitochondrial matrix. In this way, hormones and other extracellular stimuli which stimulate ATP-requiring processes through increasing the cytoplasmic concentration of Ca^{2+} may also increase intramitochondrial oxidative metabolism and hence the replenishment of ATP.

KEYWORDS

Calcium ions, mitochondria, pyruvate dehydrogenase, NAD^{+}-isocitrate dehydrogenase, oxoglutarate dehydrogenase, stimulus-response-metabolism coupling.

INTRODUCTION

Many hormones that act through cell surface receptors can influence intra-mitochondrial metabolism (Table 1). The mechanisms involved must therefore involve signal transmission not only across the plasma membrane but also across the inner-membrane of the mitochondria. This latter membrane is essentially impermeable to all charged molecules unless a specific carrier or transport system is present in this membrane. Fig. 1 summarises some of the messenger molecules which are formed in the cytoplasm of cells in response to different hormones binding to their receptors. Of the known messengers, only Ca^{2+} ions may be transferred across the inner mitochondrial membrane by a specific transport system.

The major purpose of this article will be to review the evidence that the main role of this calcium transport system is, in fact, to relay increases in the cytoplasmic concentration of Ca^{2+} brought about by hormones into the

Table 1. Examples of the regulation of intramitochondrial metabolism by hormones

Hormone	Tissue	Cytoplasmic signal	Increases in mitochondrial metabolism
Many	Many	Ca^{2+}	Oxidative metabolism
Insulin	Fat,mammary liver	(?)	Pyruvate dehydrogenase
Nor-adrenaline	Brown fat	cAMP	Uncoupling and oxidative metabolism
Glucagon	Liver	cAMP	Respiratory chain and general metabolism
Trophic	Target	(cAMP?)	Cholesterol metabolism
PTH	Kidney	(cAMP?)	$25-OHD_3$ $1,25-diOHD_3$
Thyroxine	Liver, muscle, etc.	(?)	Various

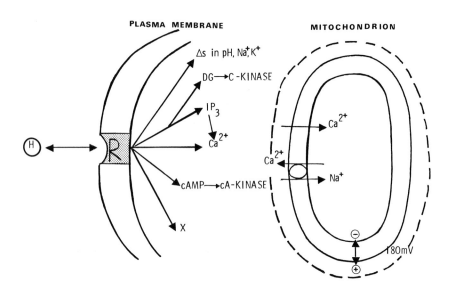

Fig. 1. Ca^{2+} may be the only cytoplasmic second messenger of hormone action which can be readily transferred across the inner-mitochondrial membrane.

mitochondrial matrix (1,2). In this way, intramitochondrial oxidative metabolism is enhanced thus resulting in the increased formation of ATP. This enhanced ATP production is required by the cytoplasmic processes such as muscle contraction and secretion which are stimulated by hormones which increase the cytoplasmic concentration of Ca^{2+}. The role of Ca^{2+} in the regulation of intramitochondrial oxidative metabolism may be considered as analagous to that of cytoplasmic Ca^{2+} on glycogen breakdown through its effects on phosphorylase kinase. Both could be said to be examples of Ca^{2+} having a role in "stimulus-response-metabolism" coupling (Fig. 2). The article will end with some comments on whether Ca^{2+} may have any role in the mitochondrial actions of the other hormones listed in Table 1.

THE CA^{2+}-SENSITIVE INTRAMITOCHONDRIAL DEHYDROGENASES AND THEIR PROPERTIES

Mitochondria from all vertebrate sources so far studied have been found to contain three important dehydrogenases which are activated several-fold by Ca^{2+} ions. The first of these to be recognised was pyruvate dehydrogenase; Ca^{2+} was found to elevate pyruvate dehydrogenase activity indirectly by causing increases in the amount of active non-phosphorylated enzyme through stimulating pyruvate dehydrogenase phosphate-phosphatase activity (3). Subsequently it was found that NAD^+-isocitrate dehydrogenase and oxoglutarate dehydrogenase were activated by Ca^{2+} more directly. In these two cases, Ca^{2+} causes substantial decreases in the K_m values for their respective substrates, threo-D_6-isocitrate and 2-oxoglutarate (4,5). Half-maximal effects of Ca^{2+} are observed with all three enzymes at about 1μM and the effective range is about 0.1 - 10μM.

These exclusively intramitochondrial dehydrogenases are without doubt important in the regulation of oxidative metabolism in vertebrate tissues. All three are also activated by increases in ADP/ATP and NAD^+/NADH ratios. We have suggested

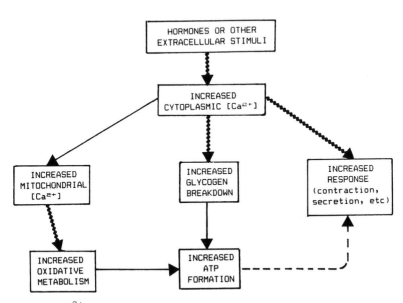

Fig. 2 Role of Ca^{2+} in stimulus-response-metabolism coupling.

that these activations should be viewed as examples of "intrinsic" control which ensures that under all circumstances changes in ATP and NADH concentrations may directly initiate appropriate changes in the provision of reducing power for the respiratory chain (1). On the other hand, the regulation by Ca^{2+} should then be considered as an example of "extrinsic" control since it is potentially a means whereby hormones and other extracellular stimuli may override the local "intrinsic" regulation. The advantage for the cell would be that oxidative metabolism and ATP synthesis is stimulated without the need to disturb the rather important ADP/ATP and NAD^+/NADH ratios and thus the ATP levels can be maintained or even increased at times when the cell needs to use more ATP.

The three dehydrogneases from non-vertebrate sources such as insects and plants appear not to be activated by Ca^{2+} but still to exhibit regulation by the nucleotide ratios (6). Since only mitochondria from vertebrate sources contain the ruthenium red inhibited transporter (see next section) which allows Ca^{2+} to be taken up into mitochondria under physiological conditions, this is powerful support for the notion that there is a close functional link between the transfer of Ca^{2+} into mitochondria and the activation of the dehydrogenases by Ca^{2+}.

THE TRANSFER OF CALCIUM ACROSS THE INNER-MEMBRANE OF MAMMALIAN MITOCHONDRIA

As indicated in Fig. 1, the calcium transport system in mammalian mitochondria has separate uptake and efflux components (2,7). Uptake of calcium occurs as Ca^{2+} via an electrophoretic uniporter mechanism which is driven by the large electrical gradient (about 180 mVolts, positive outside) across the inner-membrane. The uniporter is inhibited by the dye ruthenium red, and also by Mg^{2+} ions at concentrations (0.5-2.0mM) likely to be present in the cytoplasm of mammalian cells. The transfer of calcium out of mitochondria occurs mainly via an electroneutral exchange, or antiport, of Ca^{2+} for $2Na^+$. There is also a sodium-independent efflux pathway which may be particularly active in liver and kidney mitochondria. The $2Na^+$/Ca^{2+} antiport can be inhibited by several compounds which are better known as plasma membrane Ca^{2+}-channel blockers, the most potent being diltiazem.

There is continuous cycling of Ca^{2+} ions across the inner-membrane through the operation of the separate uptake and efflux pathways. The distribution of Ca^{2+} across the membrane will thus depend on the relative activities of the two pathways. At high concentrations of extramitochondrial Ca^{2+} the capacity of the uptake pathway can greatly exceed that of the efflux pathway. Under these conditions mitochondria can effectively diminish the extracellular concentration of Ca^{2+} to about $1\mu M$-Ca^{2+}. At this point, a steady-state is obtained at which uptake of Ca^{2+} is exactly balanced by the maximum rate of efflux of Ca^{2+}. It is important to emphasize that this ability to regulate the extramitochondrial concentration of Ca^{2+} is only observed when the free Ca^{2+} in mitochondria is sufficient to saturate the efflux pathways which is attained when the total calcium content is above 10nmoles/mg protein (corresponding to a free matrix Ca^{2+} concentration of 10-20μM). The physiological importance of this buffering action is not entirely obvious as the "set point" appears to be above the concentration of cytoplasmic Ca^{2+} in unstimulated cells. Furthermore, recent estimates of the amount of total calcium in mitochondria within intact liver and heart indicate that it is well below 10nmoles/mg protein; indeed,it is probably below 2nmoles/mg protein in unstimulated liver cells (9-11).

It has been argued (12-14), that hormones may actually trigger the release or "mobilisation" of calcium from mitochondria. For example, the initial rapid increases in the cytoplasmic concentration of Ca^{2+} observed in liver cells exposed to α-adrenergic-agonists, vasopressin or angiotensin II are undoubtedly due to the release of Ca^{2+} from intracellular stores. The conclusion that the major intracellular stores were mitochondria was based largely on apparent decreases found not only in the amount of calcium in mitochondrial fractions prepared from hormone-treated liver cells, but also that Ca^{2+} was rapidly released from liver cells on addition of uncouplers of mitochondrial respiration. However, it is now evident that these approaches have major difficulties (2). In fact, using alternative and improved techniques it has become apparent, as stated above, that the amount of calcium associated with unstimulated liver cells is less than 2nmole/mg mitochondrial protein. Moreover, this amount is increased rather than decreased when liver cells are exposed to the hormones (11). We will return to this topic later in this article.

It seems more likely, therefore, that under physiological conditions the mitochondrial calcium transport system will act with both uptake and efflux components operating well below maximum velocity. Under these conditions, an increase in the cytoplasmic concentration of Ca^{2+} will lead to an increase in the Ca^{2+} concentration within the mitochondrial matrix, with the new steady-state distribution of Ca^{2+} being acheived when the rate of uptake exactly matches the rate of efflux across the inner membrane.

STUDIES ON THE CA^{2+}-SENSITIVE DEHYDROGENASES USING INTACT ISOLATED MITOCHONDRIA.

Extensive studies have been carried out using heart, adipose tissue and liver mitochondria with the object of establishing the extent to which the dehydrogenases retain their sensitivity to Ca^{2+} when located within mitochondria (15-19).

Table 2 summarises some observations with mitochondria from rat heart and liver. In both types of mitochondria incubated under coupled conditions in media containing physiological concentrations of Na^+ and Mg^{2+}, marked increases in the activities of pyruvate dehydrogenase and oxoglutarate dehydrogenase are observed with increasing extramitochondrial concentrations of Ca^{2+}. Half-maximal effects are evident at 0.3-0.6μM. In contrast, in uncoupled mitochondria half-maximal effects are evident at about 1μM, which is similar to values obtained with the separated enzymes. It follows that the concentration of Ca^{2+} within the coupled mitochondria of intact cells is probably only 2-3 times that in the cytoplasm as long as the latter remains below the set point of about 1μM. Mg^{2+} inhibits calcium uptake into mitochondria while Na^+ stimulates calcium efflux, so in the absence of these ions a much greater gradient of Ca^{2+} across the mitochondrial inner-membrane would be predicted. In agreement with this, activation of both pyruvate dehydrogenase and oxoglutarate dehydrogenase is apparent at a rather lower range of extramitochondrial Ca^{2+} concentrations in the absence of the two ions (Table 2).

These and other studies demonstrate that the kinetic properties of the uptake and efflux components of the calcium transport system in the inner-membrane of mammalian mitochondria are such that changes in the extramitochondrial concentration of Ca^{2+} up to 1μM result in parallel changes in the activity of the intramitochondrial dehydrogenase.

EVIDENCE THAT HORMONES MAY REGULATE INTRAMITOCHONDRIAL OXIDATIVE METABOLISM THROUGH ALTERATIONS IN INTRAMITOCHONDRIAL Ca^{2+} IN HEART AND LIVER.

In the rat heart, the concentration of Ca^{2+} in the cytoplasm is increased by adrenaline (acting through β-adrenergic-receptors), glucagon and other positive

Table 2. Concentrations of extramitochondrial Ca^{2+} which give half-maximal activation of pyruvate dehydrogenase (PDH) and oxoglutarate dehydrogenase (OGDH) within uncoupled and coupled mitochondria.

Mitochondria	Condition	Ca^{2+} concentration (as nmolar) giving 50% activation of:-	
		PDH	OGDH
Rat heart	Uncoupled	980	940
	Coupled + NaCl+MgCl$_2$	468	328
	Coupled (minus Na^+/Mg^{2+})	39	21
Rat liver	Uncoupled	1059	–
	Coupled + NaCl + MgCl$_2$	484	560
	Coupled (minus Na^+/Mg^{2+})	98	121

Results from rat heart taken from Denton et al.,(15) and for rat liver from McCormack (19) from which full details can be obtained. In all cases activities in the presence of saturating extramitochondrial Ca^{2+} were at least four times that observed in the absence of Ca^{2+}. When added NaCl and MgCl$_2$ were present at 10mM and 1mM respectively.

Table 3. Effects of hormones and other treatments on the activity of pyruvate dehydrogenase in the perfused rat heart and in isolated rat hepatocytes.

Preparation	Treatment	Pyruvate dehydrogenase activity (% total) after:-		
		No further additions	Addition of Ruthenium Red	Ca-depletion of the cells
Perfused rat heart	Control	10	11	–
	Adrenaline (2μM)	41	12	–
	High calcium	42	13	–
Rat hepatocytes	Control	25	–	11
	Phenylephrine (20μM)	40	–	12
	Vasopressin (20nM)	39	–	11

Data taken from McCormack and England (23) and Oviasu and Whitton (24).

inotropic agents, whereas in the liver it is increased by α-adrenergic-agonists, vasopressin and angiotensin II. In all these circumstances the hormones increase oxygen uptake, flux through the citrate cycle and the proportion of pyruvate dehydrogenase in its active, non-phosphorylated form without detectable sustained decreases in ATP/ADP and NADH/NAD$^+$ ratios (11,20-22).

Table 3 summarises some of the observations which lend further support to the view that the activation of pyruvate dehydrogenase observed under these conditions involves changes in intramitochondrial Ca^{2+} concentration. For example, in the perfused rat heart, the effects of adrenaline can be mimicked by increasing the concentration of Ca^{2+} in the perfusing medium and can also be blocked by ruthenium red which inhibits the transfer of Ca^{2+} into mitochondria (23). With rat hepatocytes, the effects of phenylephrine (α-adrenergic-agonist) and vasopressin are no longer evident if the cells are depleted of calcium by a brief incubation in medium containing EGTA and no added calcium (24).

The most direct and convincing evidence that these hormones do act on intramitochondrial oxidative metabolism by increasing the concentration of Ca^{2+} in the mitochondrial matrix has been obtained in studies on mitochondria prepared from rat heart or liver previously exposed to an appropriate stimulating hormone. It has been found that if suitable precautions are taken there is little loss or uptake of calcium during the preparation and subsequent incubation of either liver or heart mitochondria (19,22,25). Precautions include the rapid disruption of intact tissue in cold, sodium-free buffered sucrose medium containing a high concentration of the calcium chelator EGTA. Under these conditions, activations of pyruvate dehydrogenase initiated by prior treatment of the tissues with a hormone persist not only during the preparation of the mitochondria but also during their subsequent incubation at $30^{\circ}C$ in sodium-free medium containing respiratory substrates and EGTA (Table 4). Persistent activation of oxoglutarate dehydrogenase can also be demonstrated (Table 4). However, these persistent increases in activity are quickly lost if Na$^+$ ions are added to the mitochondrial incubation medium. Since these effects of Na$^+$ are blocked by diltiazem which inhibits the sodium-dependent efflux pathway, it can be concluded that in each case the activations of pyruvate dehydrogenase and oxoglutarate dehydrogenase are the result of increases in Ca^{2+} concentration in the mitochondria from the hormone-treated tissues. Moreover, incubation of the mitochondria with sufficient Ca^{2+} to elicit a maximum stimulation of the dehydrogenases also results in the disappearance of the differences in the dehydrogenase activities in mitochondria from control and hormone-treated tissue. Finally, it has now been demonstrated directly in both heart and liver that hormones which increase the cytoplasmic concentration of Ca^{2+} result in increases in the total amount of calcium associated with mitochondrial fractions from 1-2nmole/mg protein to 2-4nmole/mg protein (11, 26).

Altogether there is now overwhelming evidence in these two contrasting tissues, in favour of the view of that when the cytoplasmic concentration of Ca^{2+} is increased by hormones or other external stimuli, the intramitochondrial concentration increases in parallel and results in the activation of the calcium-sensitive dehydrogenases. In the case of the heart, the beat-to-beat changes in the cytoplasmic concentration of Ca^{2+} probably greatly exceed the kinetic capacity of the mitochondrial calcium transport system (27,28). As a result, the changes in the intramitochondrial concentration of Ca^{2+} will reflect the "time-averaged" increases in cytosolic concentration. A similar situation may also occur in liver cells since recent studies suggest that α-adrenergic agonists may initiate a series of Ca^{2+} pulses in the cytoplasm of these cells (29).

Table 4. Persistence of the activations of pyruvate dehydrogenase (PDH) and oxoglutarate dehydrogenase (OGDH) in mitochondria prepared from intact rat heart and liver previously exposed to adrenaline.

Preparations	Additions to mitochondrial incubation medium	PDH activity (as % total) after:-		OGDH activity (as % Vmax) after:-	
		No hormone	Adrenaline	No hormone	Adrenaline
Rat heart (perfused)	None	8	20	23	35
	NaCl	8	7	25	24
	NaCl+Diltiazem	8	20	23	32
	Ca^{2+}	45	47	-	-
Rat liver (in vivo)	None	12	20	8	13
	NaCl	13	14	8	8
	NaCl+Diltiazem	13	23	9	18
	Ca^{2+}	49	51	33	35

Data taken from McCormack and Denton (25) and McCormack (22) from where full details can be obtained. After no hormone or adrenaline treatment, tissue was rapidly homogenised and mitochondria prepared. The mitochondria were then incubated at $30^{O}C$ for 5 min in KCl-based medium containing respiratory substrates, EGTA and where indicated additions of NaCl(10mM), diltiazem (300 M) and Ca^{2+} (to give a final $[Ca^{2+}]$ which maximally activated PDH.

GENERAL COMMENTS

We confidently expect that the above view of the role of Ca^{2+} in the hormonal regulation of intramitochondrial oxidative metabolism will be valid for other mammalian cells. The stimulation of the intramitochondrial dehydrogenases under conditions of increased concentrations of cytoplasmic Ca^{2+} will greatly contribute to metabolic homeostasis by allowing increased formation of NADH for the respiratory chain to occur within the stimulated cells without the need to diminish NADH/NAD$^+$ and ATP/ADP. Indeed, it seems likely that the mitochondrial NADH/NAD$^+$ ratio increases under such circumstances and that this is important in the stimulation of oxidative phosphorylation in the absence of a rise in ADP concentration. In liver cells, the hormones may also stimulate the respiratory chain (30).

The idea that the main role of the calcium transport system in mitochondria is the regulation of the intramitochondrial concentration of Ca^{2+} is not compatible with the earlier suggestions (see Refs. 12-14) that mitochondria can act as a store of calcium which can be mobilised following hormone stimulation. In any case, the evidence for these suggestions is open to alternative interpretations (2,11). The possibility still remains that mitochondria may play some role in the buffering or moderation of the cytoplasmic concentration of Ca^{2+}, if the concentration should rise above the "set-point" of about 0.5-1.0μM. However, if mitochondria act in this manner it would only be for transient periods as the concentration of Ca^{2+} under such circumstances would be sufficient to essentially saturate the Ca^{2+}-sensitive systems in both the cytoplasmic and mitochondrial compartments of the cells. Moreover, irreversible derangement of mitochondrial function may occur if excess calcium is taken up into these organelles (15,19, 27).

Although there are many examples of hormones acting on their target tissues through increases in the cytoplasmic concentration of Ca^{2+}, there are other examples of hormones which have well established effects on various facets of mitochondrial function where a role for Ca^{2+} is not so obvious (see Table 1). Nevertheless, we would like to end this article with some comments on the possibility that changes in the intramitochondrial concentration of Ca^{2+} may have some importance in these situations.

Insulin causes the activation of pyruvate dehydrogenase in adipose and other tissues and this is important in the stimulation of fatty acid synthesis (31,32). Although we first suggested that this effect of insulin may be brought about by an increase in the mitochondrial concentration of Ca^{2+}, our recent studies have shown that insulin does not, in fact, act in this manner (18). On the other hand, the effects of nor-adrenaline (via β-adrenergic-receptors) on brown adipose tissue, and glucagon on liver cells may involve a role for changes in intramitochondrial Ca^{2+} (22,30,33). However, the evidence to date is incomplete and, in any case, although increases in mitochondrial Ca^{2+} concentration may explain the observed increases in pyruvate dehydrogenase activity, they are perhaps unlikely to initiate the other mitochondrial alterations which occur (notably the uncoupling of mitochondria in brown adipose tissue and the stimulation of the respiratory chain in liver cells). Finally, it would seem important that studies are also now carried out to explore the extent to which the effects of trophic hormones, PTH and thyroxine on mitochondrial (Table 1) may be brought about by changes in the intramitochondrial concentration of Ca^{2+}. It should be noted that, at least in theory, hormones may manipulate the intramitochondrial concentration of Ca^{2+} by altering the activity of one or more components of the calcium transport system (1). Some evidence for such changes has already been obtained in both heart and liver (34-36).

ACKNOWLEDGEMENT

Work in the author's laboratories has been supported by grants from the Medical Research Council, the British Diabetic Association and the Percival Waite Salmond Bequest.

REFERENCES

1.   Denton, R.M. and McCormack, J.G. (1980). FEBS Lett.,119, 1-8.
2.   Denton, R.M. and McCormack, J.G. (1985). Amer. J. Physiol.,(In press).
3.   Denton, R.M., Randle, P.J. and Martin, B.R. (1972). Biochem. J., 128, 161-163.
4.   Denton, R.M., Richards, D.A. and Chin, J.G. (1978).Biochem. J., 176, 899-906.
5.   McCormack, J.G. and Denton, R.M. (1979). Biochem. J., 180, 533-544.
6.   McCormack, J.G. and Denton, R.M. (1981). Biochem. J., 196, 619-624.

7. Nicholls, D.G. and Akerman, K.E.O. (1982). Biochim. Biophys Acta, 683, 57-88.
8. Vaghy, P.L., Johnson, J.D., Matlib, M.A., Wang, T. and Schwartz, A.(1982). J. Biol. Chem., 257, 6000-6002.
9. Reinhart, P.H., Van der Pol, E., Taylor, W.M. and Bygrave, F.L. (1984). Biochem. J., 218, 415-420.
10. Somlyo, A.P., Bond, M. and Somlyo, A.V. (1985). Nature, 314, 622-625.
11. Assimacopoulos-Jeannet, F.D., McCormack, J.G. and Jeanrenaud, B. (1986). J. Biol. Chem. (submitted).
12. Exton, J.H. (1981). Mol. Cell. Endocrinol., 23, 233-264.
13. Williamson, J.R., Cooper, P.H. and Hoek, J.B. (1981) Biochim. Biophys. Acta, 639, 243-295.
14. Reinhart, P.H., Taylor, W.M. and Bygrave, F.L. (1984). Biochem. J., 223, 1-13.
15. Denton, R.M., McCormack, J.G. and Edgell, N.J. (1980). Biochem. J., 190, 107-117.
16. Hansford, R.G. (1981) Biochem. J. 194, 721-723.
17. Hansford, R.G. and Castro, F. (1981). Biochem. J. 198, 525-533.
18. Marshall, S.E., McCormack, J.G. and Denton, R.M. (1984). Biochem. J. 218, 249-260.
19. McCormack, J.G. (1985). Biochem. J., 231, 581-595.
20. McCormack, J.G. and Denton, R.M. (1981). Biochem. J. 194, 639-643.
21. Hems, D.A., McCormack, J.G. and Denton, R.M. (1978). Biochem. J. 176, 627-629.
22. McCormack, J.G. (1985). Biochem. J. 231, 597-608.
23. McCormack, J.G. and England, P.J. (1983). Biochem. J., 214, 581-585.
24. Oviasu, O.A. and Whitton, P.D. (1984). Biochem. J. 224, 181-186.
25. McCormack, J.G. and Denton, R.M. (1984). Biochem. J. 218, 235-247.
26. Crompton, M., Kessar, P. and Al-Nassar, I. (1983). Biochem. J. 216, 333-342.
27. Robertson, S.P., Potter, J.D. and Rouslin, W. (1982). J. Biol. Chem., 257, 1743-1748.
28. Crompton, M. and Roos, I. (1985). Biochem. Soc. Trans, 13, 667-669.
29. Woods, N.M., Cuthbertson, K.S.R. and Cobbold, P.H.(1986) Nature, (In press).
30. Halestrap, A.P., Quinlan, P.T., Armston, A.E. and Whipps, D.E. (1985). Biochem. Soc. Trans, 13, 660-664.
31. Denton, R.M. (1986). Adv. Cyclic Nucleotide and Protein Phos. Res., (In press)
32. Denton, R.M., McCormack, J.G. and Marshall, S.E. (1984). Biochem. J., 217, 441-452.
33. Gibbins, J.M., Denton, R.M. and McCormack, J.G. (1985). Biochem. J. 228, 751-755.
34. Goldstone, T.P., Duddridge,R.J. and Crompton, M.(1983). Biochem. J., 210, 463-472.
35. Kessar, P. and Crompton, M. (1981). Biochem. J., 200, 379-388.
36. Taylor, W.M., Prpic, V., Exton, J.H. and Bygrave, F.L. (1980). Biochem. J., 188, 443-450.

Résumé

On a trouvé que trois deshydrogénases clés des mitochondries de mammifères sont activées par les ions Ca^{2+} avec des effets demi-maximaux à environ 1 uM. Il s'agit de la pyruvate deshydrogénase, de la NAD^+-isocitrate deshydrogénase et de l'oxalacetate deshydrogenase. L'activation de ces enzymes peut être aussi démontrée dans la mitochondrie intacte lorsque le Ca^{2+} extramitochondrial est augmenté dans la gamme de concentration qui est considérée exister dans le cytoplasme des cellules de mammifères. On propose que le rôle principal du système de transport du calcium dans la membrane interne de la mitochondrie de mammifère est de répercuter les changements de concentration cytoplasmique de Ca^{2+} dans la matrice mitochondriale. De la sorte les hormones et autres signaux extracellulaires qui stimulent les processus exigeant de l'ATP par l'intermédiaire d'une augmentation de la concentration de Ca^{2+} cytoplasmique peuvent aussi augmenter le métabolisme oxydatif et par conséquent le rétablissement de la teneur en ATP.

New mediators

Nouveaux médiateurs

Hormones and cell regulation. Ed J. Nunez *et al.* Colloque INSERM/John Libbey Eurotext Ltd. © 1986. Vol. 139, pp. 263–273.

Role of fructose 2,6-bisphosphate in the control of glycolysis and gluconeogenesis in the liver and in lower eukaryotes

Emile Van Schaftingen

Laboratoire de Chimie Physiologique, Université Catholique de Louvain and International Institute of Cellular and Molecular Pathology, UCL 75.39, 75 avenue Hippocrate, B-1200 Brussels, Belgium.

KEYWORDS

Fructose 2,6-bisphosphate, glycolysis, gluconeogenesis, liver, Saccharomyces cerevisiae, Trypanosoma brucei

Discovered in 1980 (Van Schaftingen et al., 1980b), fructose 2,6-bisphosphate ($Fru-2,6-P_2$) appears to play an important role in the control of glycolysis and of gluconeogenesis in a variety of eukaryotic cells (reviewed by Hers and Van Schaftingen, 1982; Hers et al., 1982; Hers and Hue, 1983; Claus et al., 1984; Van Schaftingen, 1986). $Fru-2,6-P_2$ is a potent stimulator of 6-phosphofructo 1-kinase (PFK 1) in animals and in fungi. In plants, PFK 1 activity is not affected by $Fru-2,6-P_2$, but PP_i-PFK, an enzyme using inorganic pyrophosphate instead of ATP to convert $Fru-6-P$ into $Fru-1,6-P_2$, is nearly dependent for its activity on the presence of $Fru-2,6-P_2$. The sensitivity of PP_i-PFK to this phosphate ester varies with each type of plant, potato tuber containing a very sensitive enzyme, which is half-maximally saturated with a nanomolar concentration of $Fru-2,6-P_2$ under some assay conditions. This property is the basis of a sensitive assay procedure allowing one to measure less than 0.1 picomole of the novel phosphate ester (Van Schaftingen, 1984).

$Fru-2,6-P_2$ is also a potent inhibitor of fructose 1,6-bisphosphatase (FBPase 1) from animals, plants and fungi. (In plants, this effect concerns the cytosolic FBPase 1). Therefore it is able to exert a coordinate control on the interconversion between $Fru-6-P$ and $Fru-1,6-P_2$ in these organisms.

REGULATION OF THE $FRU-2,6-P_2$ CONCENTRATION IN THE LIVER

In the liver cell, $Fru-2,6-P_2$ is synthesized from $Fru-6-P$ and ATP by an enzyme called 6-phospho 2-kinase (PFK 2) and degraded to $Fru-6-P$ and P_i by a specific hydrolase, fructose 2,6-bisphosphatase (FBPase 2). One peculiarity of these two enzymic activities is that they cannot be separated from each other by any purification procedure used, including anion-exchange chromatography, hydrophobic chromatography, blue Sepharose chromatography (fig. 1) and affinity chromatography (El Maghrabi et al., 1982a; Van Schaftingen et al., 1982). This suggests that both enzymic activities are properties of a single bifunctional protein, which has now been purified to apparent homogeneity by several procedures (El Maghrabi

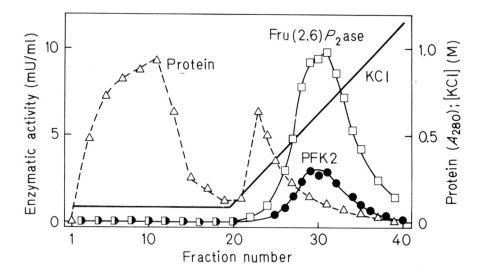

Figure 1 : Copurification of rat liver PFK 2 and FBPase 2 by Blue Sepharose
chromatography (from Van Schaftingen et al., 1982).

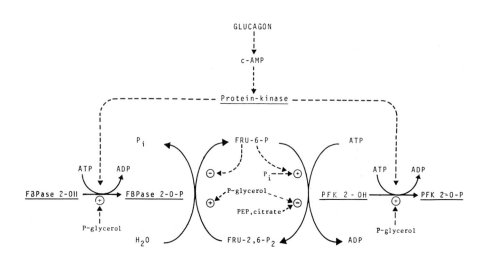

Figure 2 : Biosynthesis and biodegradation of Fru-2,6-P2 in the liver and
their control by glucagon and by metabolites. Abbreviations :
P-glycerol : glycerol 3-P; PEP : P-enolpyruvate (modified from Hers
and Van Schaftingen, 1982).

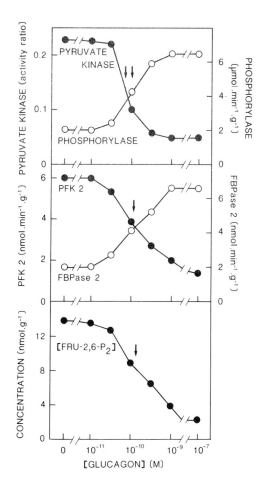

Figure 3 : Dose-response curves of the glucagon effect on the interconversion of enzymes and on the concentration of Fru-2,6-P$_2$ in hepatocytes isolated from a fed rat (from Bartrons et al., 1983).

et al., 1982a; Sakakibara et al., 1984 and unpublished results from this laboratory).

PFK 2 and FBPase 2 are regulated in a coordinate manner by covalent modification and by metabolites (Fig. 2). Indeed, liver PFK 2-FBPase 2 is a good substrate for cyclic AMP-dependent protein kinase, on which one mole of phosphate can be incorporated on a seryl residue per mole of enzyme subunit (El Maghrabi et al., 1982b). Concomitant with this phosphorylation, the PFK 2 activity is decreased, and the FBPase 2 activity is increased (El Maghrabi et al., 1982b; Van Schaftingen et al., 1981; 1982). This phosphorylation-induced change in the kinetic properties of PFK 2-FBPase 2 allows us to explain the decrease in the Fru-2,6-P$_2$ concentration induced by glucagon (fig. 3) and by several other agents acting on the liver through an increase in the concentration of cyclic

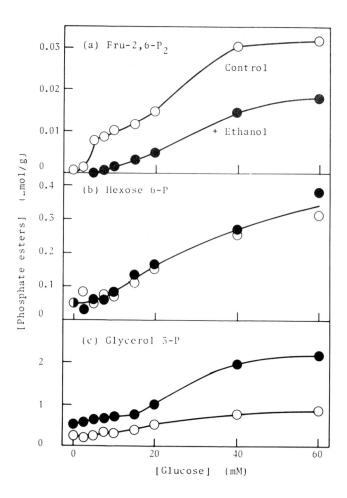

Figure 4 :   Effect of the concentration of glucose in the incubation medium on
the concentrations of Fru-2,6-P_2, hexose 6-P and glycerol 3-P in
hepatocytes isolated from a fasted rat and incubated in the absence
or in the presence of 10 mM ethanol (from Van Schaftingen *et al*.,
1984).

AMP. This decrease in the Fru-2,6-P_2 concentration is in turn responsible for
an increase in the net conversion of Fru-1,6-P_2 to Fru-6-P, which contributes
with the cAMP-mediated inactivation of pyruvate kinase to stimulate gluconeogenesis (reviewed by Engström, 1978).

The control of PFK 2 and FBPase 2 by metabolites is best exemplified by the roles
of Fru-6-P and glycerol 3-P. The former is a substrate of the kinase and a potent
inhibitor of the phosphatase. An increase in its concentration mediates the glucose effect of raising the Fru-2,6-P_2 level in isolated hepatocytes (fig. 4).
Thanks to its effect on Fru-2,6-P_2, glucose causes a stimulation of PFK 1 and
an inhibition of FBPase 1, two effects which allow us to understand the stimulation of glycolysis (Woods and Krebs, 1971) and the inhibition of gluconegenesis
(Ruderman and Herrera, 1968) exerted by the hexose in the liver. In full agreement with this view, both effects of glucose on glycolysis (Van Schaftingen *et*

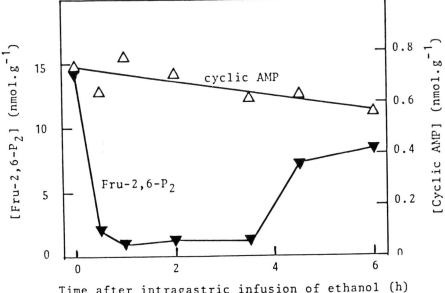

Figure 5 :   Effect of ethanol on the concentrations of Fru-2,6-P$_2$ and cAMP in
            the livers of fed rats. Ethanol (2 g/kg or a 20 % solution in saline)
            was administered by intragastric infusion.

al., 1980a) and on gluconeogenesis (Claus et al., 1975) are cancelled by glucagon
which, as indicated above, blocks the formation of Fru-2,6-P$_2$.

Glycerol 3-phosphate has effects on PFK 2 and FBPase 2 antagonistic to those of
Fru-6-P : indeed, it is a competitive inhibitor of PFK 2 (Claus et al., 1982) and
a potent deinhibitor of FBPase 2 (Van Schaftingen et al., 1982). Therefore it is
expected to play a negative role on Fru-2,6-P$_2$ concentration, an action which
allows us to understand the effect of ethanol. Ethanol causes both in vivo (fig.
5) and in vitro (isolated hepatocytes) (fig. 4) an increase in the glycerol 3-P
concentration as well as a decrease in the Fru-2,6-P$_2$ concentration (Claus et
al., 1982; Van Schaftingen et al., 1984). The effect on glycerol 3-P is due to
the rapid metabolization of ethanol through alcohol dehydrogenase and the resulting decrease in NAD/NADH ratio. That the effect on Fru-2,6-P$_2$ is secondary to
that on glycerol 3-P is indicated by the following observations :
(1) Other alcohols that are metabolized by alcohol dehydrogenase as well as glycerol, which is readily phosphorylated to glycerol 3-P, cause an increase in the
glycerol 3-P concentration and a decrease in the Fru-2,6-P$_2$ level. (2) The
two effects of ethanol on glycerol 3-P and on Fru-2,6-P$_2$ are cancelled by
4-methylpyrazole, a potent inhibitor of alcohol dehydrogenase. (3) The effect of
ethanol on Fru-2,6-P$_2$ is relieved if one increases the concentration of glucose
and, consequently, the intracellular concentration of Fru-6-P : this is expected
because of the competitive effect of Fru-6-P and of glycerol 3-P on PFK 2 and
FBPase 2 (Van Schaftingen et al., 1984).

It should be noted that ethanol causes a slow decrease in the activity of PFK 2
and an increase in the activity of FBPase 2 (Van Schaftingen et al., 1984), both

267

indicative of an increase in the phosphorylation state of PFK 2-FBPase 2. These effects are not mediated through an increase in the concentration of cAMP (see fig. 5) but are presumably also the result of the increase in the concentration of glycerol 3-P. This compound was indeed shown to accelerate the phosphorylation of PFK 2-FBPase 2 by cAMP-dependent protein kinase (Van Schaftingen et al., 1984).

REGULATION OF THE FRU-2,6-P$_2$ CONCENTRATION IN THE YEAST SACCHAROMYCES CEREVISIAE

In yeast, Fru-2,6-P$_2$ has the same role as in the liver, i.e. it stimulates glycolysis by interacting with PFK 1 and FBPase 1. It is also synthesized from Fru-6-P and ATP, but PFK 2 from Saccharomyces cerevisiae shows a striking difference when compared with liver PFK 2 : it is activated (rather than inactivated) by phosphorylation by cyclic AMP-dependent protein kinase (François et al., 1984; Yamashoji & Hess, 1984). This regulation by covalent modification comes into play in the glucose effect of stimulating glycolysis. Indeed, addition of glucose to a suspension of S. cerevisiae grown to the stationary phase elevates the concentrations of cyclic AMP, hexose 6-phosphate and Fru-2,6-P$_2$, and activates PFK 2 (fig. 6). All these effects, with the notable exception of that on hexose 6-phosphate, are almost totally cancelled by the simultaneous addition of 8-aminoacridine, which presumably inhibits adenylate cyclase and therefore blocks the cyclic AMP-mediated effects of glucose (François et al., 1984). The role of cyclic AMP in the control of the Fru-2,6-P$_2$ concentration in S. cerevisiae is further indicated by the observation that glucose has little effect on the Fru-2,6-P$_2$ level and the activation state of PFK 2 in thermosensitive mutants deficient in adenylate cyclase (François et al., 1984).

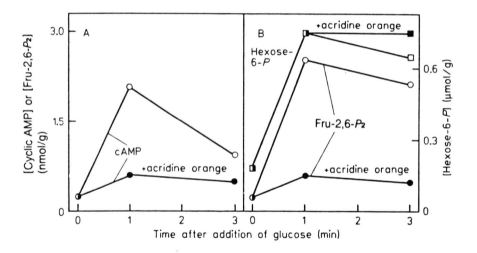

Figure 6 : Effect of glucose added with or without acridine orange on the concentrations of cAMP (A), hexose 6-P and Fru-2,6-P$_2$ (B) in Saccharomyces cerevisiae. Glucose was added at a final concentration of 0.1 M (from François et al., 1983).

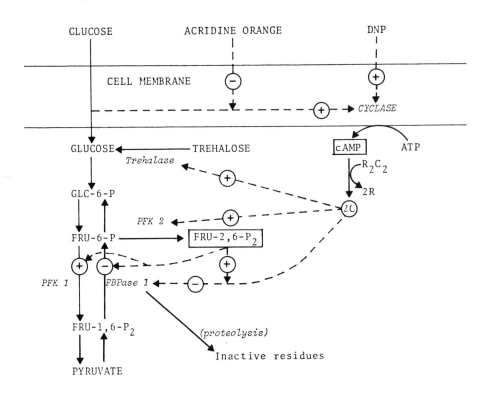

Figure 7 : Control of glycolysis and gluconeogenesis by Fru-2,6-P$_2$ and cAMP in
Saccharomyces cerevisiae (from François et al., 1983).

Fig. 7 illustrates our current understanding of the way by which glucose stimulates glycolysis in S. cerevisiae. When glucose enters the cell, it not only undergoes a rapid phosphorylation but it also stimulates the formation of cAMP by an as yet unidentified mechanism. In fungi as in animals, cyclic AMP combines with the regulatory subunit of cyclic AMP dependent protein kinase (Hixson and Krebs, 1980); the free catalytic subunit can then phosphorylate three substrates of interest for the regulation of glycolysis in yeast. One is trehalase, which is activated by phosphorylation (reviewed by Thevelein, 1984) and will then catalyse the degradation of trehalose to glucose. A second substrate is PFK 2, which, when phosphorylated, will synthesize Fru-2,6-P$_2$; this phosphate ester will in turn stimulate PFK 1 and inhibit FBPase 1. The last substrate is FBPase 1, which is partially inactivated by phosphorylation, a process accelerated by the presence of Fru-2,6-P$_2$ (Gancedo et al., 1983; Pohlig et al., 1983). The three phosphorylation events and the accelerated formation of hexose 6-P contribute to stimulate glycolysis. From this sequence of events, one can also see that cyclic AMP has the value of a glycolytic signal in S. cerevisiae, a role quite distinct from the one it plays in the liver. In contrast, the role of Fru-2,6-P$_2$ as a stimulator of glycolysis seems to have been well conserved through evolution, as further emphasized in the next section.

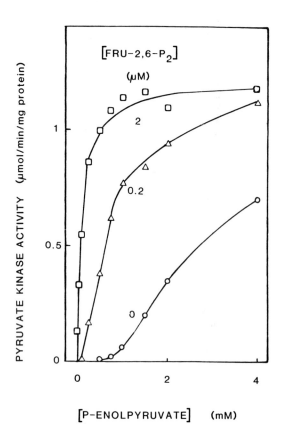

Figure 8 :   Effect of Fru-2,6-P₂ on the affinity of Trypanosoma brucei pyruvate
kinase for its substrate P-enolpyruvate.
The enzymic activity was measured with 1 mM P_i (from Van
Schaftingen et al., 1985).

STIMULATION OF PYRUVATE KINASE IN TRYPANOSOMA BRUCEI

The presence of Fru-2,6-P₂, and its role as a glycolytic signal is now well
documented in animals, plants and in fungi. Its status in protozoa is presently
uncertain and appears to greatly depend on the microorganism considered. We could
not detect this phosphate ester in Tetrahymena pyriformis but observed its
presence at micromolar concentrations in Euglena gracilis and Trypanosoma brucei.
In E. gracilis, trehalose phosphorylase was reported to be inhibited by
Fru-2,6-P₂ (Miyatake et al., 1984). We could not confirm this finding but
observed that PP_i-PFK is markedly stimulated by Fru-2,6-P₂ in this
photosynthetic protozoon (unpublished results). In T. brucei, PFK 1 activity is
known not to be affected by Fru-2,6-P₂, and PP_i-PFK and FBPase 1 are absent;
the enzyme controlled by Fru-2,6-P₂ was recently found to be pyruvate kinase
(Van Schaftingen et al., 1985) : at submicromolar concentrations, the regulatory
phosphate ester stimulates greatly this enzyme by converting the saturation curve
for phosphoenolpyruvate from a sigmoïd to a hyperbola (fig. 8). Fru-1,6-P₂ has

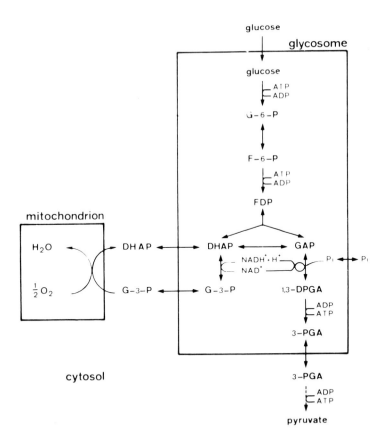

Figure 9 : Compartmentation of glycolysis in _Trypanosoma brucei_ (modified from Opperdoes 1985).

the same effect but at 4,000-fold higher concentrations. We have calculated that at physiological concentrations of substrates and of the two effectors P_i and Fru-1,6-P_2, the activity of pyruvate kinase would be too low to account for the glycolytic flux in this protozoon. It is only in the presence of micromolar concentrations of Fru-2,6-P_2 that the enzyme could be sufficiently active.

Stimulation by Fru-2,6-P_2 was not observed with pyruvate kinase from mammalian, plant or fungal origin, nor with the enzyme from _E. gracilis_. It was, however, observed on the enzyme of two other members of Trypanosomatidae family that were tested, i.e. _Leishmania major_ and _Crithidia luciliae_ (Van Schaftingen _et al._, 1985). One may wonder why Trypanosomatidae are so peculiar in having another point of control of glycolysis by Fru-2,6-P_2. The answer is not yeat clear but one has to remember that all the members of the Trypanosomatidae family have in common a unique spatial organization of glycolysis (reviewed by Opperdoes, 1985). Indeed, all enzymes catalyzing the conversion of glucose to glycerol and to 3-phosphoglycerate are enclosed in a granule called a glycosome (fig. 9) whereas phosphoglyceromutase, enolase and pyruvate kinase are in the cytosol. Pyruvate kinase is therefore the only non-equilibrium enzyme of glycolysis present in the cytosol and constitutes a privileged point of control of glycolysis in this organism.

CONCLUSION

Fru-2,6-P$_2$ is a stimulator of glycolysis in widely different cells such as the hepatocyte, the yeast Saccharomyces cerevisiae and the protozoon Trypanozoma brucei. This indicates that it had already acquired the value of a glycolytic signal early in the evolution of Eukaryotes.

ACKNOWLEDGEMENTS

This work was supported by the Fonds de la Recherche Scientifique Médicale and by the U.S. Public Health Service, Grant AM 9235.

REFERENCES

Bartrons, R., Hue, L., Van Schaftingen, E. & Hers, H.G. (1893) Biochem. J. 214, 829-837.

Claus, T.H., Pilkis, S.J. & Park, C.R. (1975) Biochim. Biophys. Acta 404, 110-123.

Claus, T.H., Schlumpf, J.R., El-Maghrabi, M.R. & Pilkis, S.J. (1982) J. Biol. Chem. 257, 7541-7548.

Claus, T.H., El-Maghrabi, M.R., Regen; D.M., Stewart, H.B., McGrane, M., Kountz, P.D., Nyfeler, F., Pilkis, J. & Pilkis, S.J. (1984) Curr. Top. Cell. Regul 23, 57-86.

El-Maghrabi, M.R., Claus, T.H., Pilkis, J., Fox, E. & Pilkis, S.J. (1982a) J. Biol. Chem. 257, 7603-7607.

El-Maghrabi, M.R., Fox, E., Pilkis, J. & Pilkis, S.J. (1982b) Biochem. Biophys. Res. Commun. 106, 794-802.

Engström, L. (1978) Curr. Top. Cell. Regul. 13, 29-51.

François, J., Van Schaftingen, E. & Hers, H.G. (1984) Eur. J. Biochem. 145, 187-193.

Gancedo, J.M., Mazon, M.J. & Gancedo, C. (1983) J. Biol. Chem. 258, 5998-5999.

Hers, H.G. & Hue, L. (1983) Ann. Rev. Biochem. 52, 617-653

Hers, H.G. & Van Schaftingen, E. (1982) Biochem. J. 206, 1-12.

Hers, H.G., Hue, L. & Van Schaftingen, E. (1982) Trends Biochem. Sci. 329-331.

Hixson, C.S. & Krebs, E.G. (1980) J. Biol. Chem. 255, 2137-2145

Hue, L., Blackmore, P.F. & Exton, J.H. (1981) J. Biol. Chem. 256, 8900-8903.

Miyatake, K., Kuramoto, Y. & Kitaoka, S. (1984) Biochem. Biophys. Res. Commun. 122, 906-911.

Opperdoes, F.R. (1985) Brit. Med. Bull. 41, 130-136.

Pohlig, G., Wingender-Drissen, R., Noda, T. & Holzer, H. (1983) Biochem. Biophys. Res. Commun. 115, 317-324.

Ruderman, N.B., Herrera, G.M. (1968) Am. J. Physiol. 214, 1346-1351.

Sakakibara, R., Kitajima, S. & Uyeda, K. (1984) J. Biol. Chem. 259, 41-46

Thevelein, J.M. (1984) Microbiol. Rev. 48, 42-59.

Van Schaftingen, E. (1984) In : Methods of Enzymatic Analysis (Bergmeyer, H.U. ed.) 3rd edn. Vol. 6, pp. 335-341, Verlag Chemie, Weinheim.

Van Schaftingen, E. (1986) In : Adv. in Enzymology, A. Meister ed., in press.

Van Schaftingen, E., Hue, L. & Hers, H.G. (1980a) Biochem. J. 192, 887-895.

Van Schaftingen, E., Hue, L. & Hers, H.G. (1980b) Biochem. J. 192, 897-901.

Van Schaftingen, E., Davies, D.R. & Hers, H.G. (1981) Biochem. Biophys. Res. Commun. 103, 362-368.

Van Schaftingen, E., Davies, D.R. & Hers, H.G. (1982) Eur. J. Biochem. 124, 143-149.

Van Schaftingen, E., Bartrons, R. & Hers, H.G. (1984) Biochem. J. 222, 511-518.
Van Schaftingen, E., Opperdoes, F.R. & Hers, H.G. (1985) Eur. J. Biochem. (in press).
Woods, H.F. & Krebs, H.A. (1971) Biochem. J. 125, 129-139.
Yamashoji, S. & Hess, B. (1984) FEBS Lett. 178, 253-256.

Résumé

Le fructose 2,6-bisphosphate est un puissant stimulateur de la 6-phosphofructo - 1-kinase d'animaux, de levures et de moisissures. Il stimule aussi la phosphofructokinase dépendante du pyrophosphate des plantes ainsi que la pyruvate kinase des Trypanosomatidés. Il est enfin un inhibiteur de la fructose 1,6-bisphosphatase d'animaux, de moisissures, de levures et de plantes. Dans le foie, le fructose 2,6-bisphosphate est synthétisé par la 6-phosphofructo 2-kinase et détruit par la fructose 2,6-bisphosphatase. Ces deux activités enzymatiques, portées par une seule protéine bifonctionnelle, sont régulées par certains métabolites, principalement le fructose 6-phosphate et le glycérol 3-phosphate, ainsi que par un phénomène de modification covalente. La protéine bifonctionnelle est, en effet, un substrat de la protéine-kinase dépendante de l'AMP cyclique : sa phosphorylation entraîne l'inactivation de la 6-phosphofructo 2-kinase et l'activation de la fructose 2,6-bisphosphatase, un mécanisme qui intervient dans l'effet du glucagon de provoquer une baisse de la concentration de fructose 2,6-bisphosphate dans le foie. La 6-phosphofructo 2-kinase de Saccharomyces cerevisiae est, elle aussi, un substrat de la protéine kinase dépendante de l'AMP cyclique mais, contrairement à l'enzyme de foie, sa phosphorylation entraîne son activation. Cet effet intervient dans le mécanisme par lequel le glucose stimule la glycolyse dans les levures.

Hormones and cell regulation. Ed J. Nunez *et al.* Colloque INSERM/John Libbey Eurotext Ltd. © 1986. Vol. 139, pp. 275–283.

PAF-acether (platelet activating factor), a mediator of cell-to-cell interaction

J. Benveniste

INSERM U.200, Univ. Paris-Sud, 32 rue des Carnets, 92140 Clamart, France.

SUMMARY

Paf-acether (platelet-activating factor, 1-O-alkyl-2-acetyl-sn-glycero-3-phosphocholine) is a potent mediator of inflammation and allergy, synthesized by various cell types, such as neutrophils, macrophages, endothelial cells and platelets. It exerts a vast array of activities, far beyond its now classical proinflammatory role. Besides activating platelets and neutrophils, it induces a cyclooxygenase-dependent acute bronchoconstriction, exhibits intense negative effect on cardiac force and coronary flow, and appears to play a pivotal role in experimental septic shock, ischemic bowel necrosis and gastric ulcer. Its precise role remains however to be delineated, probably through the use of the specific inhibitors that are now available.

KEY WORDS

Paf-acether, asthma, platelets, bronchial hyperreactivity.

DEFINITION OF PAF-ACETHER

A new substance that was capable of aggregating rabbit platelets, was described in 1972 (1). It was obtained upon specific allergen challenge of leukocytes from sensitized rabbits and was named platelet-activating factor (PAF). It is now considered as a pivotal molecule in a variety of pathophysicological conditions. More recently its molecular structure has been elucidated : 1-O-alkyl-2-acetyl-sn-glycero-3-phosphocholine (2, 3) (Fig. 1). Total synthesis yields an end product that has both the physicochemical and biological properties of the naturally occurring factor (4). Hence the name paf-acether, used as a trivial denomination emphasizing the main structural features of the molecule.

The acceptance of a biological activity, as being due to paf-acether, requires precise characterization (5). The substance should aggregate washed rabbit platelets pretreated with a cyclooxygenase inhibitor in the presence of an ADP scavenger system (creatine phosphate/creatine phosphokinase). In addition, it must migrate with synthetic paf-acether in thin layer chromatography as well as be eluted with it in high performance liquid chromatography. The substance must be inactivated by phospholipase A_2 (PLA_2), C and D and exhibit a resistance to

R. <u>arrhizus</u> lipase (6). Finally, it should induce tachyphylaxis when cross examined against synthetic paf-acether in washed rabbit platelets. The most preferred way for the detection of paf-acether is, of course, a structural analysis, but the amount of formed material is usually too low to permit the use of such a procedure (7).

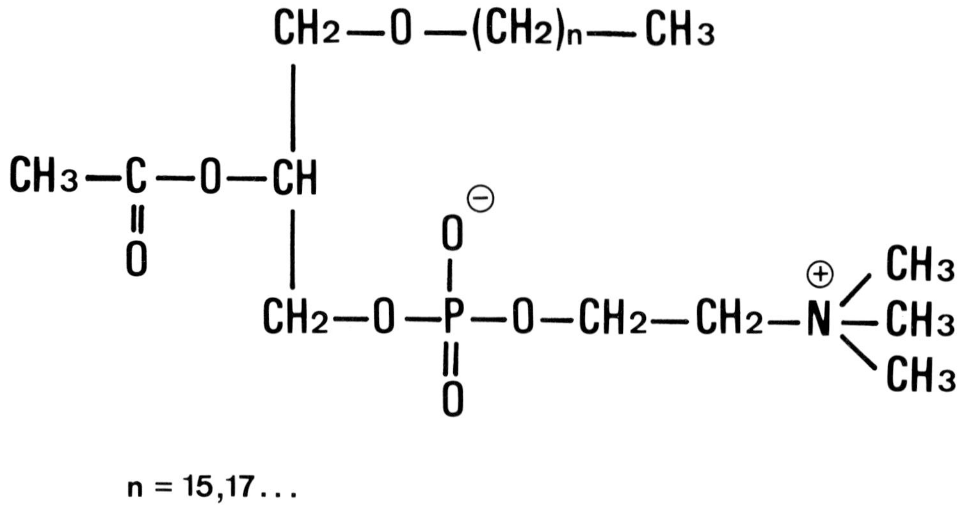

$$CH_2-O-(CH_2)_n-CH_3$$

$$CH_3-\underset{\underset{O}{\|}}{C}-O-CH$$

$$CH_2-O-\underset{\underset{O}{\|}}{\overset{\overset{O^{\ominus}}{|}}{P}}-O-CH_2-CH_2-\overset{\oplus}{N}\overset{\nearrow CH_3}{\underset{\searrow CH_3}{-CH_3}}$$

n = 15,17...

Figure 1. Structure of paf-acether.

SOURCES OF PAF-ACETHER

The presence of this compound has now been reported for a wide variety of cell types including rodent peritoneal macrophages (8), alveolar macrophages (9) and polymorphonuclear cells (10). Paf-acether was first isolated from rabbit basophils (1), yet its presence in human basophils is still controversial (11, 12). Rabbit and human platelets have also been shown to produce paf-acether (13, 14). More recently, endothelial cells and eosinophils were shown to synthesize and release large amounts of paf-acether (15, 16).

PAF-ACETHER METABOLISM

Paf-acether is formed <u>via</u> several enzymatic steps upon stimulation of various cells and organs (Rev. in 17). The substrate for the first enzyme, PLA_2, is 1-alkyl-2-acyl-sn-glycero-3-phosphocholine, an analog of phosphatidylcholine with an alkyl-ether aliphatic chain at the position 1 of the glycerol. It is noteworthy that the fatty acid esterified at the position 2 is quite often arachidonic acid. The PLA_2 is capable of hydrolyzing the alkyl-ether analog of phosphatidylcholine to form 1-alkyl-2-lyso-glycero-phosphocholine (2-lyso paf-acether) which in turn serves as a substrate for the acetyltransferase and for the formation of paf-acether. In this process, arachidonic acid can be liberated. Indeed many works have recently demonstrated the simultaneous triggering of the two main biochemical pathways leading to the generation of inflammatory mediators, namely the arachidonate-dependent and the paf-acether pathways. The inactivation of paf-acether depends on acetylhydrolase and acyl-transferase activities. The overall rate of biosynthesis of both intracellular and extracellular paf-acether depends on the balance between the various anabolic and catabolic enzymes (Fig. 2).

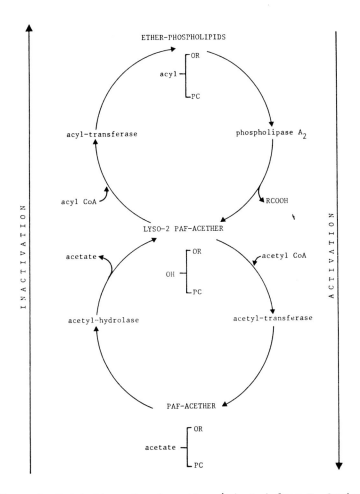

Figure 2. Metabolism of paf-acether (adapted from F. Snyder).

EFFECTS OF PAF-ACETHER

Cells

Since the original observation that paf-acether is a potent aggregatory agent on washed rabbit platelets, its activity has also been assessed on platelets from a wide variety of species, including guinea-pig, dog and man (Rev. in 18, 19). Paf-acether has been described as an activator of neutrophil aggregation (20), chemotactic activation (21) and lysosomal enzyme and oxygen radical release (22, 23).

Heart and vasculature

A hypotensive effect of an intravenous injection of paf-acether in rabbits, guinea pigs, baboons and rats has been reported. In fact, paf-acether is the same molecule as the antihypertensive polar renomedullary lipid described 25 years ago (24, 25). In dogs paf-acether induces an acute circulatory collapse (26). In all species the hypotensive effect is independent of platelet aggregation ; this is particularly clear in the rat, whose platelets are insensitive to paf-acether. The only direct vascular response to paf-acether stimulation that has been observed is in the rat portal vein preparation (27). The absence of effect in isolated vascular muscle preparations has been reported (28). We were unable to show any direct effect in isolated dog coronary arteries (R. Santamaria, unpublished observations). When injected in isolated heart Langendorff preparations, paf-acether demonstrates both a negative inotropic effect and a reduction of the flow rate (29, 30). These data would suggest an effect on the coronary circulation in the perfused organ, possibly through secondary release of arachidonate metabolites (Pretolani et al., submitted).

When intradermally administered to the rat or human, paf-acether induces edema and, in the latter case, hyperalgesia (31-34). This effect appears to be independent of platelet and neutrophil activation (33-35). A dual response - early and late - was reported in human skin (34). Whether or not this response is related to the early and late reactions observed in human asthma remains to be ascertained.

Lung and paf-acether

A direct contractile effect of paf-acether was reported on guinea pig ileum and lung tissue (36). However, we were unable to consistently reproduce these results in human and guinea-pig lung and ileal tissues (28). A small relaxation has been reported for guinea pig tracheal preparations (37). These data suggest that respiratory muscles are probably not the primary target in the pulmonary response to paf-acether.

We have recently described the presence of a specific IgE response in alveolar macrophages from asthmatic patients, resulting in the release of a lysosomal enzyme as well as paf-acether (38, 39). When alveolar macrophages were obtained from theophylline- or corticosteroid-treated patients, this release was not observed. Alveolar macrophages that can be activated by paf-acether become desensitized after repeated administration of this agent both in vitro and, by aerosol, in vivo. This is suggestive of the implication of these cells in the bronchoconstriction induced by aerosolized paf-acether (40).

Indeed, intravenous or aerosol administration of paf-acether to guinea-pigs induces an immediate increase of inspiratory pressure, reflecting a bronchoconstriction (41). This phenomenon is associated with thrombocytopenia, and pretreatment of animals with an anti-platelet serum abolished the effects of paf-acether. These mechanisms involved in the bronchoconstriction due to paf-acether have not been clearly elucidated. However there appears to be a relationship between the route of administration of this factor and the primary site of action, that is either cyclooxygenase-dependent (42) or through a lipoxygenase activity (43). The trapping of radiolabeled platelets by the lung was observed in the guinea pig 30 s after the bronchoconstriction, suggesting a relationship between the paf-acether-induced bronchoconstriction and the platelets (44). Anaphylactic effects of paf-acether have also been described in the rabbit (45).

The acute bronchomotor effect of paf-acether was demonstrated in primates, namely the baboon (46). The increase in peak inspiratory pressure after intratracheal deposition of paf-acether was associated with blood, blood gas and cell modifications, and was independent from the cyclooxygenase pathway. Similar

278

results were obtained in <u>Macaca mulatta</u> (47). The biological effects of paf-acether in comatous (cerebral death) patients has recently been documented (48 and Gateau et al., submitted). Paf-acether induced an increase in peak inspiratory pressure as well as changes in blood and blood gas values. Significant cardiac modifications, such as a severe drop in left ventricle stroke work index were also present.

More recently, paf-acether has been suspected to mediate lung diseases not so much as an acute bronchomotor agent but via increment of bronchial reactivity. This followed the initial observation by Vargaftig's group that bronchial reactivity to serotonin was greatly increased in the guinea-pig after intravenous administration of paf-acether (42, rev. in 49).

Miscellaneous effects
The spectrum of paf-acether activities is quickly expanding. Recent works have unveiled several unexpected effects of the mediator. Paf-acether induces the release of tissue-type plasminogen activator in rat, both <u>in vivo</u> and in perfused hind legs (50, 51). Paf-acether is a potent stimulator of hepatic glycogenolysis (52). It appears to be instrumental in creating ischemic bowel necrosis (53) and could very well play a pivotal role in the pathogenesis of endotoxin and septic shock (54, 55). Finally, a potent ulcerogenic action of paf-acether has been reported in the rat (56).

The variety of the sources and effects of this mediator suggests a wide patho-physiological role for it. It might be that, in fact, paf-acether acts in biological systems far from its original putative role in allergy and inflammation. It is only <u>via</u> the use of specific inhibitors that the role of paf-acether will be appreciated with certainty.

REFERENCES

1. Benveniste, J., Henson, P.M. and Cochrane, C.G. (1972): Leukocyte-dependent histamine release from rabbit platelets : the role of IgE, basophils and a platelet-activating factor. J. Exp. Med. 136, 1356-1377.
2. Benveniste, J., Tencé, M., Varenne, P., Bidault, J., Boullet, C. and Polonsky, J. (1979): Semi-synthèse et structure proposée du facteur activant les plaquettes (PAF) : paf-acéther, un alkyl éther analogue de la lyso-phosphatidylcholine. C. R. Acad. Sc. Paris 289D, 1037-1040.
3. Demopoulos, C.A., Pinckard, R.N. and Hanahan, D.J. (1979): Platelet-activating factor. Evidence for 1-O-alkyl-2-acetyl-sn-glyceryl-3-phosphorylcholine as the active component (a new class of lipid chemical mediators). J. Biol. Chem. 254, 9355-9358.
4. Godfroid, J.J., Heymans, F., Michel, E., Redeuilh, C., Steiner, E. and Benveniste, J. (1980): Platelet-activating factor (paf-acether) : total synthesis of 1-O-octadecyl 2-O-acetyl sn-glycerol-3-phosphocholine. FEBS Lett. 116, 161-164.
5. Roubin, R., Tencé, M., Mencia-Huerta, J.M., Arnoux, B., Ninio, E. and Benveniste, J. (1983): A chemically defined monokine-macrophage-defined platelet-activating factor. In Lymphokines, ed E. Pick, vol 8, pp 249-276. New York: Academic Press.
6. Benveniste, J., Le Couedic, J.P., Polonsky, J. and Tencé, M. (1977): Structural analysis of purified platelet-activating factor by lipases. Nature 269, 170-171.
7. Hanahan, D.J., Demopoulos, C.A., Liehr, J. and Tencé, M. (1977): Identification of platelet-activating factor isolated from rabbit basophils as acetyl-glyceryl-ether-phosphorylcholine. J. Biol. Chem. 255, 5514-5516.

8. Mencia-Huerta, J.M. and Benveniste, J. (1979): Platelet-activating factor (PAF) and macrophages. I. Evidence for the release from rat and mouse peritoneal macrophages and not from mastocytes. Eur. J. Immunol. 9, 409-415.
9. Arnoux, B., Duval, D. and Benveniste, J. (1980): Release of platelet-activating factor (PAF) and slow-reacting substance (SRS) from isolated perfused rat kidney. Eur. J. Clin. Invest. 10, 437-441.
10. Lotner, G.Z., Lynch, J.M., Betz, S.J. and Henson, P.M. (1980): Human neutrophil-derived platelet-activating factor. J. Immunol. 124, 676-684.
11. Betz, S.J., Lotner, G.Z. and Henson, P.M. (1980): Generation and release of platelet-activating factor (PAF) from enriched preparations of rabbit basophils ; failure of human basophils to release PAF. J. Immunol. 125, 2749-2755.
12. Camussi, G., Aglietta, M., Coda, R., Bussolino, F., Piacibello, W. and Tetta, C. (1981): Release of platelet-activating factor (PAF) and histamine. II. The cellular origin of PAF : monocytes, polymorphonuclear neutrophils and basophils. Immunology 42, 191-199.
13. Chignard, M., Le Couedic, J.P., Tencé, M., Vargaftig, B.B. and Benveniste, J. (1979): The role of platelet-activating factor in platelet aggregation. Nature 279, 799-800.
14. Chignard, M., Le Couedic, J.P., Vargaftig, B.B. and Benveniste, J. Platelet-activating factor (PAF-acether) secretion from platelets. Effect of various aggregating agent. Br. J. Haematol. 46, 455-464.
15. Camussi, G., Aglietta, M., Malavasi, F., Tetta, C., Piacibello, W., Sanavio, F. and Bussolino, F. (1983): The release of platelet-activating factor fromhuman endothelial cells in culture. J. Immunol. 131, 2387-2403.
16. Lee, T.-c., Lenihan, D.J., Malone, B., Roddy, L.L. and Wasserman, S.I. (1984): J. Biol. Chem. 259, 5526-5530.
17. Snyder, F. (1985): Chemical and biochemical aspects of platelet-activating factor : A novel class of acetylated ether-linked choline-phospholipids. Med. Res. Rev. 5, 107-140.
18. Vargaftig, B.B., Chignard, M., Benveniste, J., Lefort, J. and Wal, F. (1981): Background and present status of researchs on platelet-activating factor (PAF-acether). Ann. N.Y. Acad. Sci. 370, 119-137.
19. Pinckard, R.N., McManus, L.M. and Hanahan, D.J. (1982): Chemistry and biology of acetyl glyceryl ether phosphorylcholine (platelet-activating factor). In Advances in Inflammation Research, ed G. Weissmann, vol 4, pp 147-180. New York: Raven Press.
20. Camussi, G., Tetta, C., Bussolino, F., Caligaris Cappio, F., Coda, R., Masera, C. and Segoloni, G. (1981): Mediators of immune complex-induced aggregation of polymorphonuclear neutrophils. II. Platelet-activating factor as the effector substance of immune-induced aggregation. Int. Archs. Allergy Appl. Immun. 64, 25-41.
21. Czarnetzki, M.B. and Benveniste, J. (1981): Effect of 1-0-octadecyl-2-0-acetyl-sn-glycero-3-phosphocholine (paf-acmether) on leukocytes. I. Analysis of the in vitro migration of human neutrophils. Chem. Phys. Lipids 29, 317-326.
22. O'Flaherty, J.T., Wykle, R.L., Miller, C.H., Lewis, J.C., Waite, M., Bass, D.A., McCall, M.D. and DeChatelet, L.R. (1981): 1-0-alkyl-sn-glyceryl-3-phosphorylcholines. A novel class of neutrophil stimulants. Am. J. Pathol. 103, 70-78.
23. Poitevin, B., Roubin, R. and Benveniste, J. (1984): Paf-acether generates chemiluminescence in human neutrophils in the absence of cytochalasin B. Immunopharmacology 7, 135-144.
24. Muirhead, E.E. and Stirman, J.A. (1958): Dietary protein and hypertension in the dog : protection by ureterocaval anastomosis with a study of the kidneys so treated. Am J Pathol. 34, 561 (abs.).
25. Blank, M.L., Snyder, F., Byers, L.W., Brooks, B. and Muirhead, E.E. (1979): Antihypertensive activity of an alkyl ether analog of phosphatidylcholine. Biochem. Biophys. Res. Comm. 90, 1194-1200.

280

26. Bessin, P., Bonnet, J., Apffel, D., Soulard, C., Desgroux, L., Pelas, I. and Benveniste, J. (1983): Acute circulatory collapse caused by platelet-activating factor (paf-acether) in dogs. Eur. J. Pharmac. 86, 403-413.
27. Santamaria, R., Pourrias, B. and Benveniste, J.m (1983): Effet du paf-acéther sur la veine porte de rat. J. Pharmacol. (Paris) 14 (suppl), 83 (abs.).
28. Cerrina, J., Raffestin, B., Labat, C., Boullet, C., Bayol, A., Gateau, O. and Brink, C. (1983): Effects of paf-acether on isolated muscle preparations from the rat, guinea-pig and human lung. In Platelet-Activating Factor and Structurally-Related Ether Lipids. INSERM Symposium n° 23, eds J. Benveniste and B. Arnoux, pp 205-212. Amsterdam: Elsevier Science Publishers.
29. Benveniste, J., Boullet, C., Brink, C. and Labat, C. (1983): The actions of paf-acether (platelet-activating factor) on guinea-pig isolated heart preparations. Br. J. Pharmac. 80, 81-83.
30. Feuerstein, G., Boyd, L.M., Ezra, D. and Goldstein, R.E. (1984): Effect of platelet-activating factor on coronary circulation of the domestic pig. Am. J. Physiol. 246, H466-H471.
31. Bonnet, J., Loiseau, A.M., Orvaen, M. and Bessin, P. (1981): Platelet-activating factor (PAF-acether) involvement in acute inflammatory and pain process. In Pharmacology of Inflammation and Allergy : Lipids and Cells, eds F. Russo-Marie, B.B. Vargaftig and J. Benveniste, vol 100, pp 111-118.
32. Pinckard, R.N., Kniker, W.T., Lee, L., Hanahan, D.J. and McManus, L.M. (1980): Vasoactive properties of 1-O-alkyl-2-acetyl-sn-glyceryl-3-phosphocholine (AcGEPC) in human skin. J. Allergy Clin. Immunol. 63, 196 (abs.).
33. Pirotzky, E., Page, C.P., Roubin,R., Pfister, A., Paul, W., Bonnet, J. and Benveniste, J. (1984): Paf-acether-induced plasma exudation in rat skin is independent of platelets and neutrophils. Microcirc. Endothel. Lymph. 1, 107-122.
34. Archer, C.B., Page, C.P., Paul, W., Morley, J., McDonald, D.M. (1984): Inflammatory characteristics of platelet-activating factor (PAF-acether) in human skin. Br. J. Dermat. 110, 45-50.
35. Björk, J. and Smedegard, G. (1983): Acute microvascular effects of paf-acether, as studied by intravital microscopy. Eur. J. Pharmac. 96, 87-94.
36. Stimler, N.P., Bloor, C.M., Hugli, T.E., Wykle, R.L., McCall, C.E. and O'Flaherty, J.T. (1981): Anaphylactic actions of platelet-activating factor. Am. J. Pathol. 105, 64-69.
37. Prancan, A., Lefort, J., Barton, M. and Vargaftig, B.B. (1982): Relaxation of the guinea-pig trachea induced by platelet-activating factor and by serotonin. Eur. J. Pharmac. 80, 29-35.
38. Joseph, M., Tonnel, A.B., Torpier, G., Capron, A., Arnoux, B. and Benveniste, J. (1983): The involvement of IgE in the secretory processes of alveolar macrophages from asthmatic patients. J. Clin. Invest. 71, 221-230.
39. Arnoux, B., Simoes-Caeiro, M.H., Landes, A., Mathieu, M., Duroux, P. and Benveniste, J. (1982): Alveolar macrophages from asthmatic patients releazse platelet-activating factor (paf-acether) and lyso paf-acether when stimulated with the specific allergen. Am. Rev. Resp. Dis. 125, 70 (abs.).
40. Maridonneau-Parini, I., Lagente, V., Lefort, J., Randon, J., Russo-Marie, F. and Vargaftig, B.B. (1985): Desensitization to PAF-induced bronchoconstriction and to activation of alveolar macrophages by repeated inhalations fo PAF in the guinea-pig. Biochem. Biophys. Res. Comm. 131, 42-49.
41. Vargaftig, B.B., Lefort, J., Chignard, M. and Benveniste, J. (1980): Platelet-activating factor induces a platelet-dependent bronchoconstriction unrelated to the formation of prostaglandin derivatives. Eur. J. Pharmac. 65, 185-192.
42. Vargaftig, B.B., Lefort, J. and Rotilio, D. (1983): Route-dependent interactions between paf-acether and guinea-pig bronchopulmonary smooth muscle : relevance of cyclooxygenase mechanisms. In Platelet-Activating Factor and Structurally-Related Ether-Lipids. INSERM Symposium n°23, eds J. Benveniste and B. Arnoux, pp. 307-316. Amsterdam: Elsevier Science Publishers.

43. Bonnet, J., Thibaudeau, D. and Bessin, P. (1983): Dependency of the paf-acether-induced bronchospasm on the lipoxygenase pathway in the guinea-pig. Prostaglandins 26, 457-466.
44. Page, C.P., Paul, W., Douglas, G.J., Casals-Stenzel, J. and Morley, J. (1983): Paf-acether, platelets and bronchospasm. In Platelet-Activating Factor and Structurally-Related Ether-Lipids. INSERM Symposium n°23, eds J. Benveniste and B. Arnoux, pp. 327-334. Amsterdam: Elsevier Science Publishers.
45. McManus, L.M., Hanahan, D.J., Demopoulos, C.A. and Pinckard, R.N. (1980): Pathobiology of the intravenous infusion of acetyl-glyceryl-ether-phosphorylcholine (AGEPC), a synthetic platelet-activating factor (PAF) in the rabbit. J. Immunol. 124, 2219-2224.
46. Denjean, A., Arnoux, B., Masse, R., Lockhart, A. and Benveniste, J. (1983): Acute effect of intratracheal administration of paf-acether in baboons. J. Appl. Physiol. 55, 799-804.
47. Patterson, R. and Harris, K.E. (1983): The activity of aerosolized and intracutaneous synthetic platelet-activating factor (AGEPC) in rhesus monkey with IgE-mediated airway responses and normal monkeys. J. Lab. Clin. Med. 102, 933-938.
48. Gateau, O., Arnoux, B., Deriaz, H. and Benveniste, J. (1984): Acute effects of intratracheal administration of paf-acether (platelet-activating factor) in humans. Am. Rev. Resp. Dis. 129, A3 (abs.).
49. Morley, J., Sanjar, S. and Page, C.P. (1984): The platelet in asthma. Lancet II, 1142-1144.
50. Emeis, J.J. and Kluft, C. (1985): PAF-acether-induced release of tissue-type plasminogen activator from vessel walls. Blood 66, 86-91.
51. Klöcking, H.P., Markwardt, F. and Hoffmann, A. (1985): Release of tissue-type plasminogen activator by platelet-activating factor. Thromb. Res. 38, 413-416.
52. Buxton, D.B., Shukla, S.D., Hanahan, D.J. and Olson, M.S. (1984): Stimulation of hepatic glycogenolysis by acetylglyceryl ether phosphorylcholine. J. Biol. Chem. 259, 1468-1471.
53. Gonzalez-Crussi, F. and Hsueh, W. (1983): Experimental model of ischemic bowel necrosis. The role of platelet-activating factor and endotoxin. Am. J. Pathol. 112, 127-135.
54. Terashita, Z.-I., Imura, Y., Nishikawa, K. and Sumida, S. (1985): Is platelet-activating factor (PAF) a mediator of endotoxin shock ? Eur. J. Pharmac. 109, 257-261.
55. Inarrea, P., Gomez-Cambronero, J., Pascual, J., del Carmen Ponte, M., Hernando, L. and Sanchez-Crespo, M. (1985): Synthesis of PAF-acether and blood volume changes in Gram-negative sepsis. Immunopharmacology 9, 45-52.
56. Rosam, A.C., Wallace, J.L. and Whittle, B.J.R. (1986): Potent ulcerogenic actions of platelet-activating factor on the stomach. Nature 319, 54-56.

Résumé

Le paf-acéther (platelet-activating factor, 1-O-alkyl-2-acétyl-sn-glycéro-3-phosphocholine) est un puissant médiateur de l'inflammation et de l'allergie synthétisé par de nombreux types cellulaires (neutrophiles, macrophages, cellules endothéliales, plaquettes et organes comme le coeur et le rein). Dans tous les systèmes capables de produire ce médiateur, la même voie métabolique a été décrite : hydrolyse d'un acide gras en position 2 du glycérol d'un analogue alkyl-éther de la phosphatidylcholine. L'acide gras libéré peut être l'acide arachidonique. Le lyso composé ainsi obtenu est acétylé en présence d'acétyl-CoA par une acétyltransférase. Il est ainsi transformé en paf-acéther. Une acétyl-hydrolase est capable d'hydrolyser le paf-acéther en lyso paf-acéther et une acyltransférase transforme ce dernier en analogues alkyls de la phosphatidyl-choline. Le niveau global de biosynthèse du médiateur dépend donc de l'équilibre entre les différentes enzymes cataboliques et anaboliques. De nombreux travaux ont récemment démontré la possibilité de déclenchement simultané de la libération des deux voies métaboliques principales capables de générer des médiateurs inflammatoires, la voie dépendant de l'acide arachidonique et celle du paf-acéther.

Le paf-acéther n'est pas seulement, comme son nom l'indique, actif sur les plaquettes mais il est également un puissant activateur des neutrophiles chez lesquels il déclenche l'agrégation, une activation de type chimiotactique, la libération d'enzymes lysosomiales et de radicaux libres. Les effets du paf-acéther ont été étendus à d'autres systèmes cellulaires et à d'autres organes. C'est un puissant hypotenseur au point qu'il est considéré actuellement comme l'un des agents principaux du choc endotoxinique et du choc septique Injecté dans un système de coeur isolé de cobaye il démontre un important effet inotrope négatif et une réduction considérable du flux coronaire. C'est un des plus puissants inducteurs de l'augmentation de perméabilité vasculaire dans différentes espèces, y compris l'homme. Enfin, son rôle dans la survenue de maladies pulmonaires, notamment l'asthme, est évoqué sur de nombreux arguments expérimentaux : contraction directe du muscle lisse, il est vrai faible et variable d'un tissu à l'autre, mais surtout induction d'une puissante broncho-constriction dans différentes espèces, y compris les primates et l'homme, dont de nombreux arguments indiquent qu'elle dépend probablement de l'agrégation intrapulmonaire des plaquettes. Une autre approche de l'effet pulmonaire du paf-acéther est l'induction d'une hyperréactivité bronchique qui abaisserait le seuil de réponse à des agonistes sans effet en l'absence d'imprégnation pulmonaire par le paf-acéther.

Par ailleurs, de très nombreux effets nouveaux ont été récemment décrits : libération de l'activateur tissulaire du plasminogène, stimulation de la glycogénolyse hépatique, création de nécroses ischémiques intestinales ou d'ulcères de l'estomac chez le rat. La variété des origines et des effets de ce médiateur suggère qu'il pourrait jouer un rôle physiopathologique beaucoup plus large que suspecté à l'origine, mais c'est seulement par l'utilisation d'inhibiteurs spécifiques que ce rôle pourra être apprécié avec certitude.

Author Index
Index des auteurs

518621

3 1378 00518 6211